INORGANIC AND ORGANIC RADICALS: THEIR BIOLOGICAL AND CLINICAL RELEVANCE

INORGANIC AND ORGANIC RADICALS: THEIR BIOLOGICAL AND CLINICAL RELEVANCE

PROCEEDINGS OF
A ROYAL SOCIETY DISCUSSION MEETING
HELD ON 30 AND 31 JANUARY 1985

ORGANIZED BY
R. O. C. NORMAN, F.R.S., AND H. A. O. HILL
AND EDITED BY H. A. O. HILL

LONDON
THE ROYAL SOCIETY
1986

Printed in Great Britain for the Royal Society
by the
University Press, Cambridge

ISBN 0 85403 260 6

First published in *Philosophical Transactions of the Royal Society of London*,
series B, volume 311 (no. 1152), pages 449–689

Copyright

Published by the Royal Society
6 Carlton House Terrace, London SW1Y 5AG

PREFACE

There is now good evidence that biological systems involve radical species, whether as a result of exposure to exogenous materials or simply the consequence of normal reactions. An understanding of such radicals will not rest on one discipline but on a number acting in concert. Thus it was our intention to bring together for this Meeting those with their expertise centred on the molecular nature of radicals and those primarily concerned with biological systems, including man. There appear to be two major influences on the properties of radicals, particularly with respect to their generation and consequences of formation: photochemistry and the presence or absence of molecular oxygen. Indeed, the role of dioxygen is crucial. A diradical, or its radical relatives, can take part in a wide number of reactions either for good or for ill. Many of the contributions to this Discussion Meeting deal with the use that is made of dioxygen, or its reduced relations, in biology, or with the defence that is usually present against the production of these relations. The latter appear to occur either in the wrong place, or at the wrong time, or both. The relation between these effects, or defects, and the consequences for the whole animal, is touched upon in many of the talks. The Meeting rests on the basis provided by the opening talks, which describe methods of studying radical reactions and their properties in a wide range of environments. It continues by considering both the use organic chemists made of these reactions and their attempts to model many of the features of analogous biochemical reactions. The possible involvement of radical intermediates of other reactions, ranging from those promoted by coenzyme B_{12}-dependent enzymes to carbon tetrachloride toxicity, is dealt with by a number of speakers. The subsequent talks, which deal with the biochemical and medical features of the work, draw on the lessons learnt from working with the molecularly more simple systems applied to a variety of cells and cellular conditions ranging from leukocytes to melanomas. It is hoped that the discussion at the Meeting, both formal and informal, will encourage others to exchange information with those concerned with the biological influences of the two progenitors of most living systems.

We are very grateful to the staff of the Royal Society, particularly Miss Christine Johnson, Meetings Organizer, and Miss Sally Clinton, of the editorial department, for their help and patience.

R. O. C. Norman
H. A. O. Hill

September 1985

CONTENTS

BIOLOGICAL ASPECTS OF RADICAL CHEMISTRY

RADICALS AND THEIR PHYSIOLOGICAL AND CLINICAL RELEVANCE

Phil. Trans. R. Soc. Lond. B **311**, 451–472 (1985)
Printed in Great Britain

Inorganic radicals of relevance to biological systems

BY M. C. R. SYMONS

Department of Chemistry, The University, Leicester LE1 7RH, U.K.

There is little doubt that the most important inorganic radicals involved in biological systems are those which are intermediates in the oxygen–water redox cycle, i.e. $OH^{\cdot}$, O_2^- and $HO_2^{\cdot}$. Aspects of the structures and reactivities of these radicals are considered, together with methods of detection. In particular, the use of e.s.r. spectroscopy is outlined, including rapid-freeze and spin-trapping techniques. Attention is called to comparisons and contrasts between these radicals and corresponding sulphur-centred radicals, although these are not strictly 'inorganic'. The oxygen-centred radicals are usually generated *in vivo* by redox reactions, but they are also of importance in radiolytic processes because they are formed from water. Other radicals formed in this way whose structures and reactivities are considered include solvated electrons and hydrogen atoms.

1. BACKGROUND

Definitions are always difficult and are best avoided. Nevertheless, a working definition for the present task is that radicals are species containing one, or possibly more than one, unpaired electron; they display a tendency to form bonds to other species (molecules, ions or radicals) such that their unpaired electrons become paired. Examples of typical reactions for a radical $X^{\cdot}$ are dimerization, (1), hydrogen-atom abstraction, (2), and radical addition, (3). All but reaction (1) result in the generation of new radicals.

$$2X^{\cdot} \rightarrow X{-}X, \tag{1}$$

$$X^{\cdot} + RH \rightarrow X{-}H + R^{\cdot}, \tag{2}$$

$$X^{\cdot} + A{-}B \rightarrow X{-}A{-}B^{\cdot}. \tag{3}$$

Radicals are also frequently good electron donors (especially radical anions) or good electron acceptors (especially radical cations), but these are not properties that are in any sense unique to radicals. Many transition metal complexes contain unpaired d-electrons but, in general, these do not exhibit reactions of type (1)–(3) and hence, with some exceptions, do not qualify for the description 'radical'.

A few inorganic radicals are stable with respect to dimerization. These include $NO^{\cdot}$, which forms a weakly bound dimer at low temperatures; O_2^-, which probably does not dimerize (but can form a weak complex with O_2 to give O_4^-; Lindsay *et al.* 1983); $^{\cdot}NO_2$, which forms N_2O_4 reversibly at room temperature; O_3^-, the oxonide ion which has no well defined dimer; $^{\cdot}NF_2$, which is formed reversibly from N_2F_4; $^{\cdot}SO_2^-$, formed in solution from dithionite ions; $S_2O_4^{2-}$, and $^{\cdot}ClO_2$, which shows no tendency to dimerize. All but $^{\cdot}NO_2$ are π^*-radicals in that their SOMOs (semi-occupied molecular orbitals) are π^* in nature and are strongly delocalized over the molecular frames. Other radicals of interest in the present context will dimerize irreversibly if no other mode of reaction presents itself. Thus, for example, $OH^{\cdot}$ radicals give H_2O_2. In contrast, $HO_2^{\cdot}$ radicals, which in principle can give HO_4H dimers, undergo a preferable disproportionation to give H_2O+O_2.

After the beautifully unifying concepts of nucleophilic and electrophilic reaction mechanisms were developed in the thirties (Ingold 1953), there was considerable resistance to the idea that homolytic mechanisms might apply to reactions in solvents, especially in aqueous solutions. There was no easy way of establishing the involvement of radicals until M. G. Evans and his coworkers, who included Polanyi, Baxendale and Uri, showed that free-radical polymerization was a sensitive method of trapping radicals, greatly magnifying their effect via long chain processes (Baxendale *et al.* 1946). For example, if a dilute aqueous solution of acrylonitrile is treated with a drop of a radical-generating solution, the whole solution becomes filled with precipitated polymer in a most dramatic fashion.

In particular, these studies strongly supported the concept that induced oxidations from Fenton's reagent and related systems proceeded via the formation of $OH^{\cdot}$ radicals (4).

$$Fe^{2+}+H_2O_2 \rightarrow Fe^{3+}OH^{-}+OH^{\cdot}. \tag{4}$$

In the presence of most organic compounds, the highly reactive $OH^{\cdot}$ radicals react to give induced oxidations rather than reacting further with the iron complexes or with H_2O_2.

My own initiation into this area of chemistry arose when I noticed that permanganate ions reacted smoothly in strongly alkaline aqueous solutions to give manganate and oxygen, and that this system could induce the oxidation of organic compounds (R—H) which did not react with MnO_4^- directly (Symons 1953 *a*, *b*). I suggested that reactions (5) and (6) were responsible:

$$MnO_4^- + OH^- \rightleftharpoons MnO_4^{2-} + {}^{\cdot}OH, \tag{5}$$

$${}^{\cdot}OH + RH \rightarrow H_2O + R^{\cdot}. \tag{6}$$

Subsequent studies seemed to give strong support to the idea (Symons 1953 *a*, *b*, 1954; Lott & Symons 1960; Carrington & Symons 1963). This reaction hinges on the fact that the electron-donating ability of OH^- increases steadily as its solvation by water is reduced. Thus the reaction is only of use in strongly alkaline media and has not been used very widely.

Because of the prevalence of transition metals, particular iron, in biological systems, it is inevitable that reactions with H_2O_2 will be potential sources of $OH^{\cdot}$ radicals and of O_2^-–$HO_2^{\cdot}$ radicals, while oxygen is always a source of O_2^-–$HO_2^{\cdot}$ radicals. (It is necessary to consider both O_2^- and $HO_2^{\cdot}$ in this context, because the pK_a of $HO_2^{\cdot}$ is 4.9, so that formation of O_2^- is always accompanied by some $HO_2^{\cdot}$. This is important because of the very different reactivities of these species.) Sometimes the system is designed to produce these radicals in abundance, because of their high reactivities, but otherwise it is necessary to keep their concentrations low, and Nature has developed many efficient ways for accomplishing this task.

My aim is to make some general comments about these inorganic radicals. Three aspects are considered: electronic structures; reactivities; and methods of detection. These topics are closely interlinked, because reactivities stem largely from the nature of the SOMOs of the radicals and of the neighbouring filled and empty levels. It is these SOMOs that are so well probed by e.s.r. spectroscopy.

2. Detection of radicals

At this stage I consider, briefly, various methods of detection of radicals before considering their structures, because I can then use spectroscopic evidence in support of my statements.

(a) *Chemical methods*

These are legion, and include the induced oxidation and polymerization tests mentioned above. Some of the best are enzymatic, such as the use of superoxide dismutase to remove O_2^- radicals selectively. The spin-trapping technique outlined in §2*d* is an excellent example of a chemical test coupled with e.s.r. spectroscopy. My concern in this review is, however, not with chemical methods and their many limitations, but with spectroscopic techniques.

(b) *Optical methods*

Ultraviolet and infrared absorption, ultraviolet emission and Raman or resonance Raman scattering techniques have all been used effectively in the study of radicals. Of these, infrared and normal Raman techniques are relatively insensitive, so that high concentrations of radicals are required. Resonance Raman is far more sensitive, but it can at present only be used for radicals having well defined, low-lying allowed electronic transitions. The use of tunable u.v. lasers promises to open up this technique, but the problem may then be that systems of interest may absorb the incident light too strongly to permit the extraction of useful information. Ultraviolet spectroscopy has proven to be the most powerful of these techniques, but firm identification of the absorbing species is sometimes difficult, because the information relates to the difference between ground and excited states and is only a good 'fingerprint' if prior knowledge of the spectrum is available. Furthermore, many radicals of interest have poorly defined u.v. spectra, or are transparent in the normal range.

(c) *Electron spin resonance*

This technique is uniquely applicable to radicals and to paramagnetic transition metal complexes because it detects transitions between electron spin levels, and these are confined to systems containing unpaired electrons. There are many books and reviews which outline the theory (Wertz & Bolton 1972; Atherton 1973) and chemical applications (Symons 1978) of this technique, and one which is solely concerned with inorganic radicals (Atkins & Symons 1967). I confine my attention here to a brief overview of the method, to clarify later comments. This is most readily accomplished by examining figures 1–7 and by reference to the closely related and far more widely known technique of n.m.r. spectroscopy.

Figure 1 shows the act of absorption, namely the electron spin-flip transition induced by microwave radiation where the degeneracy of the $S = \pm\frac{1}{2}$ levels has been lifted by a static magnetic field (B_z). Instrumentation is very similar to that for n.m.r. spectroscopy, except that it is not possible to sweep the microwave field, so it is normal practice to sweep the applied magnetic field. The effect of this on the energy levels is shown in figure 2. This shows that the rate of divergence of the electron energy levels, which is governed by the g value ($h\nu = g\mu_\beta B$, when ν is the microwave frequency and B the applied magnetic field), can vary. If there is no orbital motion of the unpaired electron, $g = 2.0023$. This can be thought of as the 'free-spin' value and taken as an effective zero. Orbital motion leading to an orbital magnetic moment occurs when the SOMO is close in energy to empty (virtual) or filled orbitals with which it

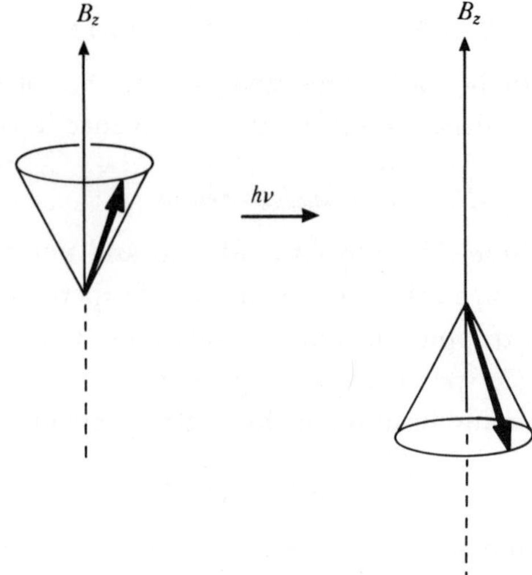

FIGURE 1. The e.s.r. transition induced by the microwave field ($h\nu$). The electron is pictured as a small magnet (arrow) precessing about the direction of the applied magnetic field B_z.

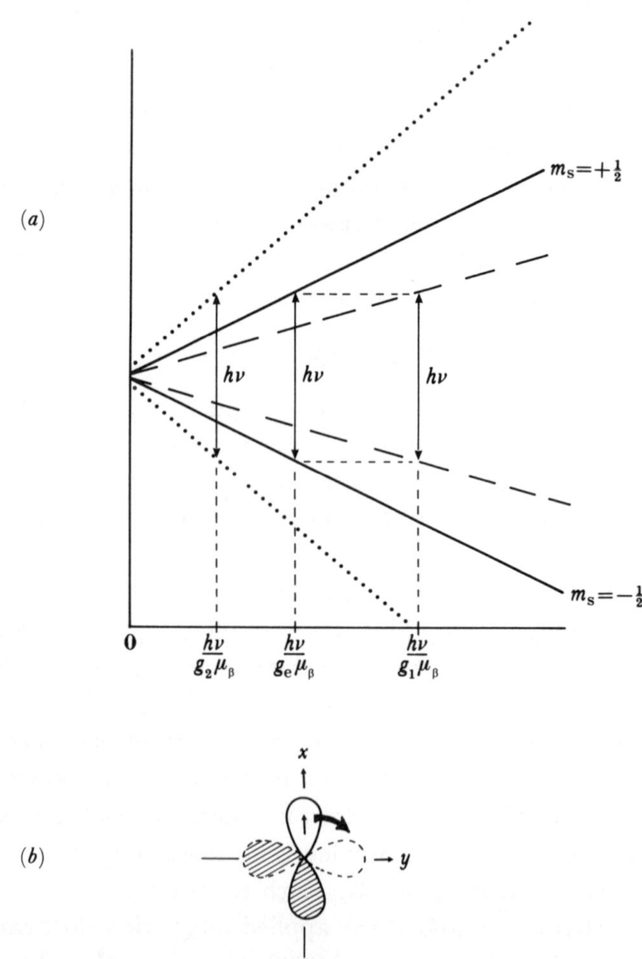

FIGURE 2. (a) The effect of an applied static magnetic field (B) on the $M_S = \pm\frac{1}{2}$ levels of an unpaired electron. These are degenerate at zero field. The thick lines represent the spin-only behaviour, the transition ($h\nu$) occurring at a field corresponding to $g_e = 2.0023$. The broken lines are for a radical having a low-lying vacant excited state, giving $g_1 < 2.0023$, and the dotted lines are for a radical having a neighbouring filled level, giving $g_2 > 2.0023$. (Taken from Symons 1978.) (b) Movement of the unpaired electron from one p-orbital (p_x) into another (p_y) induced by a magnetic field (B_z) gives rise to orbital motion and hence a shift in the g value.

can be magnetically coupled. As indicated in figure 2, Δg is negative (shift to high field) if the coupled level is empty and positive if it is filled.

For most organic radicals, shifts in g are small and are not of major diagnostic value. However, for many transition metal complexes, and for several of the oxygen- and sulphur-centred radicals discussed herein, Δg can be large, and controls the form of the e.s.r. spectra.

The other major factor controlling the form of the e.s.r. spectrum is electron–nuclear hyperfine coupling, which can be compared with nuclear–nuclear spin-coupling in n.m.r. spectroscopy.

This leads to a splitting of the $M_S = \pm\frac{1}{2}$ states as indicated in figure 3, the resulting pattern of lines being characteristic of the type, and number of coupled nuclei. Some examples are shown in figures 4–6. In particular, for the much studied nitroxide radicals, $R_2\dot{N}O$, mentioned below, coupling to the ^{14}N nucleus results in three equally intense lines (figure 5*a*) because $I = 1$. Thus, $\frac{1}{3}$ of the radicals have $M_I = +1$, $\frac{1}{3}$ have $M_I = 0$ and $\frac{1}{3}$ have $M_I = -1$. In favourable cases, hyperfine coupling to other nuclei, including ^1H and ^{13}C is also observed in the spectra of $R_2\dot{N}O$ radicals (figures 7 and 8).

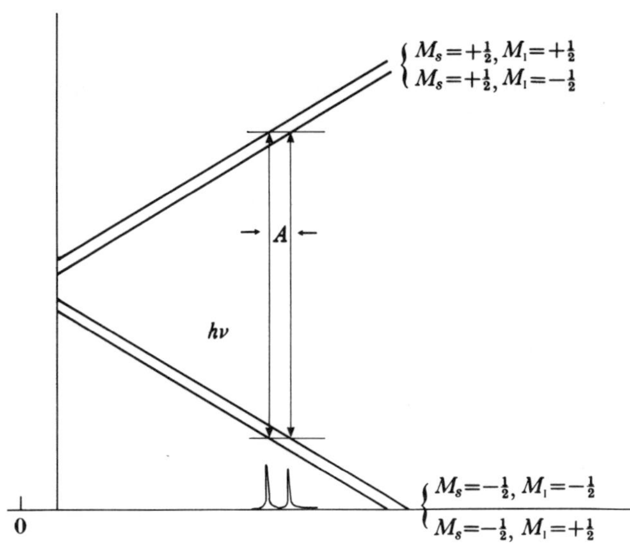

FIGURE 3. Divergence with field of the $M_S = \pm\frac{1}{2}$ levels in the presence of a single nucleus having $I = \frac{1}{2}$, in the high-field approximation. Note the two allowed transitions involve no change in M_I.

As with n.m.r. spectra, variation in linewidths can be used to extract kinetic information about the radicals, and changes in the hyperfine parameters can be used to estimate the degree of solvation and hence the environment of the radicals. It is important to note that both g and A are, in general, anisotropic, having x, y and z components. Liquid-phase spectra are averages, and give only the isotropic parts of these tensors, and the rate of averaging may make important contributions to the linewidths.

In some cases, the effects of anisotropy and other factors are so great that e.s.r. signals are too broad to detect in the liquid phase. One approach is then to freeze the solution and measure the solid-state e.s.r. spectrum. This is a less sensitive procedure because the resulting 'powder' spectra may cover a wide field range, as illustrated in figure 9 for a solution containing O_2^-.

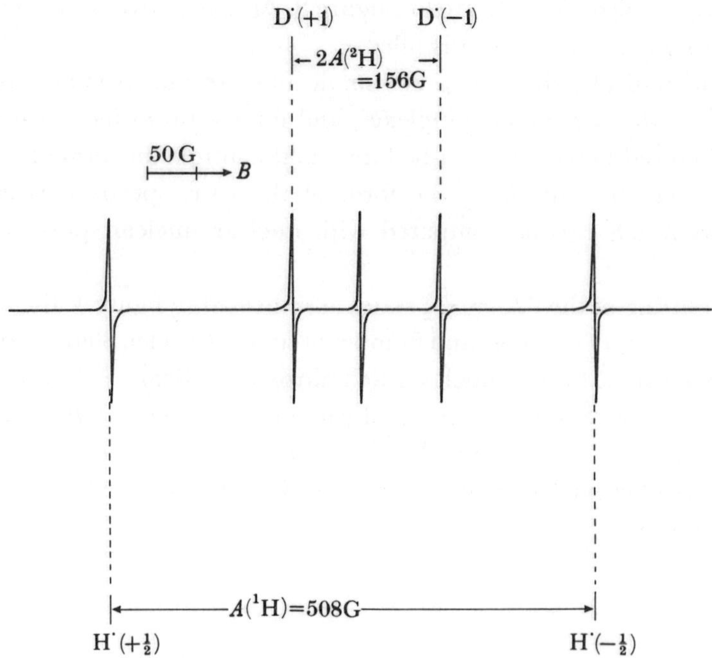

FIGURE 4. First derivative e.s.r. spectrum showing features for trapped hydrogen and deuterium atoms.

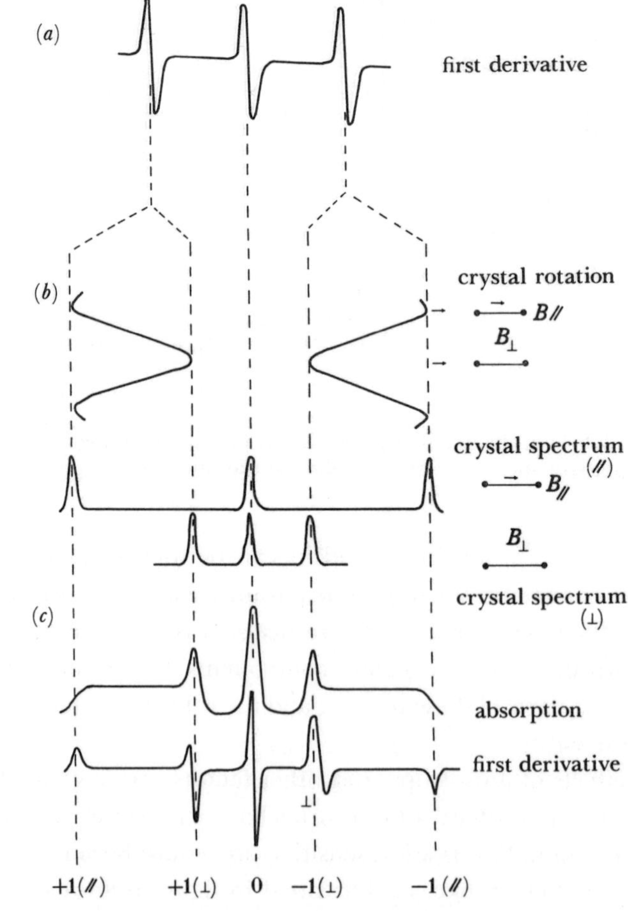

FIGURE 5. This figure illustrates the connection between a typical solution spectrum for a radical with a single coupled nucleus with $I = 1$, (*a*), and the changes on rotation observed for a single crystal. (*b*) The rotation is selected such that the parallel and perpendicular orientations are traversed. An isotropic *g* value has been assumed. The corresponding powder spectrum in absorption and as the first derivative is given in (*c*).

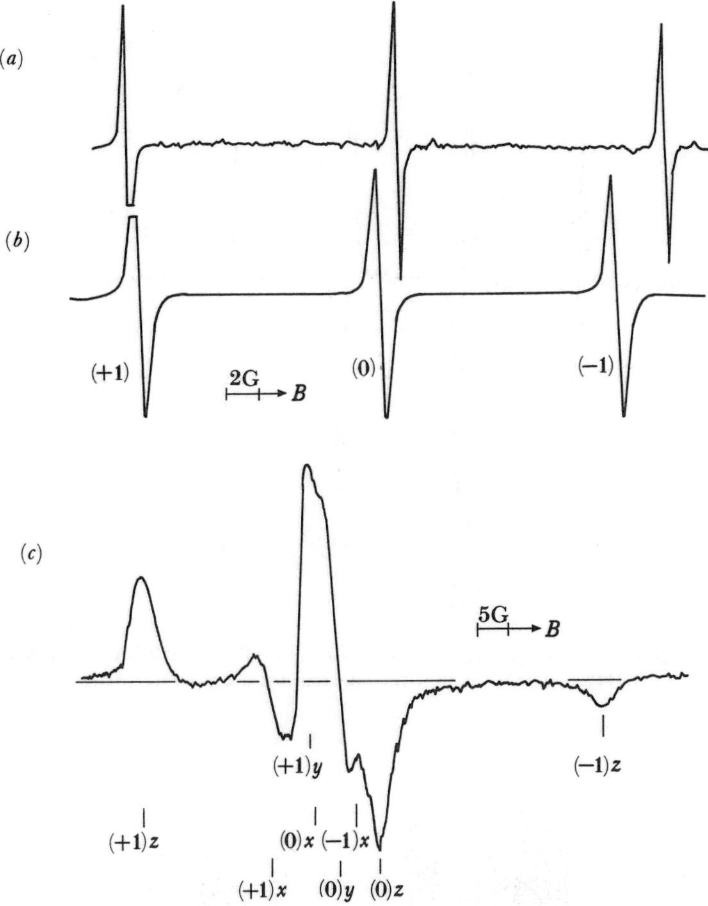

FIGURE 6. First derivative X-band e.s.r. spectra for $(Me_3C)_2NO^{\cdot}$ radicals (*a*) in water at 20 °C; (*b*) in dodecane at 20 °C and (*c*) in CD_3OD at 77 K. In (*c*) $(CD_3)_3CNO^{\cdot}$ radicals were used to give better definition.

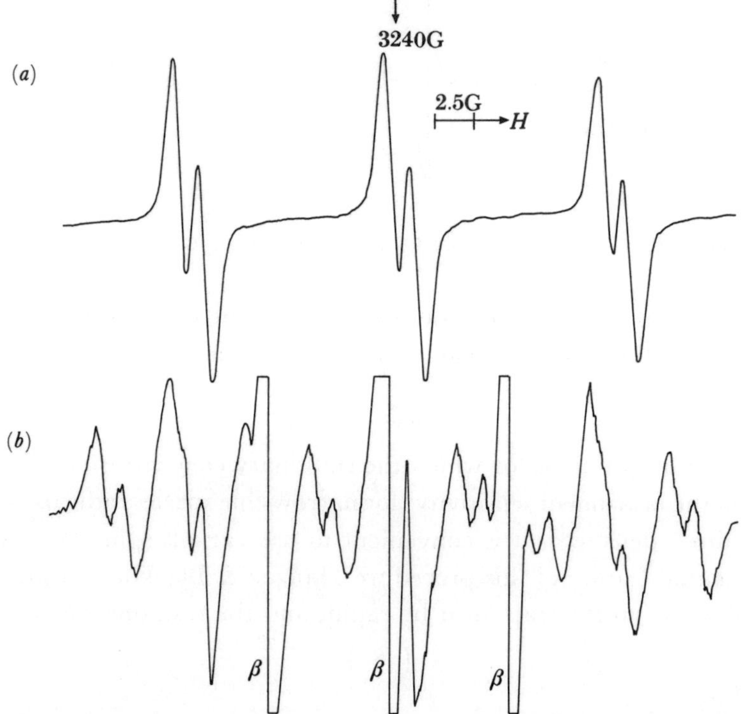

FIGURE 7. First derivative X-band e.s.r. spectra for (*a*) PBN and $^{\cdot}[^{12}C]Cl_3$; (*b*) PBN + $^{\cdot}[^{13}C]Cl_3$.

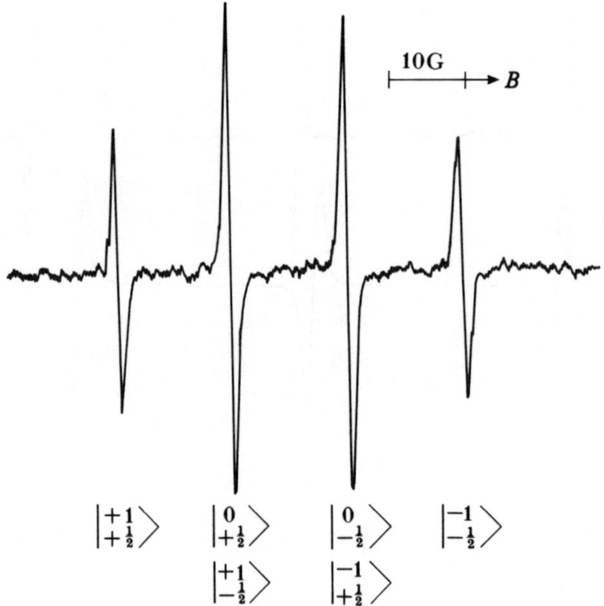

FIGURE 8. First derivative X-band e.s.r. spectrum assigned to the OH˙ radical adduct of DMPO. Note that $A(^1H) \approx A(^{14}N)$ so that the spectrum is a quartet.

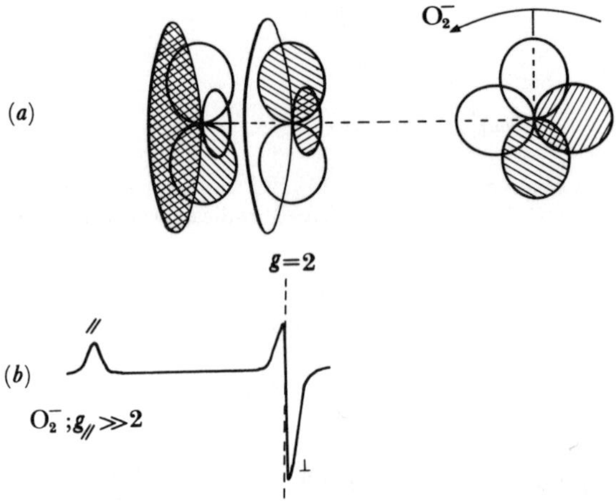

FIGURE 9. (*a*) The π-orbitals for O_2^- and (*b*) the resulting e.s.r. spectrum.

Also, for aqueous systems, phase separation giving ice as the dominant phase almost always occurs, and this can lead to spin-exchange effects which often modify the e.s.r. spectra.

(*d*) *Spin trapping*

In these cases, and in situations for which the stationary concentration of reactive radicals is below the level of the instrument sensitivity (for narrow-line species radicals at concentrations of *ca.* 10^{-7} mol^{-1} dm^{-3} detected), it is convenient to use various spin traps, as illustrated in scheme 1. The essential feature of this procedure (Janzen & Blackburn 1968; Chalfont *et al.* 1968) are that additions to the trap must be rapid, and the resulting nitroxide radical must

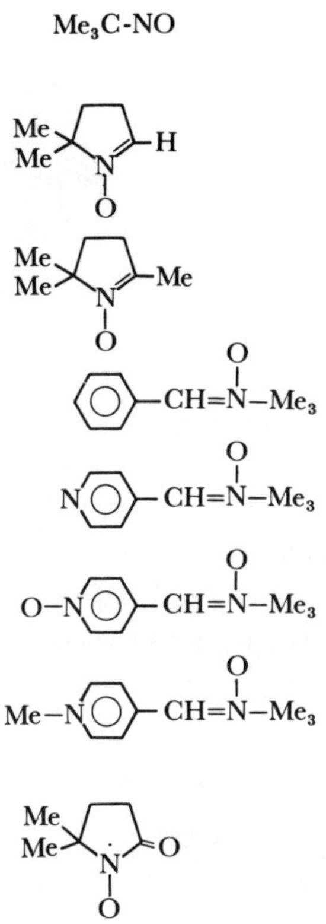

Structure	Abbreviation
Me_3C-NO	NTB
(Me, Me, N, O, H)	DMPO
(Me, Me, N, O, Me)	TMPO
C_6H_5–CH=N(O)–Me_3	PBN
N(ring)–CH=N(O)–Me_3	4-PyBN
O–N(ring)–CH=N(O)–Me_3	4-POBN
Me–N(ring)–CH=N(O)–Me_3	4-MePyBN
(Me, Me, N, O, =O)	DMPOX

SCHEME 1. Examples of commonly used spin traps.

be far more stable than the parent radical, so that relatively high concentrations can accumulate.

Although this is a good method for proving the participation of radicals, identification and attempts to quantify yields are fraught with difficulties (Janzen 1971). Some of these are briefly indicated below.

Identification frequently rests upon a fingerprinting process, in which an established spectrum is used as a gauge. This presupposes that all radical adducts of a given trap have unique spectra, which may sometimes be a dangerous concept. The following considerations are pertinent.

(i) $A(^{14}N)$ varies slightly with the nature of R$^{\cdot}$. This is especially the case for RNO traps.

(ii) For nitrone traps, the unique β-proton gives a variable coupling. This is determined by the degree of overlap between the π*-orbital containing the unpaired electron and the C—H σ-orbital (θ in figure 10; the proton coupling is proportional to $\cos^2\theta$). The nature of R, including its bulk, will govern the equilibrium value of θ. Thus $A(^1H)$ is a strongly variable parameter and hence a useful tool for identification.

(iii) The most convincing datum, if it can be seen, is direct hyperfine splitting from one or more nuclei of the added radical, R$^{\cdot}$. A nice example of this is the doublet splitting from ^{13}C for $^{\cdot}[^{13}C]Cl_3$ radicals, shown in figure 7 (Symons *et al.* 1982). This provides unambiguous evidence that $^{\cdot}CCl_3$ radicals were being trapped. Hyperfine coupling to protons in R$^{\cdot}$ is usually small, and lost in the linewidths. However, such coupling can sometimes be detected by using resolution enhancement techniques, which comprise computer processing of the data, or better, by using ENDOR (electron–nuclear double resonance) methods.

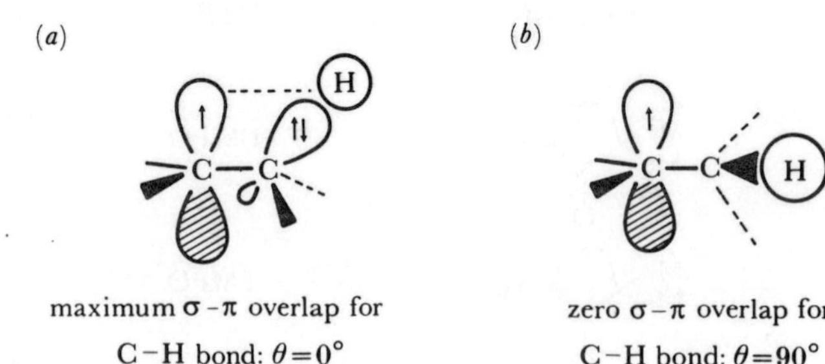

FIGURE 10. Overlap (σ–π or hyperconjugative) for a C—H bond in a radical $R_2\dot{C}—CHR'_2$ showing how this depends upon θ, the angle between this bond and the radical plane.

Thus, ideally, oxygen-centred radicals should be studied by using ^{17}O enrichment. This nucleus has $I = \frac{5}{2}$ and hence splitting of each component into sextets would be observed. I am not aware of any such studies for the oxygen-centred radicals discussed herein.

I now turn to a consideration of the structures and reactivities of various oxygen-centred radicals, because these are by far the most important inorganic radicals involved in biological systems.

3. Dioxygen derivatives

(a) *Dioxygen*

Dioxygen has a ground-state triplet level, the two unpaired electrons being accommodated, formally, in the degenerate pair of antibonding π-orbitals, $\pi_x^*(\uparrow), \pi_y^*(\uparrow)$. It is noteworthy that, because of strong coupling to rotational levels, the e.s.r. spectrum for O_2 as a low-pressure gas comprises sets of many narrow lines spread over a wide field range. Unfortunately, these are so extensively broadened for O_2 in solution that no resonance is detectable. However, oxygen is an important source of internal fluctuating magnetic fields, which may broaden the e.s.r. features of other radicals.

It is important to note that there are several low-lying excited states for dioxygen, probably the most important being the $^1\Delta$ state in which the electrons are paired in one of the π^*-orbitals, leaving the other vacant. Again, the $(\pi_x^*)^2$, $(\pi_y^*)^0$ description is not strictly correct for the gas-phase molecule, but I suggest that it may be suitable for $^1\Delta O_2$ in aqueous solution. This possibly significant suggestion stems from our knowledge of the asymmetric solvation of O_2^-, in which hydrogen bonding is directed towards the filled π^*-orbital only (cf. figure 11). I think some elements of this solvation would also occur for aqueous solutions of the $^1\Delta$ form of dioxygen. This would serve to distinguish between π_x and π_y. It might also have a measurable effect on the emission spectrum for $^1\Delta$ as it falls to the ground state. Such an effect could be important because this spectrum is used for identification of singlet dioxygen (Kanofsky 1983, 1984).

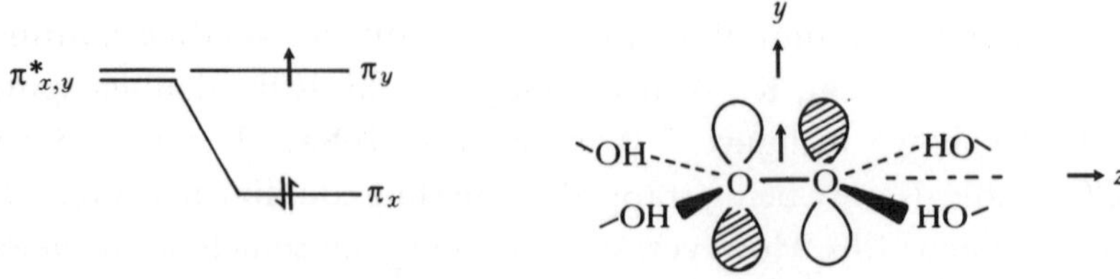

FIGURE 11. Effect of solvation (hydrogen bonding) on the π_x^*- and π_y^*-orbitals for the superoxide ion. Bonding is largely confined to the x–z plane.

The reactivity of $^1\Delta O_2$ is clearly different from that for the ground, $^3\Sigma$, state. The latter reacts as a bi-radical, and it is a good radical scavenger, whereas the former can react both as a nucleophile and as an electrophile. Because O_2 is thought to be produced in its $^1\Delta$ form in certain reactions, and because the spin-flip transition to the ground state is only weakly allowed, this reactivity difference may be important.

(*b*) *The superoxide ion*

Oxygen has a high electron affinity, the excess electrons being accommodated in the antibonding $\pi_{x,y}$ level. Again there is strong spin–orbit coupling, but for O_2^- in solution or in its salts the degeneracy of the π_x^*- and π_y^*-orbitals is lifted by hydrogen bonding or crystal-field effects, as indicated in figure 11. This has the effect of largely quenching orbital motion but, nevertheless, when a magnetic field, B, lies along the molecular axis (z) orbital motion is encouraged and increases linearly with field, thus increasing the rate of divergence of the spin-doublet levels (g_z) (figure 2). This facile circulation cannot occur for B_x or B_y so, in first order, g_x and g_y remain close to the normal free-spin value.

I stress that the magnitude of the down-field shift for g_z is not so much a property of the O_2^- radical as of the strength of the electric field governing Δ. This must be precise if a well defined z feature is to be observed in the solid-state e.s.r. spectrum. In the liquid phase there are two uncertainties. One is that the signal is located at the weighted average of the g values, and the other is that the formal g_z value will itself fluctuate as the solvation is modulated by the normal motionally induced changes in solvation. The result is a line too broad for detection by e.s.r.

Some of these effects are nicely illustrated by our e.s.r. results for the growth of solvation for O_2^- radicals generated by radiation-induced electron addition at 4 K (Eastland & Symons 1977; Symons & Stephenson 1981). Initially, the oxygen molecules in solvents such as ethanol exist in solvent cages, being surrounded by ROH molecules hydrogen bonded to themselves, not to oxygen. When these glassy solutions are exposed to ionizing radiation at 4 K, electron addition occurs (as shown by a decreased yield of trapped electrons), but no e.s.r. signals characteristic of O_2^- are observable. As the temperature is raised, a broad, very low-field g_z feature appears, which narrows and moves to higher fields in discrete stages as the temperature is raised. At *ca.* 90 K this feature has become identical with that observed from normal frozen solutions of NaO_2 in ethanol. We conclude that we are observing the progressive addition of EtOH molecules to O_2^- by hydrogen bonding, the changes indicating that at equilibrium there are four such bonds. To explain the consequent splitting of the π_x^*- and π_y^*-orbitals, this solvation must be confined to a planar arrangement. These arguments have been strongly supported by recent electron spin-echo studies (Narayama & Kevan 1980).

The reactivity of the superoxide ion is relatively low. It forms stable salts, and shows little tendency to dimerize. Its reactivity is twofold: it is a one-electron donor, and it is a base (or nucleophile). Its electron-donating power is a function of solvation: the anion is strongly stabilized in aqueous solution, and hence will have a lower electron-donating power than in an aprotic environment.

Protonation occurs in water with a pK_a of *ca.* 4.9. This means that there will still be significant protonation in neutral solutions, and because the reactivity of $HO_2^{\cdot}$ is far greater than that of O_2^- in certain respects, some reactions exhibited by aqueous superoxide solutions may well be due to $HO_2^{\cdot}$.

The superoxide ion also acts as a nucleophile in some of its reactions, especially towards carbonyl groups, as in its reactions with esters, (7) (Galliani & Rindone 1981).

$$O_2^- + R-C(=O)OR' \rightleftharpoons R-C(OO^{\cdot})(OR)-O^-. \qquad (7)$$

Subsequent reactions are not yet well established (Johlman *et al.* 1983), but such nucleophilic addition has received strong support from the observation of a radical having g values close to those for normal $ROO^{\cdot}$ radicals (see below) when superoxide ions react with dimethylformamide (Green *et al.* 1979; Symons *et al.* 1980). It seems that subsequent reactions are less facile in anhydrous solutions: addition of water leads to re-formation of solvated O_2^- ions, rather than to break-down products as with the ester adducts.

Note that in this addition the e.s.r. results support the concept that O_2^- adds to the C=O group via electron-pair donation; as a nucleophile not a radical. So the SOMO remains a π^* oxygen orbital just as it does on protonation. In contrast, addition of O_2^- to the C=N double bond of a nitrone involves the SOMO of the anion, so the negative charge is retained by oxygen and the SOMO becomes the normal N—O π^*-orbital (8).

$$(H)(R')C=N(O)-R + O_2^- \rightarrow H-C(OO^-)(R')-\dot{N}(O)-R. \qquad (8)$$

This dichotomy in reactivity appears to be well defined. However, it may not always be clear which mechanistic role is involved. Thus, for example, in its addition to $>C=O$ to give $>C(OO^{\cdot})(O^-)$ adducts, the obvious route involves direct nucleophilic attack (7). However, the alternative radical mode, comparable with (8), gives $>C(OO^-)(O^{\cdot})$, but this would surely undergo internal electron-transfer to give the former species very rapidly. Thus the distinction is somewhat esoteric. However, for bimolecular displacements, such as with alkyl halides, the distinction should be clear because as a nucleophile the products would be $ROO^{\cdot}$ radicals and halide ions, whereas as a radical they would be RO_2^- anions and halogen atoms. The former would be relatively more favoured for chlorides, but the latter would be relatively more favoured for iodides.

Presumably the radical $O^{-\cdot}$ should show a similar dichotomy in its reactivity. Both mechanisms lead to the same product for addition to $>C=O$ units, but the products would be distinct for displacement reactions with alkyl halides.

(*c*) $HO_2^{\cdot}$ *and* $RO_2^{\cdot}$ *radicals*

Apart from the proton splitting for the former, these two radicals have very similar e.s.r. spectra (table 1). The $HO_2^{\cdot}$ radical has never been detected in fluid solution, but is readily

Table 1. Typical e.s.r. parameters for various inorganic radicals

radical	matrix	*g* values				nucleus	hyperfine coupling constants/G[a]			
		g_x	g_y	g_z	g_{av}		A_x	A_y	A_z	A_{iso}
$O^{-\cdot}$	D_2O/NaOD[b]	2.070	2.002	2.002	2.025	—	—	—	—	—
$OH^{\cdot}$	H_2O[c]	2.059	2.009	2.0027	—	^{1}H	0	45	28.5	24.5
O_2^-	MeOH + H_2O[d]	2.078	2.000	2.000	2.026	—	—	—	—	—
$HO_2^{\cdot}$	$H_2O + H_2O_2$[e]	2.0353	2.0086	2.0042	2.0160	^{1}H	13.8	3.6	15.7	11.0
$O_3^{-\cdot}$	$KClO_3$[f]	2.0025	2.0174	2.0113	2.0104	—	—	—	—	—
$Cl_2^{-\cdot}$	H_2O[g]	2.000	2.040	2.040	2.027	^{35}Cl	102	10	10	40.8
$ClOH^{-\cdot}$	H_2O[gh]	2.004	2.017	2.017	2.013	^{35}Cl	59	−16	−16	9
$ClOO^{\cdot}$	$KClO_3$[f]	1.9983	2.0017	2.0130	2.0043	^{35}Cl	5.3	7.2	14.9	9.1
$ClO^{\cdot}$	CO_2[i]	1.889	1.899	2.66	2.149	^{35}Cl	30	*ca.* 0	*ca.* 0	*ca.* 10
$^{\cdot}NO_2$	H_2O[j]	2.0066	1.9920	2.0022	2.0003	^{14}N	50.6	49.84	70.21	56.88
$^{\cdot}SO_2^-$	H_2O[kl]	—	—	—	2.0057	^{33}S	—	—	—	14.2
		2.0026	2.0026	2.0026	2.0026	^{33}S	153	115.5	115.5	128
$^{\cdot}SO_3^-$	H_2O[m]	—	—	—	2.0028					

[a] $1\ G = 10^{-4}\ T$.
[b] Blandamer *et al.* 1964.
[c] Brivati *et al.* 1967.
[d] Symons & Stephenson 1981.
[e] Wyard *et al.* 1968.
[f] Eachus *et al.* 1968.
[g] Catton & Symons 1969.
[h] Ginns & Symons 1972.
[i] Trainer *et al.* 1983.
[j] Atkins *et al.* 1962.
[k] Atkins *et al.* 1964.
[l] Chantry *et al.* 1962.
[m] Ozawa & Kwan 1983.

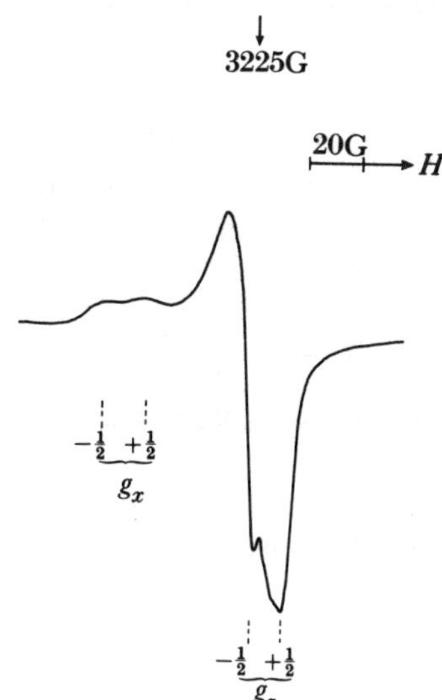

Figure 12. First derivative X-band e.s.r. spectrum for $H_2O + H_2O_2$ after exposure to ^{60}CO γ-rays at 77 K and on cooling to *ca.* 150 K, showing features assigned to $HO_2^{\cdot}$ radicals.

characterized in glassy or crystalline media (figure 12) (Catton & Symons 1969*b*). However, $RO_2^{\cdot}$ radicals such as $Me_3COO^{\cdot}$ are readily detectable in fluid solution although the spectra are always broad unresolved singlets (Ingold 1969; Kalyanaraman *et al.* 1983). Two factors contribute to this width: at low temperatures, uncertainty in the averaging of the *g* anisotropy is the major factor, while at higher temperatures incipient rotational quantization, usually called 'spin rotation', causes an alternative broadening. Thus there are optimum temperatures at which fairly narrow lines can be produced. For $HO_2^{\cdot}$ in water, strong hydrogen bonding will inhibit spin-rotation broadening, but fluctuations in solvation will undoubtedly contribute to line-broadening. However, the g values are now primarily determined by the structure of the radicals, which are markedly bent. Thus one of the degenerate π-orbitals of O_2^- is involved in covalent bonding to H or R, the SOMO being strongly confined to the other π^*-orbital. Even so, coupling can still occur, as evidenced by the well defined shift in g_z (*ca.* 2.034). Unfortunately, hyperfine coupling to protons in the alkyl units is not resolved, so it is not possible to identify specific $RO_2^{\cdot}$ radicals, though, in principle, this might be achieved by using ENDOR or electron spin–echo techniques.

As stressed above, $HO_2^{\cdot}$ radicals are of importance in solutions containing O_2^- ions. Their reactivity, apart from being weak acids, is centred largely on the unpaired electron. They are less reactive than $OH^{\cdot}$ radicals, because the SOMO is delocalized, but they do dimerize reversibly to give unstable tetroxides, RO_4R. They can add to double bonds, as, for example, in their addition to the C═N bonds of nitrones, their reactivity being far greater in this respect than that of O_2^-. Thus it may well be $HO_2^{\cdot}$ that is spin trapped by nitrones rather than O_2^- in neutral aqueous solutions.

A major switch in redox behaviour occurs when O_2^- is protonated, because $HO_2^{\cdot}$ is a very poor electron donor, but a good electron acceptor. The resulting HO_2^- (or RO_2^-) anions are quite strongly basic, and protonate reversibly to give H_2O_2 or ROOH molecules. This again greatly modifies the redox properties, because HO_2^- is a good electron donor, while H_2O_2 is a good acceptor (9).

$$H_2O_2 + e^- \rightarrow OH^{\cdot} + OH^-. \tag{9}$$

This alternation in electron-donating and accepting behaviour on proton gain or loss is, of course, of vital importance in the chain of events connecting O_2 and H_2O.

These radicals ($HO_2^{\cdot}$ and $RO_2^{\cdot}$) are also able to abstract hydrogen from C—H bonds, particularly if these are activated. A very important example is step (10) in the reaction chain that leads to efficient lipid peroxidation:

$$R^{\cdot} + O_2 \rightleftharpoons RO_2^{\cdot}, \tag{10}$$

$$RO_2^{\cdot} + \text{—CH=CH—CH}_2\text{—} \rightarrow RO_2H + \text{—}\overline{\dot{C}H\text{—}CH\text{—}\dot{C}H}\text{—}. \tag{11}$$

Reaction (10), which occurs with high efficiency, is an important example of the reactivity of dioxygen, and is of major significance because it converts a relatively stable radical into one with greater reactivity.

4. The hydroxyl radical

In many ways, this is the most important of the oxygen-centred radicals, and it is also probably the radical that is most frequently misassigned. Formation from water requires very high energy, and is normally accomplished by ionizing radiation. It is probably the most reactive of the radical species formed in the radiolysis of aqueous solutions, the other major species being solvated electrons and hydrogen atoms (12)–(14).

$$H_2O \rightarrow H_2O^+ + e^-, \tag{12}$$

$$H_2O + H_2O^+ \rightarrow OH^{\cdot} + H_3O^+, \tag{13}$$

$$e^- + H_3O^+ \rightarrow H^{\cdot} + H_2O. \tag{14}$$

It can be formed chemically from OH^- ions, especially if these are only weakly solvated, when they become good electron donors, as stressed above. Generation from H_2O_2 is the most common chemical method, with the use of ultraviolet light, which ruptures the O—O bond, or certain transition metal complexes which are good electron donors, such as Fe^{II} or Ti^{III} (15).

$$H_2O_2 + Fe^{2+}_{aq} \rightleftharpoons FeOH^{2+}_{aq} + OH^{\cdot}. \quad (15)$$

Many ferrous complexes are equally able to generate $OH^{\cdot}$ radicals from H_2O_2, so this combination can be biologically dangerous.

The pK_a of $OH^{\cdot}$ radicals is close to that of water, so $O^{-\cdot}$ anions are unlikely to be of importance in biochemistry. (The latter are formed readily in the radiolysis of aqueous alkaline glasses, and have an e.s.r. spectrum again determined by the nature of the solvation of the ion (Blandamer *et al.* 1964; Symons 1982*a*, *b*). One important chemical property is their ability to react with dioxygen to form the ozonide radical ion, $O_3^{-\cdot}$ (Atkins *et al.* 1962), which, because of its relatively high stability, may be of some importance.)

The SOMO for $OH^{\cdot}$ is similar to that for O_2^-, and again, e.s.r. spectra are controlled by hydrogen bonding for aqueous systems. When formed by radiolysis of ice crystals, well defined e.s.r. spectra are obtained (Symons 1982*a*, *b*; Brivati *et al.* 1967), with the expected large positive shift in g_z, and a large hyperfine coupling to the proton ($A_{iso} \approx (-)$ 27 G). However, in frozen glassy systems this 'parallel' (g_z) feature is broadened, often so strongly that it is completely lost (Riederer *et al.* 1983). This result is important because it underlines the control exerted by solvation. As with O_2^-, and for the same reasons, no spectrum for $OH^{\cdot}$ radicals in the liquid phase has ever been detected.

It is therefore helpful to use a spin trap to establish the presence of $OH^{\cdot}$ radicals in solution, especially as the rapid-freeze technique has not proved satisfactory. One major problem is the high reactivity of $OH^{\cdot}$ radicals. They are expected to react indiscriminately at close to diffusion-controlled rates, so high concentrations of traps are required to compete with other substrates. Furthermore, these secondary radicals are expected to react with the traps to give other nitroxide radicals whose spectra may obscure that for the $OH^{\cdot}$ adducts. Fortunately, if $OH^{\cdot}$ radicals react with large biopolymers, the resulting bio-radicals may not react efficiently with the traps, and the adducts, once formed, may give very broad-lined spectra because of slow tumbling rates. If several nitroxide radicals are formed concurrently, and are sufficiently stable, they can be separated by using high-pressure liquid chromatography (Moriya *et al.* 1982; Makino & Riesz 1982), greatly facilitating spectral analysis.

There are many examples of the use of spin traps to establish the presence of $OH^{\cdot}$ radicals, and some attempts at quantification. A few selected examples are given in the references (Finkelstein *et al.* 1980; Marriott *et al.* 1980; Green *et al.* 1979; Floyd 1982; Berlin & Haseltine 1981; Komiyama *et al.* 1982). It seems certain that the best trap so far investigated is DMPO (scheme 1), whose hydroxy adduct has a distinctive four-line spectrum (figure 8) and is relatively long-lived.

There are, however, at least two potential dangers in concluding that detection of this adduct provides clear proof that $OH^{\cdot}$ radicals were the precursors. One is that nitrones have relatively low ionization potentials, and might become oxidized to the radical cations. These would be expected to react rapidly with water to give the —OH adduct. Another is that certain hydroxylating agents are thought to be capable of hydroxylating the traps without actual release of free $OH^{\cdot}$ radicals (Lown & Chen 1981) and further chemical tests are needed to distinguish between such possibilities. Thus pitfalls abound: nevertheless, there can be little doubt that in several cases free $OH^{\cdot}$ radicals are genuinely involved and can be identified by spin trapping. However, I am personally dubious about some examples in which relatively large, delocalized radicals are invoked as being precursors to the formation of free $OH^{\cdot}$ radicals (Floyd 1982; Berlin & Haseltine 1981; Komiyama *et al.* 1982).

The $OH^{\cdot}$ radical is often described as being 'electrophilic'. Indeed, it is common practice to classify radicals as 'nucleophilic' or 'electrophilic'. This does not relate to the contrast outlined above, in which radicals such as O_2^- can react as a radical or as a nucleophile. Rather, it relates to the tendency for the radical to donate its unpaired electron to the substrate, or to accept an electron from the substrate, as the transition state is approached. The situation is shown schematically in figure 13, for two extreme types of radicals. One, which I prefer to classify as an electron-donor radical, tends to transfer its electron into the LUMO (lowest unoccupied, or virtual molecular orbital) while the other, an electron-acceptor radical, tends to accept an electron-pair from the HOMO (highest occupied molecular orbital) of the substrate while donating its own electron simultaneously, as shown in figure 13*b*. Radicals such as methyl commonly fall between these extremes, while $OH^{\cdot}$ radicals, or radicals such as $^{\cdot}CF_3$, can be classified as electron-acceptor species. In contrast, the radical $^{\cdot}BH_3^-$ is a good electron-donor radical.

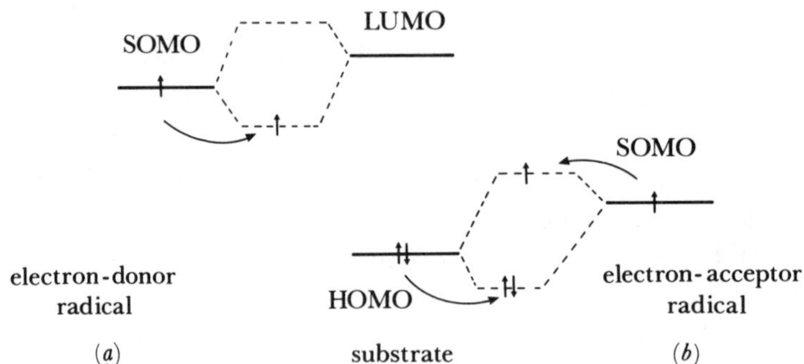

FIGURE 13. Schematic representation of the interaction between the LUMO of a substrate with an electron-donor radical (*a*), and the HOMO of a substrate with an electron-acceptor radical (*b*).

The hydroxyl radical is clearly an 'electron-acceptor' radical. In one limit, thought by some to make a major contribution towards its reactivity, a reaction such as hydrogen-atom abstraction can be thought of as proceeding via an initial electron transfer, followed by a very rapid proton transfer, as in (16).

$$OH^{\cdot}+A{-}H \overset{(a)}{\rightleftharpoons} OH^{-}+{}^{\cdot}AH^{+} \overset{(b)}{\rightarrow} H_2O+A^{\cdot}. \qquad (16)$$

Both (16*a*) and (16*b*) are often favourable processes, but to what extent there is any real division between stages (*a*) and (*b*) is unknown.

5. One-electron and two-electron processes

It has been realized for very many years that reactions such as the conversion of dioxygen into hydrogen peroxide may proceed by one- or two-electron routes. Biomolecules are well equipped to react by either route. For example, haem-iron can be present in the two-, three-, or four-valent states, and molybdenum in xanthine oxidase moves from Mo^{VI} to Mo^{IV} without the *necessary* involvement of Mo^{V}. However, there are clearly occasions when reactive radicals, especially $OH^{\cdot}$, are useful for rapid, indiscriminate, attack on unwanted species. Thus both routes need to be available, and controls are needed which can call on one while suppressing the other, to suit demand.

6. Reactions with transition metal complexes

When $OH^{\cdot}$ radicals react directly with the metal centre, rather than with one of the ligands, they normally lose all radical character, the SOMO becoming part of the metal d-orbital system, as in (17).

$$Fe^{2+} + OH^{\cdot} \rightarrow Fe^{3+}—OH^{-}. \tag{17}$$

In contrast, reactions with O_2^- may result in a complex in which the SOMO remains largely on the ligand, which therefore retains considerable radical character. This depends intimately on orbital availability and on relative electronegativities. For example, e.s.r. studies have shown that for the $Fe^{III}–O_2^-$ haemoglobin derivative, the SOMO is extensively delocalized onto iron, whereas for the isoelectronic cobalt derivative it is largely confined to dioxygen (Symons & Petersen 1968; Dickinson & Chien 1973; Dickinson & Symons 1983).

7. Other inorganic radicals

(*a*) *Solvated electrons*

In general, excess electrons are accommodated in chemical systems within antibonding orbitals of electron-affinic molecules. However, if these are not immediately available, they may become 'solvated'. This means that they become quite strongly localized in a stationary orbital which is probably unique in having no *central* nucleus. These species are closely related to F-centres in alkali halide crystals, whose structures have been unambiguously defined by e.s.r. spectroscopy (Delbecq 1963). A first-order description of an F-centre can be simply viewed as one in which a single halide ion is removed from the crystal lattice, being replaced by an electron that remains quite strongly localized in an s-type orbital within the anion site (Claxton *et al.* 1966). Similarly, my view of an aquated electron is obtained by imagining the removal of, say, a chloride ion from water, and replacing it by an electron, leaving the local structure approximately unchanged (Symons 1976, 1981). This simple model has received extensive experimental support, but is not universally accepted (Golden & Tuttle 1978; Stradowskii & Hamill 1976).

It is unlikely that solvated electrons are ever involved in biological reactions except when exposed to ionizing radiation. They are not well characterized by e.s.r. spectroscopy, because rapid exchange processes completely remove all hyperfine coupling to solvent nuclei. They are, however, characterized by an intense visible absorption band, by which their reactions can be monitored in pulse-radiolysis studies.

(*b*) *Hydrogen atoms*

These are also not normally invoked as intermediates in biological reactions. As stressed above, they are an important intermediate in the radiolysis of water, and are well characterized in e.s.r. spectroscopy by a narrow 508 G doublet. There is an interesting chemical reaction that relates hydrogen atoms and electrons, namely (18) (Baxendale & Hughes 1958).

$$H^{\cdot} + OH^{-}_{solv} \rightleftharpoons H_2O + e^{-}_{solv}. \tag{18}$$

Although we have direct e.s.r. evidence in favour of this equilibrium (Symons & Zimmerman 1976), I still find it surprising!

(c) *Chlorine-centred radicals*

Hydroxyl radicals are able to form weak adducts with chloride ions (19) (Catton & Symons 1969*a*; Ginns & Symons 1972).

$$OH^{\cdot} + Cl^{-} \rightleftharpoons HO \dot{-} Cl^{-}. \tag{19}$$

These are classified as σ^* radicals because the SOMO is the antibonding O—Cl σ-orbital. I have no idea to what extent their formation may modify reactions of $OH^{\cdot}$ radicals in biological systems. They readily react with a second chloride ion to give the more stable Cl_2^- radical, which also has a σ^* SOMO (20).

$$Cl^{-} + HOCl^{-} \rightleftharpoons Cl \dot{-} Cl^{-} + OH^{-}. \tag{20}$$

Both are potential chlorine-atom donors and may chlorinate rather than hydroxylate. $HOCl^-$ is an electron donor, giving the biologically important HOCl molecule and ClO^- anion, as well as being an electron acceptor.

Electron donation by ClO^- may also be a reaction of some biological significance. Unfortunately, the $ClO^{\cdot}$ radical, like the isostructural O_2^- radical has a degenerate π^* SOMO and hence has never been detected in the liquid phase. I know of no spin-trapping work on $ClO^{\cdot}$ radicals. Addition could occur via Cl or O addition: the former should give a resolved quartet hyperfine coupling from chlorine, but the latter is more probable.

A radical formed by photolysis of chlorine dioxide in acidic glasses, originally thought to be $ClO^{\cdot}$ (Atkins *et al.* 1962) was later shown to be the rearranged dioxide, $ClOO^{\cdot}$ (Eachus *et al.* 1968). The e.s.r. spectrum for this radical exhibits a well defined hyperfine coupling to chlorine nuclei and only a small *g* value variation, showing that it is strongly bent, as expected. A species showing very large *g* shift and a greater coupling to chlorine has recently been identified as genuine $ClO^{\cdot}$ (Trainer *et al.* 1983).

(d) $NO^{\cdot}$ *and* $^{\cdot}NO_2$ *radicals*

Both $NO^{\cdot}$ and $^{\cdot}NO_2$ are relatively stable inorganic radicals which might be expected to form in biological systems under certain circumstances. The $NO^{\cdot}$ radical is of particular importance in its ability to act as a ligand, especially to haem iron. Thus, for example, deoxymyoglobin and haemoglobin readily react with NO to give the nitrosyl derivatives. As with dioxygen complexes, there is appreciable bonding, and the SOMO moves extensively into metal 3d-orbitals. However, the resulting e.s.r. spectra are characterized by triplet splitting from the ^{14}N nucleus [$I = 1$], showing that delocalization is extensive (Rein *et al.* 1972). It is interesting to note that e.s.r. signals from nitrosyl haemproteins were detected in early studies of tumours, when it was hoped that e.s.r. spectroscopy might be an important tool in studies of cancers. Initially, it was hoped that these signals established the importance of radicals (Vithayathil *et al.* 1965), but later the signals were recognized as being characteristic of the Fe—NO unit (Chiang *et al.* 1972), formed from fortuitously added nitrate or nitrite ions.

(e) *The* $^{\bullet}SO_2^-$ *and* $^{\bullet}SO_3^-$ *radicals*

The former radical is probably the reactive species in reactions of dithionites, because it is formed reversibly from $S_2O_4^{2-}$ in aqueous solutions at room temperature, giving rise to a well defined e.s.r. spectrum (Atkins *et al.* 1964). The latter radical has been implicated in reactions involving sulphite ions (or HSO_3^-) and prostaglandin synthase (Mottley *et al.* 1982). In this work, it was detected indirectly by using a DMPO trap. In fact, $^{\bullet}SO_3^-$ radicals have a remarkably narrow solid-state e.s.r. feature, and give rise to a narrow liquid-phase singlet that is readily detectable (Ozawa & Kwan 1983). Hence, this is a radical that might well prove to be directly detectable by e.s.r. spectroscopy if it were formed in significant yields.

8. Comparison of oxygen- and sulphur-centred radicals

Although purely inorganic sulphur analogues, such as $S_2^{-\bullet}$, $HS_2^{\bullet}$ and $HS^{\bullet}$ are not, so far as I am aware, thought to be of importance in most biological systems, various alkyl derivatives are almost certainly involved. It may therefore be of interest to draw a few comparisons between the two groups.

First, there is a major difference in reactivity between $OH^{\bullet}$ and $^{\bullet}SH$ or $^{\bullet}OR$ and $^{\bullet}SR$ radicals. Returning to reaction (16), the $^{\bullet}SH$ radical has a lower electron affinity and SH^- has a lower proton affinity than OH^-. Hence $^{\bullet}SH$ or $^{\bullet}SR$ radicals are generally less reactive than $OH^{\bullet}$ radicals. Conversely, RSH molecules are good hydrogen-atom donors, and it seems probable that they play an important 'healing' role in biological systems where biomolecules, especially DNA, have suffered attack by radicals. Not only are $RS^{\bullet}$ radicals relatively stable, they also have a mechanism of protecting themselves that is not apparently available to oxygen-centred radicals.

(a) σ* *radicals*

A major contrast between the two groups is the relative prevalence of σ^* sulphur species such as $RS\dot{-}SR^-$, $R_2S\dot{-}SR_2^+$ and probably $RS\dot{-}SR_2$ radicals. All these, like the radicals $HO\dot{-}Cl^-$ and $Cl\dot{-}Cl^-$ mentioned above, have σ^* SOMOs. This situation is illustrated in figure 14. (In this figure, the normal strong σ-bonding is contrasted with the weakening that results on electron gain into the σ^*-orbital, and with loss of one electron from the σ-orbital. Although radicals of the latter type are known (Wang & Williams 1981; Symons 1982), I know of no examples in which they are involved in biological systems.)

It is curious that the analogous oxygen-centred σ^* radicals do not appear to have any chemical significance. Some current theories support the idea that oxygen-centred σ^* radicals are likely to be far less stable than corresponding sulphur-centred σ^* radicals, but others suggest the reverse. This means that there are not likely to be any obvious explanations.

These σ^* radicals can be formed by oxidative or reductive mechanisms, as in the reactions (21)+(22) and (23).

$$RS^- \rightarrow RS^{\bullet} + e^-, \tag{21}$$

$$RS^{\bullet} + RS^- \rightleftharpoons RS\dot{-}SR^-, \tag{22}$$

$$RS{-}SR + e^- \rightarrow RS\dot{-}SR^-, \tag{23}$$

$$RS^{\bullet} + R_2S \rightleftharpoons RS\dot{-}SR_2. \tag{24}$$

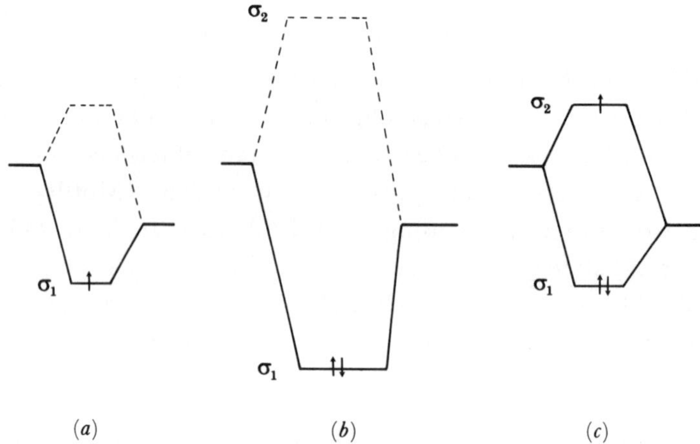

FIGURE 14. Qualitative energy level diagram for σ_1^1 cations (*a*) and $\sigma_1^2\sigma_2^1$ anion-radicals (*c*) relative to that for the parent molecule (*b*).

Comparing reactions of OH$^{\cdot}$ or RO$^{\cdot}$ radicals with those of RS$^{\cdot}$ radicals, the former readily extract hydrogen from organic compounds, whereas the latter tend to react as in (22) or (24) to form σ^* radicals, which are greatly stabilized, and are unlikely to cause damage to biomolecules.

There is considerable controversy regarding the significance of radicals formulated as $RS\dot{-}SR_2$ above (Symons 1974; Nelson *et al.* 1977, 1978). The e.s.r. spectra of intermediates widely encountered in sulphur-radical studies, characterized by features at $g \approx 2.060$, $g \approx 2.025$ and $g \approx 2.00$, were originally thought to be a result of RS$^{\cdot}$ radicals (Symons 1974). However, it is now clear that this contains two inequivalent sulphur nuclei (Hadley & Gordy 1974). The idea that this species is the expected $RS\dot{-}SR_2$ σ^* radical is attractive mechanistically (Symons 1974; Nelson *et al.* 1977, 1978), but there are strong reasons for believing that RSS$^{\cdot}$ radicals may be responsible (Hadley & Gordy 1974). The current situation has recently been discussed (Symons 1985). It may be that by a quirk of nature, both types of radical have very similar spectra.

Nature profits from these differences by using RSH compounds to heal the wounds caused by radicals such as OH$^{\cdot}$. The RSH molecules act as hydrogen donors, and provided the RS$^{\cdot}$ radicals are stabilized as in (22) or (24) before they react with oxygen, trouble is averted. Reaction with oxygen, to give RS—OO$^{\cdot}$ radicals, is thought to be of significance although present evidence is not compelling (Copeland 1975). If it does occur, it converts a relatively stable radical into a more reactive species which can no longer acquire added stability by forming a σ^* species.

I thank Professor R. O. C. Norman and Dr H. A. O. Hill for inviting me to make this contribution.

REFERENCES

Atherton, N. M. 1973 *Electron spin resonance*. London: Halsted Press.
Atkins, P. W., Brivati, J. A., Keen, N., Symons, M. C. R. & Trevalion, P. A. 1962 *J. chem. Soc.*, pp. 4785–4789.
Atkins, P. W., Horsfield, A. & Symons, M. C. R. 1964 *J. chem. Soc.*, pp. 5220–5225.
Atkins, P. W. & Symons, M. C. R. 1967 *The structure of inorganic radicals*. Amsterdam: Elsevier.
Baxendale, J. H., Evans, M. G. & Park, M. 1946 *Trans. Faraday Soc.* **42**, 155–166.
Baxendale, J. H. & Hughes, G. 1958 *Z. phys. Chem.* (*Frankf. Ausg.*) **14**, 323–326.

Berlin, V. & Haseltine, W. A. 1981 *J. biol. Chem.* **256**, 4747–4758.
Blandamer, M. J., Shields, L. & Symons, M. C. R. 1964 *J. chem. Soc.*, pp. 4352–4357.
Brivati, J. A., Symons, M. C. R., Tinling, D. J. A., Wardale, H. W. & Williams, D. O. 1967 *Trans. Faraday Soc.* **63**, 2112–2119.
Carrington, A. & Symons, M. C. R. 1963 *Chem. Rev.* **63**, 443–460.
Catton, R. C. & Symons, M. C. R. 1969*a* *J. chem. Soc.* A, pp. 446–451.
Catton, R. C. & Symons, M. C. R. 1969*b* *J. chem. Soc.* A, pp. 1393–1395.
Chalfont, G. R., Perkins, M. J. & Horsfield, A. 1968 *J. Am. chem. Soc.* **90**, 7141–7144.
Chiang, R. W., Woolum, J. C. & Commoner, B. 1972 *Biochim. biophys. Acta* **257**, 452–460.
Chien, J. C. W. 1969 *J. chem. Phys.* **51**, 4220–4230.
Claxton, T. A., Greenslade, D. J., Root, K. D. J. & Symons, M. C. R. 1966 *Trans. Faraday Soc.* **62**, 2050–2056.
Copeland, E. S. 1975 *J. magn. Reson.* **20**, 124–129.
Delbecq, C. J. 1963 *Z. Phys.* **171**, 560–566.
Dickinson, L. C. & Chien, J. C. W. 1973 *Biochem. biophys. Res. Commun.* **51**, 587–590.
Dickinson, L. C. & Symons, M. C. R. 1983 *Chem. Soc. Rev.* **12**, 387–414.
Eachus, R. S., Edwards, P. R., Subramanian, S. & Symons, M. C. R. 1968 *J. chem. Soc.* A, pp. 1704–1711.
Eastland, G. W. & Symons, M. C. R. 1977 *J. phys. Chem.* **81**, 1502–1510.
Finkelstein, E., Rosen, G. M. & Rauckman, E. J. 1980 *Archs Biochem. Biophys.* **200**, 1–9.
Floyd, R. A. 1982 *Can. J. Chem.* **60**, 1577–1585.
Galliani, G. & Rindone, B. 1981 *Tetrahedron* **37**, 2313–2320.
Ginns, I. S. & Symons, M. C. R. 1972 *J. chem. Soc. Dalton Trans.*, pp. 143–147.
Golden, S. & Tuttle, T. R. 1978 *J. phys. Chem.* **82**, 944–947.
Green, M. R., Hill, H. A. O., Okolow-Zubkowska, M. J. & Segal, A. W. 1979 *FEBS Lett.* **100**, 23–25.
Green, M. R., Hill, H. A. O. & Turner, D. R. 1979 *FEBS Lett.* **103**, 176–180.
Hadley, J. H. & Gordy, W. 1974 *Proc. natn Acad. Sci. U.S.A.* **71**, 3106–3110.
Ingold, C. K. 1953 *Structure and mechanism in organic chemistry*. London: Bell & Sons Ltd.
Ingold, K. U. 1969 *Acct. chem. Res.* **2**, 1–9.
Janzen, E. G. 1971 *Acct. chem. Res.* **4**, 31–43.
Janzen, E. G. & Blackburn, B. J. 1968 *J. Am. chem. Soc.* **90**, 5909–5912.
Johlman, C. L., White, R. L., Sawyer, D. T. & Wilkins, C. L. 1983 *J. Am. chem. Soc.* **105**, 2091–2092.
Kalyanaraman, B., Mottley, C. & Mason, R. P. 1983 *J. biol. Chem.* **258**, 3855–3858.
Kanofsky, J. R. 1983 *J. biol. Chem.* **258**, 5991–5993.
Kanofsky, J. R. 1984 *J. Am. chem. Soc.* **106**, 4277–4278.
Komiyama, T., Kuchi, T. K. & Sugiura, Y. 1982 *Biochem. Pharmac.* **31**, 3651–3655.
Lindsay, D. M., Herschbach, D. R. & Kwiram, A. L. 1983 *J. phys. Chem.* **87**, 2113–2120.
Lott, K. A. K. & Symons, M. C. R. 1960 *Discuss. Faraday Soc.* **29**, 205–216.
Lown, J. W. & Chen, H. H. 1981 *Can. J. Chem.* **59**, 390–394.
Makino, K. & Riesz, P. 1982 *Can. J. Chem.* **60**, 1480–1488.
Marriott, P. R., Perkins, M. J. & Griller, D. 1980 *Can. J. Chem.* **58**, 803–807.
Moriya, F., Makino, K., Suzuki, N., Rokushika, S. & Hatano, H. 1982 *J. Am. chem. Soc.* **104**, 830–835.
Mottley, C., Mason, R. P., Chignell, C. F., Sivarajah, K. & Eling, T. E. 1982 *J. biol. Chem.* **257**, 5050–5056.
Narayana, M. & Kevan, L. 1980 *J. chem. Phys.* **72**, 2891–2894.
Nelson, D. J., Petersen, R. L. & Symons, M. C. R. 1977 *J. chem. Soc. Perkin Trans.* II, pp. 2005–2015.
Nelson, D. J., Petersen, R. L. & Symons, M. C. R. 1978 *J. chem. Soc. Perkin Trans.* II, pp. 225–231.
Ozawa, T. & Kwan, T. 1983 *J. chem. Soc. chem. Commun.*, pp. 1295–1296.
Rein, H., Ristan, O. & Scheks, W. 1972 *FEBS Lett.* **24**, 24–27.
Riederer, H., Hüttermann, J., Boon, P. J. & Symons, M. C. R. 1983 *J. magn. Reson.* **54**, 54–66.
Stradowskii, Cz. & Hamill, W. H. 1976 *J. phys. Chem.* **80**, 1054–1058.
Symons, M. C. R. 1953*a* *Research VI*, pp. 5S–8S.
Symons, M. C. R. 1953*b* *J. chem. Soc.*, pp. 3956–3965.
Symons, M. C. R. 1954 *J. chem. Soc.*, pp. 3676–3679.
Symons, M. C. R. 1974 *J. chem. Soc. Perkin Trans.* II, pp. 1618–1622.
Symons, M. C. R. 1976 *Chem. Soc. Rev.* **5**, 337–358.
Symons, M. C. R. 1978 *Chemical and biochemical aspects of electron spin resonance spectroscopy*. Berkshire: Van Nostrand Reinhold Co. Ltd.
Symons, M. C. R. 1981 *Radiat. Phys. Chem.* **17**, 425–429.
Symons, M. C. R. 1982*a* *J. chem. Soc. chem. Commun.*, pp. 869–871.
Symons, M. C. R. 1982*b* *J. chem. Soc. Faraday Trans.* I **78**, 1953–1959.
Symons, M. C. R. 1985 *Int. J. radiat. Oncology*. (In the press.)
Symons, M. C. R., Albano, E., Slater, T. F. & Tomasi, A. 1982 *J. chem. Soc. Faraday Trans.* I **78**, 2205–2214.
Symons, M. C. R., Eastland, G. W. & Denny, L. R. 1980 *J. chem. Soc. Faraday Trans.* I **76**, 1868–1874.
Symons, M. C. R. & Petersen, R. L. 1978 *Proc. R. Soc. Lond.* B **201**, 285–300.
Symons, M. C. R. & Stephenson, J. M. 1981 *J. chem. Soc. Faraday Trans.* I **77**, 1579–1583.

Symons, M. C. R. & Zimmerman, D. N. 1976 *Int. J. radiat. Phys. Chem.* **8**, 395–396.
Trainer, M., Helten, M. & Knapska, D. 1983 *J. chem. Phys.* **79**, 3648–3655.
Vithayathil, A. J., Ternberg, J. L. & Commoner, B. 1965 *Nature, Lond.* **207**, 1246–1249.
Wang, J. T. & Williams, F. 1981 *J. chem. Soc. chem. Commun.*, pp. 176–178.
Wertz, J. F. & Bolton, J. R. 1972 *Electron spin resonance*. New York: McGraw-Hill.

Discussion

H. Sies (*Institut für Physiologische Chemie I, Universität Dusseldorf, F.R.G.*). I was interested by Dr Symons's remark about the different chemical reactivity of singlet molecular oxygen in aprotic solvents, and that singlet oxygen might be more soluble in non-aqueous phases than triplet oxygen. Would he care to speculate on the biological significance?

M. C. R. Symons. My comment is purely speculative and centres on the idea that solvated singlet O_2 might have an asymmetric distribution of charge with, say, $\delta+$ in π_x and $\delta-$ in π_y. Thus, singlet O_2 may be better solvated by water than normal triplet O_2. Thus, in fact, I am suggesting that the singlet form should be more soluble in the aqueous phase than triplet oxygen.

I am afraid I know of no clear examples of this predicted situation as yet.

Phil. Trans. R. Soc. Lond. B **311**, 473–482 (1985)
Printed in Great Britain

Fast kinetic studies of dioxygen-derived species and their metal complexes

BY B. H. J. BIELSKI

Chemistry Department, Brookhaven National Laboratory, Upton, New York 11973, *U.S.A.*

The role of metals in the reactivity of HO_2/O_2^- with compounds of biological interest is discussed. A scheme that illustrates the various reactions that a transition metal complex can undergo when reacting with HO_2/O_2^- is presented in terms of ligand and pH effects.

The decomposition of hydrogen peroxide catalysed by ferrous ion is reviewed in terms of new rate data for the reactions of ferric ion with perhydroxyl (HO_2) and superoxide (O_2^-) radicals. The new results support a mechanism proposed by Barb and his coworkers (W. G. Barb, J. H. Baxendale, P. George & K. R. Hargrave, *Trans. Faraday Soc.* **47**, 462–500 (1951)) and negates the occurrence of the Haber–Weiss reaction in this system.

In the presence of Mn^{II} complexes, O_2^- reacts to form MnO_2^+ transients and Mn^{III} complexes. Their reactivities with ascorbate, Trolox (6-hydroxy-2,5,7,8-tetramethylchroman-2-carboxylic acid) and NADH–NADPH is discussed.

INTRODUCTION

While the chemical and physical properties of the superoxide radical (O_2^-) and its conjugate acid, the perhydroxyl radical (HO_2), were first studied solely to elucidate their roles in the radiolysis of oxygen-containing aqueous solutions, the discovery in the late 1960s that O_2^- plays an important role in living systems (McCord & Fridovich 1969) resulted in renewed research efforts and a shift towards the investigation of the reactivity of these oxy species with biological compounds. A major question that arose at the time concerned the nature and origin of various deleterious effects observed when O_2^- and H_2O_2 are present together in a biological system. The suppression of such effects by addition of catalase, which destroys hydrogen peroxide as follows:

$$2\,H_2O_2 \xrightarrow{\text{catalase}} 2\,H_2O + O_2, \tag{1}$$

or superoxide dismutase (SOD), a class of enzymes that accelerates the disproportionation of superoxide radicals,

$$2\,O_2^- + 2\,H_2O \xrightarrow{\text{SOD}} H_2O_2 + 2\,OH^- + O_2, \tag{2}$$

or both catalase and SOD led to the belief that powerful oxidizing species such as the hydroxyl radical (OH) or singlet molecular oxygen ($^1O_2$) were generated from peroxide and O_2^- by the so-called 'Haber–Weiss' reaction (Haber & Weiss 1934)

$$HO_2 + H_2O_2 \rightarrow {}^{\cdot}OH + H_2O + O_2 \tag{3}$$

and by the disproportionation of HO_2/O_2^- respectively. A number of reported rate measurements indicate a very low reactivity between HO_2/O_2^- and H_2O_2 (k values range from 10 to 10^{-4} $dm^3\,mol^{-1}\,s^{-1}$; Weinstein & Bielski (1979) and references therein), thus putting in question reaction (3) as an efficient source of OH radicals under physiological conditions. Recent studies on $^1O_2$ formation in the disproportionation of HO_2/O_2^- have also shown that this species is not formed (Arudi *et al.* 1984).

As the reported toxic effects of superoxide radicals are unquestionably real and because the Haber–Weiss reaction and singlet molecular oxygen production were shown to be ineffective only under extremely clean conditions, the role of metal catalysis became the overriding question. The possibility that metal catalysis may play a major role in the deleterious effects caused by HO_2/O_2^- would not be surprising in view of the ubiquity of metals in living cells.

The approach in this laboratory to these queries is twofold. The first aspect involves basic research on the reactivities of metals with HO_2/O_2^- in isolation, to determine if metal–HO_2/O_2^- complexes are formed and, if they are formed, to determine their chemical properties. Secondly, we have been measuring the reactivity of such metal–oxy species and metals themselves with compounds of biological interest.

Metal–HO_2/O_2^- complexes

Metal–dioxygen complexes can be formed by two well-documented methods (Buxton & Sellers 1977); the reaction between a reduced metal ion and molecular oxygen,

$$M^{n+} + e^-_{aq} \rightarrow M^{(n-1)+}, \tag{4}$$

$$M^{(n-1)+} + O_2 \rightarrow MO_2^{(n-1)+}; \tag{5}$$

or the reaction between a metal ion in a stable oxidation state and an active oxygen species,

$$HO_2 \rightleftharpoons O_2^- + H^+, \tag{6}$$

$$M^{n+} + HO_2 \rightarrow MOOH^{n+}, \tag{7}$$

$$M^{n+} + O_2^- \rightleftharpoons MO_2^{(n-1)+}. \tag{8}$$

The first thorough investigation of reaction (7) involved the reaction of HO_2 with ferrous ion near pH 1 (Jayson *et al.* 1969, 1973). Despite the complexity of the system, they succeeded in resolving the spectra of transient complexes and showed that their kinetic measurements were consistent with the following reaction mechanism (k values are given at 25 °C):

$$HO_2 + Fe^{2+} \rightarrow Fe^{3+}HO_2^-; \quad k_9 = 1.2 \times 10^6\ dm^3\ mol^{-1}\ s^{-1}, \tag{9}$$

$$Fe^{3+}HO_2^- + Fe^{2+} \rightleftharpoons Fe^{3+}HO_2^-Fe^{2+}; \quad K_{10} = 27 \pm 2\ M^{-1}, \tag{10}$$

$$Fe^{3+}HO_2^- \rightarrow Fe^{3+} + HO_2^-; \quad k_{11} = 1.8 \times 10^3\ s^{-1}, \tag{11}$$

$$Fe^{3+}HO_2^-Fe^{2+} \rightleftharpoons Fe^{3+} + Fe^{2+}HO_2^-; \quad k_{12} = 2.5 \times 10^4\ s^{-1}, \tag{12}$$

$$H^+ + Fe^{2+}HO_2^- \rightarrow Fe^{2+} + H_2O_2. \tag{13}$$

To investigate further the reactivity of HO_2/O_2^- radicals with the Fe^{II}/Fe^{III} system, experiments were designed (Rush & Bielski 1985) that used a 100 ms time window (Stuglik & Zagorski 1981) that exists under certain pulse-radiolytic conditions and represents the timespan between formation and onset of precipitation of Fe^{III}. Because of the rapid oxidation of Fe^{II}_{aq} near neutrality, oxygen-containing ferrous sulphate solutions were prepared by rapid mixing of anaerobic ferrous sulphate solutions with oxygen-saturated formate solutions. The solutions exiting from the jet mixer were transferred to the radiation cell, where they were pulse-irradiated within five seconds of mixing. The ferrous iron concentration and radiation doses were adjusted so that $[O_2^-] > [Fe^{2+}]$. By taking into consideration that Fe^{III} is present in a hydrolysed form at pH 7.2, where these experiments were performed, the mechanism can be described as follows:

$$H_2O \xrightarrow{O_2,\ HCOONa} O_2^-, \qquad (14)$$

$$Fe^{2+}_{aq} + O_2^- \xrightarrow{H^+} Fe(OH)^{2+}_{aq} + H_2O_2; \quad k_{15} = 10^7\ dm^3\ mol^{-1}\ s^{-1}, \text{ (Rush \& Bielski 1985)}, \qquad (15)$$

$$FeOH^{2+}_{aq} \rightleftharpoons Fe(OH)^+_{aq} + H^+; \quad pK = 3.3\ mol, \text{ (Schneider 1984)} \qquad (16)$$

$$Fe(OH)^+_{2aq} + O_2^- \longrightarrow Fe^{2+}_{aq} + O_2 + 2OH^-; \quad k_{17} = 1.5 \times 10^8\ dm^3\ mol^{-1}\ s^{-1} \text{ (Rush \& Bielski 1985)}. \qquad (17)$$

Competition studies of HO_2/O_2^- with Fe^{III}/Ce^{III} in the acid range led to the determination of the rates of reaction between HO_2/O_2^- and $FeSO_4^+$.

$$FeSO_4^+ + HO_2 \rightarrow Fe^{2+} + O_2 + HSO_4^-; \quad k \approx 10^3\ dm^3\ mol^{-1}\ s^{-1}, \text{ (Rush \& Bielski 1985)} \qquad (18)$$

$$FeSO_4^+ + O_2^- \rightarrow Fe^{2+} + O_2 + SO_4^{2-}; \quad k = 1.5 \times 10^8\ dm^3\ mol^{-1}\ s^{-1} \text{ (Rush \& Bielski 1985)}. \qquad (19)$$

The exceptionally high rate at which O_2^- reduced $FeSO_4^+$ makes this the predominant pathway by which the HO_2/O_2^- radical pair reacts with Fe^{III} even at pH = 0.

Haber & Weiss (1934) found in their classic study of the decomposition of hydrogen peroxide catalysed by ferrous ion that at high H_2O_2 concentrations the amount of peroxide consumed and oxygen formed was much greater than the amount of Fe^{II} oxidized. They suggested that this stoichiometric imbalance might result from reaction (3). Although they were unaware at the time that reaction (3) is insignificant, Barb *et al.* (1951) concluded after a careful reinvestigation of this system that the overall mechanism can be described by the following equations.

$$Fe^{2+} + H_2O_2 \rightarrow Fe^{3+} + OH^\cdot + OH^-; \quad k_{20} = 53.0 \pm 0.7\ dm^3\ mol^{-1}\ s^{-1}, \text{ (Barb et al. 1951)} \qquad (20)$$

$$Fe^{2+} + OH^\cdot \rightarrow Fe^{3+} + OH^-; \quad k_{15} = 3.2 \times 10^8\ dm^3\ mol^{-1}\ s^{-1}, \text{ (Stuglik \& Zagorski 1981)} \qquad (15)$$

$$H_2O_2 + OH^\cdot \rightarrow H_2O + HO_2; \quad k_{21} = 4.5 \times 10^7\ dm^3\ mol^{-1}\ s^{-1}, \text{ (Schwarz 1962)} \qquad (21)$$

$$Fe^{2+} + HO_2 \rightarrow Fe^{3+} + HO_2^-; \quad k_9 = 1.2 \times 10^6\ dm^3\ mol^{-1}\ s^{-1}, \text{ (Jayson et al. 1973)} \qquad (9)$$

$$Fe^{3+} + O_2^- \rightarrow Fe^{2+} + O_2; \quad k_{19} = 1.5 \times 10^8\ dm^3\ mol^{-1}\ s^{-1} \text{ (Rush \& Bielski 1985)}. \qquad (17)$$

This early study examined the competition of Fe^{II} with Fe^{III} for HO_2/O_2^-, and measured the ratio of these reactions over the pH range from 1 to 2.65. By using their ratio and the more recently obtained values of $K_{HO_2} = 1.6 \times 10^{-5}$ mol (Bielski *et al.* 1985) and $k_9 = 1.2 \times 10^6$ dm³ mol⁻¹ s⁻¹ (Jayson *et al.* 1975) the rate of reaction between Fe^{III} and O_2^- can be calculated. This gives $k_{19} = 2.5 \times 10^8$ cm^3 mol^{-1} s^{-1}. The close agreement of this rate to $k_{19} = 1.5 \times 10^8$ dm^3 mol^{-1} s^{-1} (see below) corroborates the correctness of their mechanism. It should be noted that if the ratio of H_2O_2/Fe^{II} is initially high, this mechanism accounts for the results observed by Haber & Weiss (1934) without involving the direct interaction of HO_2 with hydrogen peroxide. This leaves the Fenton reaction (reaction (20)) as the sole $OH^{\cdot}$ radical source in this system.

Studies of the reactions between Mn^{II} complexes and HO_2/O_2^- radicals (Cabelli & Bielski 1984*a*, *b* and references therein), taken in conjunction with the aforementioned Fe^{II} system, led to the formulation of an overall mechanism (scheme of reactions) that illustrates the various reactions a transition metal complex can undergo when reacting with HO_2 or O_2^-.

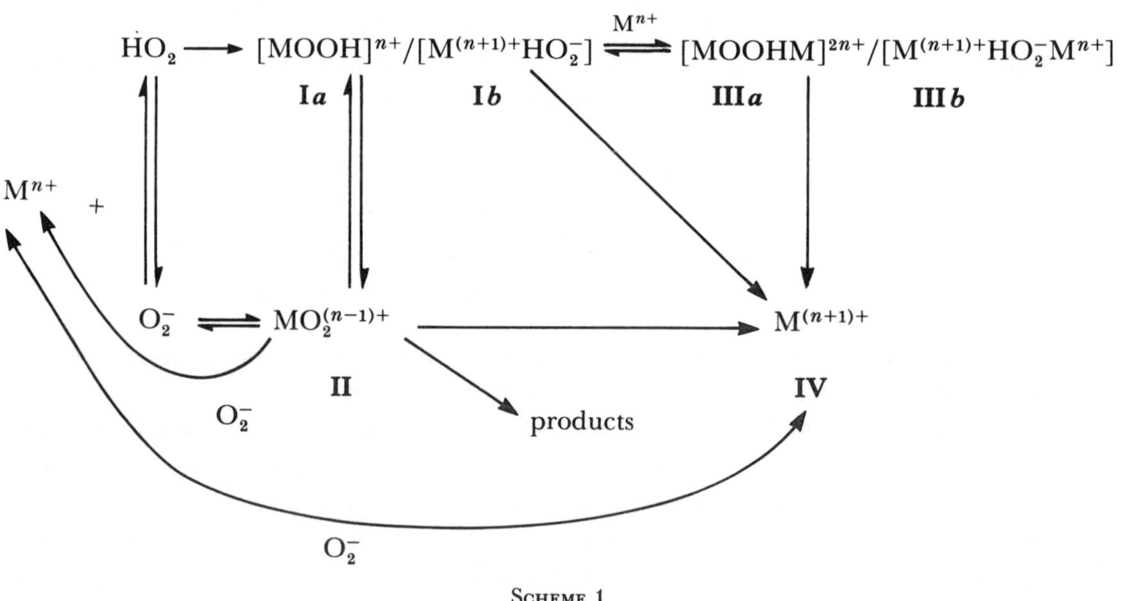

SCHEME 1

The various reactions and equilibria shown in scheme 1 represent competitive pathways. The reaction mechanism of a particular metal complex with HO_2/O_2^- is affected by the nature of its ligands and the degree by which one pathway predominates over another. Superimposed upon these effects is the pH effect, which not only controls the distribution of HO_2/O_2^- but also the form(s) of the metal complex in the medium.

As illustrated in scheme 1, the reactions can be divided into those involving predominantly protonated species and those involving $MO_2^{(n-1)+}$ and O_2^- itself. The $pK_{HO_2} = 4.8$ is well established and experimental evidence indicates that upon formation of metal–HO_2/O_2^- complexes a shift towards a lower pK occurs (p$K \approx 2.0$–3.5). The protonated complexes **I*a*/I*b*** and **III*a*/III*b*** have been discussed for both Mn^{II} and Fe^{II} reactions with HO_2. For Fe^{II}, intermediates were described as outer sphere complexes (**I*b*/III*b***) where the electron had already been transferred and the transients were Fe^{III}–peroxy species, a conclusion reached on the basis of spectral evidence (Jayson *et al.* 1969, 1973). Studies of HO_2 with Mn^{II}

complexes suggest, however, that although the reaction pathway is analogous to that of the Fe^{II} system, the transients appear to be inner sphere perhydroxyl complexes, again on the basis of spectral evidence (Cabelli & Bielski 1984*a*, *b*). The species formed in the more physiologically relevant pH range include the unprotonated metal–O_2^- complexes. Such $MO_2^{(n-1)+}$ species have been reported previously for the following metals and complexes: Fe^{2+} (Jayson *et al.* 1973); Fe^{II}–EDTA (Halliwell 1975; Butler & Halliwell 1982; Ilan & Czapski 1977; McClune *et al.* 1977; Bull *et al.* 1982); Mn^{2+} (Pick-Kaplan & Rabani 1976; Bielski & Chan 1978; Götz & Lengfelder 1983; Cabelli & Bielski 1984*a*, *b*); Mn^{II}–EDTA and Mn^{II}–NTA (Lati & Meyerstein 1978); Ba^{2+}/Ca^{2+} (Bray *et al.* 1977); Ni^{2+}, Cr^{2+} (Sellers & Simic 1976; Ilan *et al.* 1975; V^{5+}, Ti^{4+}, Ce^{3+}, Zr^{4+}, U^{6+}, Mo^{6+} (Samuni & Czapski 1970). A study of Mn^{II} complexes (sulphate, formate, phosphate, pyrophosphate) with O_2^- showed that MnO_2^+ formation and disappearance is very sensitive to the nature of the ligand. In the presence of sulphate and formate, the transient disappears without ever forming Mn^{III} while both Mn^{II} phosphate and pyrophosphate ultimately give the corresponding Mn^{III} complexes.

As the reactivity of metals and metal–O_2^- complexes with biological compounds is of great interest and because the MnO_2^+/Mn^{III} system is well characterized, we have some preliminary results on the reactivity of these species with ascorbic acid, Trolox (6-hydroxy-2,5,7,8-tetramethylchroman-2-carboxylic acid) and NADPH/NADH.

Reactivity of HO_2/O_2^- *and* MnO_2^+/Mn^{III}) *with ascorbic acid and Trolox*

Ascorbic acid is not only an excellent HO_2/O_2^- scavenger (Cabelli & Bielski 1983), but also a well-known synergistic agent for vitamin E. As these compounds may well play important roles in the protection of membranes against oxy-radical attack, they have been the subject of intensive research.

Although the reactivity of ascorbic acid with numerous transition metals has been studied at great length in the acid pH range, fewer such studies have been performed near neutrality because of the instability of some metal ions (Fe^{2+}/Fe^{3+}, Mn^{2+}/Mn^{3+}) in such media. In particular, the reactions between Mn^{III} and ascorbic acid have been studied in very acidic solutions where Mn^{III} is stable in the presence of Mn^{II}. In 0.5–2.0 mol dm^{-3} perchloric acid, ascorbic acid reacts with Mn^{3+}_{aq} and $Mn(OH)^{2+}_{aq}$ at rates $k = 6 \times 10^3$ and 5.3×10^4 dm mol^{-1} s^{-1} respectively (Pelizzetti *et al.* 1978). As illustrated in table 1, the various Mn^{III} and ascorbate (AH^-) complexes, with the exception of Mn^{III}–pyrophosphate, are much more reactive than HO_2 and O_2^-. The Mn^{III} complexes were studied in isolation in N_2O-saturated solutions, where they were generated pulse radiolytically from the corresponding Mn^{II} complexes. Similarly the HO_2/O_2^- radicals and MnO_2^+–formate complex were generated in solutions containing, in addition to ascorbate, appropriate amounts of formate, oxygen and Mn^{II}.

Earlier studies of the reactivity of HO_2/O_2^- with Trolox (Bielski 1983) had shown that while HO_2 reacts with this vitamin E analogue at a rate of 2×10^5 dm^3 mol^{-1} s^{-1}, the superoxide radical reacts with a rate constant $k < 0.1$ dm^3 mol^{-1} s^{-1} (see table 1 and figure 1).

The reactivity of MnO_2^+–formate with Trolox was studied in a stopped-flow spectrophotometer equipped with two jet mixers in tandem and an O_2^--generating plasma lamp. The MnO_2^+–formate complex that was generated from O_2^- and Mn^{II}–formate in the first mixer was then mixed with Trolox solutions in the second mixer and scanned for absorbance changes at 390 nm. All experiments were done in 60% ethanol and at 23.5 °C, where the spontaneous decay of the

Table 1. The reactivity of HO_2/O_2^- and MnO_2^+/Mn^{III} with ascorbic acid/ascorbate (AH_2/AH^-), Trolox (TH, T^-) and NADH/NADPH at 23.5 °C

reactants	pH	$k/(dm^3\ mol^{-1}\ s^{-1})$	references
HO_2/AH_2	2.0	1.6×10^4	Cabelli & Bielski (1983)
O_2^-/AH^-	7.0	5.0×10^4	Cabelli & Bielski (1983)
MnO_2^+–formate/AH^-	7.4	3.5×10^5	Cabelli & Bielski (1985)
Mn_{aq}^{3+}/AH_2	−0.3	6.0×10^3	Pelizzetti *et al.* (1977)
$MnOH_{aq}^{2+}/AH_2$	0.3	5.3×10^4	Pelizzetti *et al.* (1977)
Mn^{III}–sulphate/AH^-	5.6	1.8×10^6	Cabelli & Bielski (1985)
Mn^{III}–phosphate/AH_2/AH^-	4.7	1.4×10^6	Cabelli & Bielski (1985)
Mn^{III}–pyrophosphate/AH^-	7.0	1.4×10^4	Cabelli & Bielski (1985)
HO_2/TH	2.0	2.0×10^5	Bielski (1983)
O_2^-/TH/T^-	10.0	0.1	Bielski (1983)
MnO_2^+–formate/TH	5.4	1.1×10^5	this paper
Mn^{III}–pyrophosphate/TH	2.0	2.0×10^5	this paper
Mn^{III}–pyrophosphate/TH	7.8	5.0×10^1	this paper
HO_2/NADH	5.0	1.3×10^5	Nadezhdin & Dunford (1979)
O_2^-/NADH	8.6	27.0	Land & Swallow (1971)
Mn^{3+}–pyrophosphate/NADH	5.0	0.8	Bielski (1984)
Mn^{3+}–pyrophosphate/NADPH	5.0	0.3	Bielski (1984)
Mn^{3+}–sulphate/NADPH	5.4	3.1×10^6	this paper
HO_2/LDH–NADH	5.0	2.0×10^6	Bielski & Chan (1976)
O_2^-/LDH–NADH	8.0	10^5	Bielski & Chan (1976)

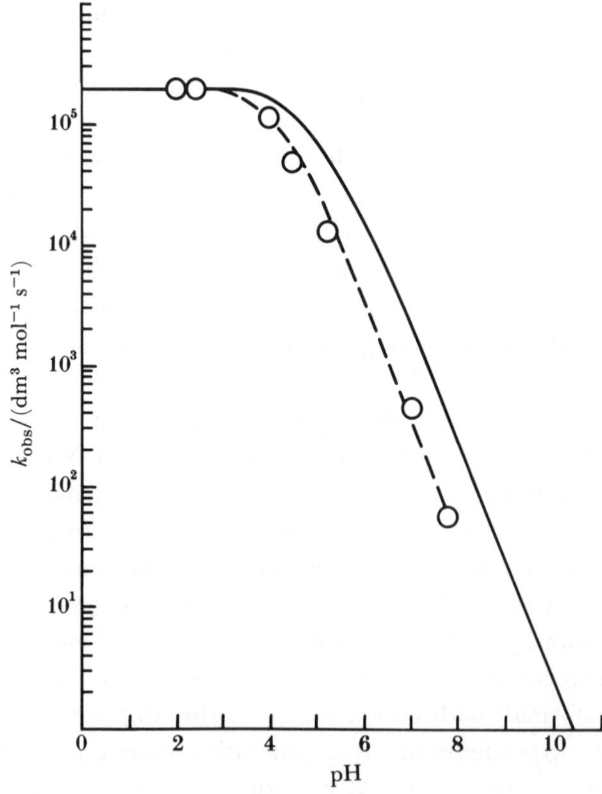

Figure 1. The pH profile for the reactions of Trolox with HO_2 (——) and Mn^{3+}–pyrophosphate (– – –).

MnO_2^+–formate complex is identical to its decay in aqueous solutions. The observed rate for

$$MnO_2^+\text{–formate} + \text{Trolox} \rightarrow Mn^{II}\text{–formate} + \text{T-semiquinone} + O_2 \quad (22)$$

at pH 5.4 is $k_{22} = (1.1 \pm 0.2) \times 10^4\ dm^3\ mol^{-1}\ s^{-1}$.

The oxidation of Trolox by Mn^{III}–pyrophosphate was studied in 60% ethanolic solutions as a function of pH under pseudo-first-order conditions by the stopped-flow method.

$$Mn^{III}\text{–pyrophosphate} + \text{Trolox} \rightarrow \text{T-semiquinone} + H^+ + Mn^{II}\text{–pyrophosphate}. \quad (23)$$

The corresponding rate constants were calculated from observed optical changes at 390 nm, the absorption maximum for Trolox semiquinone. The results are shown in figure 1 (broken line) where the plateau region represents reaction (23); $k_{23} = (2.0 \pm 0.4) \times 10^5\ dm^3\ mol^{-1}\ s^{-1}$ at 23.5 °C. The profile of the curve is described by the simple kinetic equation $k_{obs} = k_{23}/(1 + K_c/H^+)$, where

$$K_c = [Mn(HP_2O_7)_2^{3-}]\,[H_4P_2O_7]/[Mn(H_2P_2O_7)_3^{3-}] = 5.63 \times 10^{-5}\ mol; \quad \text{(Davies 1969).}$$

The reactivity of HO_2/O_2^- *and* Mn^{III}) *complexes with* NADH/NADPH

The involvement of the NADH/NADPH cofactors in the metabolic chemistry of O_2^- is well documented. For example, it has been observed (Mayewsky 1974, 1975) that animals exposed to hyperbaric oxygen show first an immediate increase in O_2^- level that is followed by a drop in cellular NADPH, suggesting that some NADPH depletion mechanism is operational. Also, it is believed (Babior 1981) that during phagocytosis the crucial reaction is the reduction of dioxygen to O_2^- at the expense of NADH/NADPH.

In vitro experiments have shown that an NADH-depletion mechanism can be initiated by O_2^- in presence of NADH-activating enzymes (Bielski & Chan 1976) or certain metal complexes (Curnutte *et al.* 1976). In an effort to elucidate the basic mechanism for the Mn^{II}-catalysed chain oxidation of NADPH by O_2^- proposed by Curnutte *et al.* (1976), pulse-radiolysis and stopped-flow experiments have been done in the presence of different Mn^{II} complexes. As can be seen from table 1, while Mn^{III}–pyrophosphate is unreactive towards NADPH, Mn^{III}–sulphate reacts at a rate of $k_{26} = 3.1 \times 10^6\ dm^3\ mol^{-1}\ s^{-1}$. The latter value was obtained in a pulse radiolysis study in which Mn^{III}–sulphate was generated from the Mn^{II} salt by $OH^{\bullet}$ radical oxidation (pH 5.4) as follows.

$$H_2O \overset{N_2O}{\rightsquigarrow} OH^{\bullet}, \quad (24)$$

$$OH^{\bullet} + Mn^{II}\text{–sulphate} \rightarrow Mn^{III}\text{–sulphate} + OH^-, \quad (25)$$

$$Mn^{III}\text{–sulphate} + NADPH \rightarrow Mn^{II}\text{–sulphate} + NADP^{\bullet}, \quad (26)$$

$$2\ NADP^{\bullet} \rightarrow (NADP)_2. \quad (27)$$

It should be noted that in spite of the fast rate of reaction (26) this mechanism does not involve a catalytic cycle.

Preliminary pulse-radiolysis studies of the Mn^{II}–phosphate/NADH system in presence of dioxygen, led to the observation of a chain reaction that corroborates the Curnutte *et al.* (1976) mechanism. An additional step observed in this study, which was beyond the time resolution of the original experiments, shows that a short-lived transient MnO_2^+–phosphate complex is formed which yields Mn^{III}–phosphate that reacts with NADH as follows.

$$H_2O \xrightarrow[HCOO^-,\, O_2^-]{Mn^{II}\text{–phosphate}} O_2^-, \tag{28}$$

$$O_2^- + Mn^{II}\text{–phosphate} \rightarrow MnO_2^+\text{–phosphate}, \tag{29}$$

$$MnO_2^+\text{–phosphate} \rightarrow Mn^{III}\text{–phosphate}, \tag{30}$$

$$Mn^{III}\text{–phosphate} + NADH \rightarrow Mn^{II}\text{–phosphate} + NAD^{\cdot} + H^+, \tag{31}$$

$$NAD^{\cdot} + O_2 \rightarrow NAD^+ + O_2^-. \tag{32}$$

The kinetics of this system have not yet been completely resolved because of complex reactions that involve, in addition to NADH, the various complexes NADH forms with Mn^{II}.

As is apparent from these results the overall reaction schemes for such systems are strongly dependent upon the ligands present in the medium and the nature of the various metal complexes.

Conclusions

Our kinetic results taken in conjunction with the mechanism proposed by Barb *et al.* (1951) led to the following conclusions. First, because the only source of OH radicals in an aqueous solution containing $Fe^{2+}/Fe^{3+}/H_2O_2/HO_2/O_2^-$ is the Fenton reaction, there is a high probability that the Haber–Weiss reaction does not occur in aqueous solutions under any conditions. Secondly, a metal-mediated generation of OH radicals from hydrogen peroxide by the superoxide radical appears possible if, as for ferric ion, the O_2^- radical can transform the metal ion or complex to an oxidation state in which it will react with H_2O_2 by a Fenton-type reaction.

Although the research on the Mn^{II}–HO_2/O_2^- systems is at a much more preliminary state, two basic trends have been observed. If O_2^- is allowed to react with Mn^{II}–complexes to yield either MnO_2^+–transients or, ultimately, Mn^{III} complexes, the reactivity of these metal species towards some biological compounds is greatly enhanced. Also, because Mn^{III}–complexes have much longer lifetimes than O_2^- in the pH range 4–7, they could have potentially much more deleterious effects even if their reactivity towards some organic compounds is not increased from that of free O_2^-.

This research was supported by NIH Grant R01 GM23656-08 and done at Brookhaven National Laboratory, which is operated under contract DE-AC02-76CH00016 with the U.S. Department of Energy.

References

Arudi, R. L., Bielski, B. H. J. & Allen, A. O. 1984 Search for singlet oxygen luminescence in the disproportionation of HO_2/O_2^-. *Photochem. Photobiol.* **39**, 703–706.

Babior, B. M. 1981 The superoxide-forming enzyme from human neutrophils. In *Membranes, molecules, toxins and cells* (ed. K. Bloch, L. Bolis & D. C. Tosteson), pp. 147–153. London: John Wright.

Barb, W. G., Baxendale, J. H., George, P. & Hargrave, K. R. 1951 Reactions of ferrous and ferric ions with hydrogen peroxide. I. The ferrous ion reaction. *Trans. Faraday Soc.* **47**, 462–500.

Bielski, B. H. J. & Chan, P. C. 1976 Reevaluation of the kinetics of lactate dehydrogenase-catalyzed chain oxidation of nicotinamide adenine dinucleotide by superoxide radicals in the presence of EDTA. *J. biol. Chem.* **251**, 3841–3844.

Bielski, B. H. J. & Chan, P. C. 1978 Products of the reaction of superoxide and hydroxyl radicals with Mn(II) cation. *J. Am. chem. Soc.* **100**, 1920–1921.

Bielski, B. H. J. 1983 Evaluation of the reactivities of HO_2/O_2^- with compounds of biological interest. In *Oxy radicals and their scavenger systems*, vol. I (ed. G. Cohen & R. E. Greenwald), pp. 1–7. Amsterdam: Elsevier.

Bielski, B. H. J., Arudi, R. L. & Cabelli, D. E. 1984 Spectral and kinetic properties of the products resulting from reactions of HO_2/O_2^- with manganese(II) complexes. In *Oxygen radicals in chemistry and biology* (ed. W. Bors, M. Saran & D. Tait), pp. 1–15. Berlin, New York: Walter de Gruyter.

Bielski, B. H. J., Cabelli, D. E., Arudi, R. L. & Ross, A. B. 1985 Rate constants for reactions of HO_2/O_2^- in aqueous solutions. *J. Phys. Chem. Ref. Data* (In the press.)

Bray, R. C., Mautner, G. N., Fielden, E. M. & Carle, C. I. 1977 Studies on the superoxide ion: complex formation with barium and calcium ions detected by e.p.r. spectroscopy and kinetically. In *Superoxide and superoxide dismutases* (ed. A. M. Michelson, J. M. McCord & I. Fridovich), pp. 61–75. London: Academic Press.

Bull, C., Fee, J. A., O'Neill, P. & Fielden, E. M. 1982 Iron-ethylenediaminetetraacetic acid (EDTA)-catalyzed superoxide dismutation revisited. *Archs Biochem. Biophys.* **215**, 551–555.

Butler, J. & Halliwell, B. 1982 Reactions of iron-EDTA chelates with superoxide radical. *Archs Biochem. Biophys.* **218**, 174–178.

Buxton, G. V. & Sellers, R. M. 1977 The radiation chemistry of metal ions in aqueous solutions. *Coord. Chem. Rev.* **22**, 195–274.

Cabelli, D. E. & Bielski, B. H. J. 1983 Kinetics and mechanism for the oxidation of ascorbic acid/ascorbate by HO_2/O_2^- radicals. A pulse radiolysis and stopped-flow photolysis study. *J. phys. Chem.* **87**, 1809–1812.

Cabelli, D. E. & Bielski, B. H. J. 1984*a* Pulse radiolysis study of the kinetics and mechanisms of the reactions between manganese(II) complexes and HO_2/O_2^- radicals. 1. Sulfate, formate, and pyrophosphate complexes. *J. phys. Chem.* **88**, 3111–3115.

Cabelli, D. E. & Bielski, B. H. J. 1984*b* Pulse radiolysis study of the kinetics and mechanisms of the reactions between manganese(II) complexes and HO_2/O_2^- radicals. 2. The phosphate complex and an overview. *J. phys. Chem.* **88**, 6291–6294.

Cabelli, D. E. & Bielski, B. H. J. 1985 The interaction between Mn^{3+}/MnO_2^+ complexes and ascorbates. A pulse radiolysis and stopped-flow photolysis study. *J. phys. Chem.* (Submitted.)

Curnette, J. T., Karnovsky, M. L. & Babior, B. M. 1976 Manganese-dependent NADPH oxidation by granulocyte particles. *J. clin. Invest.* **57**, 1059–1067.

Davies, G. 1969 Some aspects of the chemistry of manganese (III) in aqueous solution. *Coord. Chem. Rev.* **4**, 199–224.

Götz, F. & Lengfelder, E. 1983 On the mechanism of the catalytic scavenging of superoxide radical by manganese pyrophosphate: a pulse radiolysis study. In *Oxy radicals and their scavenger systems*, vol. I (ed. G. Cohen & R. A. Greenwald), pp. 228–233. Amsterdam: Elsevier Biomedical.

Haber, F. & Weiss, J. 1934 The catalytic decomposition of hydrogen peroxide by iron salts. *Proc. R. Soc. Lond.* A **147**, 332–351.

Halliwell, B. 1975 The superoxide dismutase activity of iron complexes. *FEBS Lett.* **56**, 34–38.

Ilan, Y. A., Czapski, G. & Ardon, M. 1975 The formation of CrO_2^{2+} in the reaction of $Cr^{2+} + O_2$ in aqueous acid solutions. *Israel J. Chem.* **13**, 15–21.

Ilan, Y. A. & Czapski, G. 1977 The reaction of superoxide radical with iron complexes of EDTA studied by pulse radiolysis. *Biochim. biophys. Acta* **498**, 386–394.

Jayson, G. G., Keene, J. P., Stirling, D. A. & Swallow, A. J. 1969 Pulse-radiolysis study of some unstable complexes of iron. *Trans. Faraday Soc.* **65**, 2453–2464.

Jayson, G. G., Parsons, B. J. & Swallow, A. J. 1973 Oxidation of ferrous ions by perhydroxyl radicals. *J. chem. Soc. Faraday Trans. I*, pp. 236–242.

Land, E. J. & Swallow, A. J. 1971 One-electron reactions in biochemical systems as studied by pulse radiolysis. IV. Oxidation of dihydronicotinamide-adenine dinucleotide. *Biochim. biophys. Acta* **234**, 34–42.

Lati, J. & Meyerstein, D. 1978 Oxidation of first-row bivalent transition-metal complexes containing ethylenediaminetetraacetate and nitrilotriacetate ligands by free radicals: a pulse radiolysis study. *J. chem. Soc. Dalton Trans.* pp. 1105–1118.

Mayevsky, A., Jamieson, D. & Chance, B. 1974 Oxygen poisoning in the unanesthetized brain: correlations of the oxidation-reduction state of pyridine nucleotide with electrical activity. *Brain Res.* **76**, 481–491.

Mayevsky, A. 1975 The effect of trimethadione on brain energy metabolism and EEG activity of the conscious rat exposed to HPO. *J. Neurosci. Res.* **1**, 131–142.

Meisel, D., Ilan, T. A. & Czaposki, G. 1974 Hydroperoxyl radical reactions. III. Pulse radiolytic study of the reaction of the hydroperoxyl radical with some metal ions. *J. phys. Chem.* **78**, 2330–2334.

McClune, G. J., Fee, J. A., McClusky, G. A. & Groves, J. T. 1977 Catalysis of superoxide dismutation of the reaction and evidence for the direct formation of an iron(III)-ethylenediaminetetraacetic acid peroxo complex from the reaction of superoxide with iron(II)-ethylenediaminetetraacetic acid. *J. Am. chem. Soc.* **99**, 5220–5222.

McCord, J. M. & Fridovich, I. 1969 Superoxide dismutase – an enzymic function for erythrocuprein (hemocuprein). *Biol. Chem.* **244**, 6049–6055.

Nadezhdin, A. & Dunford, H. B. 1979 Oxidation of nicotinamide adenine dinucleotide by hydroperoxyl radical. A flash photolysis study. *J. phys. Chem.* **83**, 1957–1961.

Pelizzetti, E., Mentasti, E. & Pramauro, E. 1978 Outer-sphere oxidation of ascorbic acid. *Inorg. Chem.* **17**, 1181–1186.

Pick-Kaplan, M. & Rabani, J. 1976 Pulse radiolytic studies of aq. $Mn(ClO_4)_2$ solutions. *J. phys. Chem.* **80**, 1840–1843.

Rush, J. D. & Bielski, B. H. J. 1985 Kinetic studies of the reactions of HO_2/O_2^- with ferrous/ferri ions. *J. phys. Chem.* (Submitted.)

Samuni, A. & Czapski, G. 1970 Complexes of peroxy radical with transition metal ions. *J. phys. Chem.* **74**, 4592–4594.

Schneider, W. 1984 Hydrolysis of Fe(III) – Chaotic olation versus nucleation. *Comments Inorg. Chem.* **4**, 205–223.

Schwarz, H. 1962 A determination of some rate constants for the radical processes in the radiation chemistry of water. *J. phys. Chem.* **66**, 255–262.

Sellers, R. M. & Simic, M. G. 1976 Pulse radiolysis study of reactions of some reduced metal ions with molecular oxygen in aqueous solution. *J. Am. chem. Soc.* **98**, 6145–6150.

Stuglik, Z. & Zagorski, Z. P. 1981 Pulse radiolysis of neutral iron(II) solutions: oxidation of ferrous ions by OH radicals. *Radiat. Phys. Chem.* **17**, 229–233.

Weinstein, J. & Bielski, B. H. J. 1979 Kinetics of the interaction of HO_2 and O_2^- radicals with hydrogen peroxide. The Haber–Weiss Reaction. *J. Am. chem. Soc.* **101**, 58–62.

Discussion

R. L. Willson (*Department of Biochemistry, Brunel University, Uxbridge, Middlesex*). I am very interested in Dr Bielski's studies on O_2^- and $HO_2^{\cdot}$ with NADH in the absence and presence of the dehydrogenase enzyme. When he published this work some time ago he attributed the observed increase in the rate constant to a change in the conformation of NADH when it is bound to the enzyme. Is this still totally his view, or is there any possibility that a local environment in the enzyme can cause the equilibrium $HO_2^{\cdot}/O_2^-$ to be shifted in favour of $HO_2^{\cdot}$?

B. H. J. Bielski. Although I have no evidence that the increased rate of NADH oxidation in the presence of dehydrogenases is a result of a conformational change in the cofactor, the magnitude of the observed rate constants suggests enzyme activation of NADH; $k(HO_2+NADH) = 1.8\times10^5\ dm^3\ mol^{-1}\ s^{-1}$ (Nadezhdin & Dunford 1979); $k(HO_2+LDH\text{–}NADH) = 2.0\times10^6\ dm^3\ mol^{-1}\ s^{-1}$ and $k(HO_2+GAPDH\text{–}NADH) = 2.0\times10^7\ dm^3\ mol^{-1}\ s^{-1}$ (Bielski & Chan 1976, 1980). Of course it is quite possible that the observed increase in the rate of reaction is a result of a combination of enzyme activation of NADH and a local pH shift in favour of HO_2.

References

Bielski, B. H. J. & Chan, P. C. 1976 Re-evaluation of the kinetics of lactate dehydrogenase-catalyzed chain oxidation of nicotinamide adenine dinucleotide by superoxide radicals in the presence of ethylenediaminetetra-acetate. *J. biol. Chem.* **251**, 3841–3844.

Bielski, B. H. J. & Chan, P. C. 1980 Glyceraldehyde-3-phosphate dehydrogenase-catalyzed chain oxidation of reduced nicotinamide adenine dinucleotide by perhydroxyl radicals. *J. biol. Chem.* **255**, 874–876.

Nadezhdin, A. & Dunford, H. B. 1979 Oxidation of nicotinamide adenine dinucleotide by hydroperoxyl radical. A flash photolysis study. *J. phys. Chem.* **93**, 1957–1961.

R. L. Willson. Has Dr Bielski studied the reaction of organic peroxy radicals, $RO_2^{\cdot}$, with NADH? Unlike $HO_2^{\cdot}$, in many instances peroxy radicals are unlikely to be dissociated at pH 7 and are therefore more likely to react with NADH. Has he any rate data concerning such organic peroxy radical reactions or any evidence that they do occur?

B. H. J. Bielski. Unfortunately I never studied the reactivity of $RO_2^{\cdot}$ radicals with NADH.

Phil. Trans. R. Soc. Lond. B **311**, 483–503 (1985)
Printed in Great Britain

Reactivity and activation of dioxygen-derived species in aprotic media (a model matrix for biomembranes)

By D. T. Sawyer, J. L. Roberts, Jr, T. S. Calderwood, H. Sugimoto and M. S. McDowell

Department of Chemistry, University of California, Riverside, California 92521, *U.S.A.*

In aprotic media the electrochemical reduction of dioxygen yields superoxide ion (O_2^-), which is an effective Brønsted base, nucleophile, one-electron reductant, and one-electron oxidant of reduced transition metal ions. With electrophilic substrates (organic halides and carbonyl carbons) O_2^- displaces a leaving group to form a peroxy radical ($ROO^{\cdot}$) in the primary process. Superoxide ion oxidizes the activated hydrogen atoms of ascorbic acid, catechols, hydrophenazines and hydroflavins. Combination of O_2^- with 1,2-diphenylhydrazine yields the anion radical of azobenzene, which reacts with O_2 to give azobenzene and O_2^- (an example of O_2^--induced autoxidation). With phenylhydrazine, O_2^- produces phenyl radicals. The *in situ* formation of $HO_2^{\cdot}$ (O_2^- plus a proton source) results in H-atom abstraction from allylic and other groups with weak heteroatom—H bonds (binding energy (b.e.) less than 335 kJ). This is a competitive process with the facile second-order disproportionation of $HO_2^{\cdot}$ to H_2O_2 and O_2 ($k_{bi} \approx 10^4\ mol^{-1}\ s^{-1}$ in Me_2SO).

Addition of $[Fe^{II}(MeCN)_4]\ (ClO_4)_2$ to solutions of hydrogen peroxide in dry acetonitrile catalyses a rapid disproportionation of H_2O_2 via the initial formation of an adduct $[Fe^{II}(H_2O_2)^{2+} \leftrightarrow Fe(O)(H_2O)^{2+}]$, which oxidizes a second H_2O_2 to oxygen. In the presence of organic substrates such as 1,4-cyclohexadiene, 1,2-diphenylhydrazine, catechols and thiols the Fe^{II}–H_2O_2/MeCN system yields dehydrogenated products; with alcohols, aldehydes, methylstyrene, thioethers, sulphoxides, and phosphines, the $Fe^{II}(H_2O_2)^{2+}$ adduct promotes their monoxygenation. The product from the FeO^{2+}–H_2O_2 reaction, $[Fe^{II}(H_2O_2)_2^{2+}]$, exhibits chemistry that is closely similar to that for singlet oxygen ($^1O_2$), which has been confirmed by the stoichiometric dioxygenation of diphenylisobenzofuran, 9,10-diphenylanthracene, rubrene and electron-rich unsaturated carbon–carbon bonds ($Ph_2C{=}CPh_2$, $PhC{\equiv}CPh$ and *cis*-$PhCH{=}CHPh$). In dry ligand-free acetonitrile (MeCN), anhydrous ferric chloride ($Fe^{III}Cl_3$) activates hydrogen peroxide for the efficient epoxidation of alkenes. The $Fe^{III}Cl_3$ further catalyses the dimerization of the resulting epoxides to dioxanes. These observations indicate that strong Lewis acids that are coordinatively unsaturated, $[Fe^{II}(MeCN)_4]^{2+}$ and $[Fe^{III}Cl_3]$, activate H_2O_2 to form an effective oxygenation and dehydrogenation agent.

When catalytic quantities of superoxide ion are introduced into a dry acetonitrile solution that contains excess substrate (Ph_2SO or $PhCH_2OH$), ambient air, 1,2-diphenylhydrazine and iron$^{(II)}$, the substrate is rapidly and efficiently monoxygenated, the combination provides a catalytic system for the autoxidation of organic substrates via reaction cycles that closely mimic cytochrome P_{450} monoxygenase enzymes.

1. Introduction

(*a*) *Aprotic solvents as models for the chemical environment in biological membranes*

The elucidation of the detailed reaction mechanisms for chemical reactions in biological systems is difficult. An organism or an intact organelle is the biological equivalent of a black box with known inputs and outputs, but whose inner processes are complicated and not understood.

Thus, analysis of the internal reactions of a biological system generally must proceed by isolating and studying each hypothetical reaction inside the black box.

A question then arises as to whether these biochemical reactions take place in an essentially aqueous or in a non-aqueous hydrophobic environment. Without a definitive answer to this question the reactions have been studied in both aqueous and aprotic media. A fundamental justification for the use of aprotic solvents is that they allow the study of species, particularly anion radicals, that are too reactive to study in solvents that have labile protons.

In addition, the properties of oxy ions and metal ions may be significantly different in aprotic solvents and water. Water is such a strong ligand that it displaces weaker ligands, and is both a moderate acid and a weak base. In an aprotic solvent the redox properties of ions, especially anions, are significantly affected and the delicate balance of ligand–metal interactions that is associated with metal ion catalysis is dramatically changed. Thus, studies in aprotic solvents may reveal reaction pathways that are not observed in water and reaction rates that are enhanced by several orders of magnitude. Another important factor is the enhanced solubility of O_2 in aprotic media (at a partial pressure of 1 atm† its concentration is 1 mM in H_2O, 5 mM in dimethylformamide and 8 mM in acetonitrile (Sawyer *et al.* 1982*a*).

A further justification for studies in aprotic solvents is that they provide an environment that closely parallels that of a lipid bilayer membrane with embedded proteins (low proton availability with a polar hydrophilic character). Even the solution in the interior of the cell (the cytosol) may have less protic character than pure water because most of the water in the concentrated cytosol is bound to inorganic ions, proteins, other biomolecules and the membrane surface. If this reasoning is valid, then the chemistry for dioxygen species (O_2^{-}, $HO_2^{\cdot}$, H_2O_2, HO_2^{-}) in aprotic solvents is likely to be analogous to that in cells, on the surface of cell membranes, and within biomembranes.

For several reasons both aqueous or aprotic solutions are inadequate model matrices for biological processes. Reaction volumes in biological systems are small and contain many solid organelles surrounded by membranes of large surface area. Thus, the interior of a cell has many solid–solution interfaces with charged double layers that present substantial voltage gradients. Cellular membranes also act as barriers for species whose transport is limited by their size, shape and net charge, and thereby create large concentration gradients. Moreover, the identity and concentration of the reactive intermediates that exist inside a cell are difficult to determine. These factors contribute considerable uncertainty as to the relevance of the reaction pathways in homogeneous aqueous and non-aqueous systems to biological systems.

(*b*) *Previous studies of* O_2*-activation via one-electron reduction*

The role of oxygen in aerobic metabolism has occupied the attention of chemists since the time of Lavoisier. Triplet ground-state dioxygen, $^3O_2$, is a diradical and reacts slowly with most spin-paired organic compounds because the direct reaction of a triplet molecule to give singlet products is a spin-forbidden process (Taube 1965; Hamilton 1974). Therefore, much attention has been focused on how $^3O_2$ is able to react with singlet-state organic compounds in the presence of catalysts and cofactors, in other words, the problem of oxygen activation. A number of modes of activation have been discovered. These include (*a*) the introduction of free-radical

† 1 atm = 101325 Pa.

initiators that promote reactions with triplet dioxygen by a radical chain mechanism such as lipid peroxidation (Simic & Karel 1980; Pryor 1976); (*b*) the excitation of $^3O_2$ to a singlet state (Krinsky 1979; Foote 1976); (*c*) the binding of triplet-state dioxygen to a transition metal ion that has unpaired electron spins (Spiro 1980) and (*d*) the successive addition of electrons to $^3O_2$ to form active intermediates, the first being the superoxide ion, O_2^- (Frimer 1983; Roberts & Sawyer 1983).

The discovery that superoxide ion is produced in biological systems, albeit as a by-product, and the further discovery of a family of enzymes that catalyse superoxide ion dismutation has provided an important stimulus for research on superoxide ion chemistry (Fridovich 1982).

In aqueous solution at pH 7, superoxide ion reacts rapidly in three characteristic modes. First, as a weak base, for example protonation followed by rapid disproportionation (Bielski 1978); second, as a one-electron reductant of easily reducible substrates, for example Fe^{III} cytochrome *c*, quinones, and oxidized transition metal complexes of Fe^{III}, Mn^{III}, Cu^{II}, Ru^{III}, and Mo^{VI} (Sawyer & Valentine 1981); and third, as a one-electron oxidant, for example, oxidative addition to Fe^{II} EDTA to form a ferric–peroxo complex, Fe^{III} (O_2^{2-}) (Bull *et al.* 1983). Without the stabilization of the peroxide ion afforded by a metal cation or proton, superoxide cannot act as a one-electron oxidant (Sawyer *et al.* 1978).

In aprotic solvents, reduction of dioxygen in the presence of $Zn^{II}(bipy)_3^{2+}$, $Cu^I(MeCN)_4^+$ and $Fe^{II}TPP$ produces, respectively $Zn^{II}(bipy)_3O_2$, $Cu^{II}(O_2)$, and the side-on bonded peroxo complex, $O_2Fe^{III}(TPP)^-$, where bipy represents 2,2′-bipyridine and TPP represents tetrapheylporphyrin (Sawyer *et al.* 1984*a*). The latter complex also is formed by the reaction of O_2^- with $Fe^{II}TPP$ (McCandlish *et al.* 1980). Similarly, the addition of four moles of O_2^- per mole of molybdenum(VI)3,5-di-t-butylcatecholate dimer, $[Mo^{VI}(O)(DTBC)_2]_2$, yields two moles of the peroxide adduct, $[Mo^{VI}(O)(O_2)(DTBC)_2]^{2-}$ plus two moles of dioxygen. This represents a catalysed disproportionation, $4O_2^- \rightarrow 2O_2^{2-} + 2O_2$ (Lim & Sawyer 1982). These net reactions resemble the oxidative additions observed in aqueous solution. To date, these complexes are essentially inert as reactants or catalysts for the oxygenation of organic substrates, but their chemistry has not been fully explored.

Bielski (1978) has reported refined values for the acid–base, spectral and kinetic properties of O_2^- and its conjugate acid, $HO_2^{\cdot}$ (the perhydroxyl radical), in aqueous solutions. Superoxide ion is a weak base (the pK_a for $HO_2^{\cdot}$ is 4.9) such that at pH 7, *ca.* 1% of the total superoxide exists as $HO_2^{\cdot}$. Both $HO_2^{\cdot}$ and O_2^- spontaneously disproportionate via a series of reactions.

$$HO_2^{\cdot} + HO_2^{\cdot} \xrightarrow{k\,=\,8.6\times10^5\ \mathrm{mol^{-1}\ s^{-1}}} H_2O_2 + O_2; \tag{1}$$

$$HO_2^{\cdot} + O_2^- \xrightarrow{k\,=\,1\times10^8\ \mathrm{mol^{-1}\ s^{-1}}} HO_2^- + O_2; \tag{2}$$

$$O_2^- + O_2^- \xrightarrow{k\,<\,0.35\ \mathrm{mol^{-1}\ s^{-1}}} O_2^{2-} + O_2. \tag{3}$$

The tendency of superoxide ion to disproportionate enhances its effective basicity.

$$2O_2^- + H_2O \rightleftharpoons O_2 + HO_2^- + OH^-; \tag{4}$$

$$2O_2^- + HB \rightarrow O_2 + HO_2^- + B^-. \tag{5}$$

This effect is most clearly seen in aprotic solvents, where O_2^- is long-lived (Chin *et al.* 1982). Sawyer & Gibian (1979) estimated an equilibrium constant of 2.5×10^8 for reaction (4) in water. Thus, even acids much weaker than water (up to $pK_a \approx 23$) can be deprotonated by O_2^- when proton dissociation is controlled by thermodynamic rather than kinetic factors; for example, protons on oxygen and nitrogen acids are generally more labile than those on carbon acids (Crooks 1975).

There are significant differences in the acid–base, redox, and nucleophilic properties of O_2^- in water and in dipolar aprotic solvents such as dimethyl sulphoxide (Me_2SO), dimethylformamide (DMF), and acetonitrile (MeCN). Superoxide ion in water is a weaker base, a weaker one-electron reducing agent and a weaker nucleophile than it is in aprotic solvents (Roberts & Sawyer 1983). These effects are mainly attributable to the strong solvation of O_2^- by water (Koppenol 1983) in contrast to that by aprotic solvents. In water, O_2^- displays a notable lack of reactivity with molecules of biological interest (Bielski 1983; Gebicki & Bielski 1981), except for those that contain acidic protons (ascorbic acid, thiols) or compounds that are easily reducible by single-electron transfer (quinones, Fe^{III} cytochrome *c*). However, $HO_2^{\cdot}$ is an effective one-electron oxidant and can initiate lipid autoxidation (Gebicki & Bielski 1981); i.e. $HO_2^{\cdot}$ is a 'hotter' radical and a more effective hydrogen-atom abstractor than O_2^- (Valentine 1979; Frimer 1983; Sawyer & Roberts 1983). This has shifted the focus of attention from O_2^- to $HO_2^{\cdot}$ in the search for species responsible for the cytotoxicity that accompanies the generation of superoxide ion in biological systems. A significant aspect of this cytotoxicity is the destruction of membranes by autoxidation of their lipid components; this has prompted a re-examination of diene fatty acid autoxidation (Porter 1984).

The initiator of lipid autoxidation most often invoked is the hydroxyl radical ($OH^{\cdot}$), which can be derived from hydrogen peroxide (produced by disproportionation of superoxide) via several routes. The first is the well studied Fenton reaction and its analogues (Green & Hill 1984; Walling 1975).

$$H_2O_2 + M^{n+} \rightarrow M^{(n+1)+} + OH^- + OH^{\cdot} \tag{6}$$

$$O_2^- + M^{(n+1)+} \rightarrow O_2 + M^{n+} \tag{7}$$

Another route, demonstrated in aprotic solvents, is the base-induced decomposition of hydrogen peroxide (Roberts *et al.* 1978):

$$2H_2O_2 + OH^- \rightarrow O_2^- + 2H_2O + OH^{\cdot}. \tag{8}$$

In aprotic solvents the uncatalysed disproportionation of superoxide ion is slow and, in the absence of strong solvation, the 'naked' anion is a much better nucleophile. This behaviour parallels that of another small non-polarizable anion, fluoride ion, which is a weak base and poor nucleophile in water, but a strong base and good nucleophile in aprotic solvents (Valentine 1979; Clark 1980).

2. Results and discussion

(a) *New aspects of O_2^- reactivity in aprotic solvents*

This section summarizes recent studies of superoxide ion reactions with polyhalogenated hydrocarbons and halogenated alkenes, carbonyl compounds and substrates with activated secondary amine functions (substituted hydrazines, hydrophenazines and hydroflavins). The reactions of O_2^- with organic substrates in aprotic solvents yield activated oxygen intermediates (peroxy radicals, peroxides and H_2O_2) that represent a significant biological hazard, especially if they are formed within biomembranes.

(i) *Oxygenation of polyhalogenated hydrocarbons*

The stoichiometries and kinetics for the reaction of O_2^- with polyhalogenated alkanes and alkenes are summarized in table 1 (Roberts *et al.* 1983; Calderwood *et al.* 1983; Calderwood & Sawyer 1984). The normalized first-order rate constants, $k_1/[S]$, that are reported in tables 1–4 were determined by the rotated ring-disk electrode method (Roberts *et al.* 1983) under pseudo-first-order conditions ($[substrate] \gg [O_2^-]$).

The nucleophilicity of O_2^- toward primary alkyl halides (scheme 1) results in an S_N2 displacement of halide ion from the carbon centre. The normal reactivity order, primary

Table 1. Stoichiometries and kinetics for the reaction of 0.1–5 mm O_2^- with polyhalogenated hydrocarbons in dimethylformamide (0.1 m tetraethylammonium perchlorate) at 25 °C[a]

substrate concentration, [S] = 1–10 mm	O_2^-/S	products/S	$k_1/[S]$ ($mol^{-1}\ s^{-1}$)[b]
CCl_4	5	$HOC(O)O^-$, $4Cl^-$, $3.3O_2$	3800.0
$FCCl_4$	5	$HOC(O)O^-$, $3Cl^-$, $2.5O_2$	4.0
$HCCl_3$	4	$HOC(O)O^-$, $3Cl^-$, $2O_2$	0.4
CF_3CCl_3	4	$CF_3C(O)O^-$, $3Cl^-$, $2.4O_2$	400.0
$PhCCl_3$	4	$PhC(O)O^-$ (70%), $PhC(O)OO^-$ (30%) $3Cl^-$, $2.4O_2$	50.0
$MeCCl_3$	—		< 0.1
$HOCH_2CCl_3$	4	$HOCH_2C(O)O^-$, $3Cl^-$	47.0
$(p\text{-ClPh})_2CHCCl_3$ (DDT)	1	$(p\text{-ClPh})_2C{=}CCl_2$, Cl^-	100.0
$(p\text{-MeOPh})_2CHCCl_3$ (Methoxychlor)	1	$(p\text{-MeOPh})_2C{=}CCl_2$, Cl^-	10.0
$(p\text{-ClPh})_2CFCCl_3$ (F-DDT)	1	$(p\text{-ClPh})_2C{=}CCl_2$, Cl^-	170.0
PhCHBrCHBrPh	2	2PhCH(O), $2Br^-$, O_2	1000.0
MeCHBrCHBrMe	2	2MeCH(O), $2Br^-$, O_2	160.0
CH_2BrCH_2Br (EDB)	2	$2CH_2(O)$, $2Br^-$, O_2	2000.0
$CH_2BrCHBrCH_2Cl$ (DBCP)	5	$2CH_2(O)$, $HOC(O)O^-$, $2Br^-$, Cl^-, $2O_2$	4000.0
n-BuBr	1	Br^-	960.0
CH_2ClCH_2Cl	2	$2CH_2(O)$, $2Cl^-$, O_2	24.0
cis-CHCl=CHCl	4	$2HOC(O)O^-$, $2Cl^-$	10.0
$CH_2{=}CCl_2$	3	$HOC(O)O^-$, $2Cl^-$, O_2	2.0
$CHCl{=}CCl_2$	5	$2HOC(O)O^-$, $3Cl^-$, $1.5O_2$	9.0
$CCl_2{=}CCl_2$	6	$2HOC(O)O^-$, $4Cl^-$, $3O_2$	15.0
$(p\text{-ClPh})_2C{=}CCl_2$ (DDE)	3	$HOC(O)O^-$, $(p\text{-ClPh})_2C{=}O$, $2Cl^-$, O_2	2.0

[a] Stoichiometries determined for O_2^--substrate reactions by titration of excess $(Me_4N)O_2$ (with voltammetric detection); for released Br^- and Cl^- by titration with $AgNO_3$; for released O_2 by negative-scan voltammetry; and for organic products by ether extraction and capillary-column gas chromatography.

[b] Pseudo-first-order rate constants, k_1 (normalized to unit substrate concentration [S]), were determined from measurements with a glassy carbon–glassy carbon ring-disc electrode that was rotated at 900 rev min^{-1}.

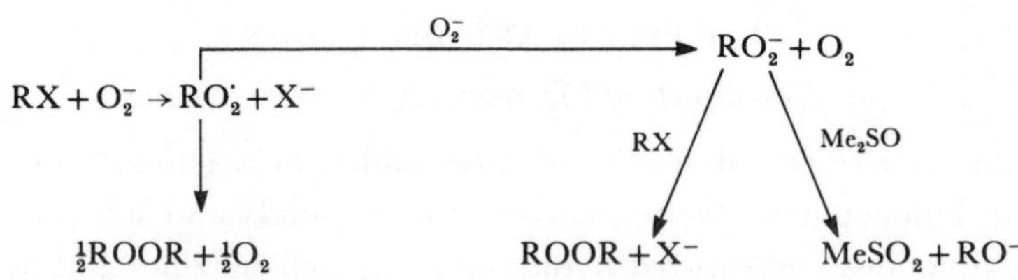

Scheme 1

> secondary > tertiary and leaving-group order I > Br > OTs > Cl are observed, as are the expected stereoselectivity and inversion at the carbon centre. In dimethylformamide the final product is the dialkyl peroxide. The peroxy radical ($ROO^{\cdot}$), which is produced in the primary step and has been detected by spin trapping (Merritt & Johnson 1977), is an oxidant that is readily reduced by O_2^- to form the peroxy anion (ROO^-). Because the latter can oxygenate Me_2SO to its sulphone, the main product in this solvent is the alcohol (ROH) rather than the dialkylperoxide.

Although formation of the dialkyl peroxide is shown in the prototype reaction (scheme 1), hydroperoxides, alcohols, aldehydes and acids have also been isolated. The extent of these secondary paths depends on the choice of solvent and reaction conditions. Secondary and tertiary halides also give substantial quantities of alkene elimination products.

The reaction of O_2^- with CCl_4 and $RCCl_3$ compounds almost certainly cannot occur via an S_N2 mechanism because the carbon-atom centre is inaccessible. Rather, superoxide ion appears to attack a chlorine atom with a net result that is equivalent to an electron transfer from O_2^- to chlorine (scheme 2). This step is analogous to the 'single-electron transfer' (s.e.t.) mechanism that has been proposed for many nucleophilic reactions; an initial transfer of an electron followed by collapse of a radical pair (Eberson 1982).

$$E^+ + Nu^- \rightarrow [E^{\cdot}\ {}^{\cdot}Nu] \rightarrow E\text{—}Nu \qquad (9)$$

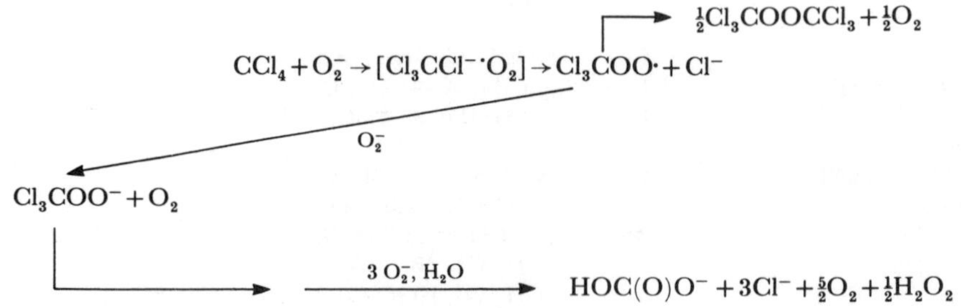

$$CCl_4 + O_2^- \rightarrow [Cl_3CCl^{-}\,{}^{\cdot}O_2] \rightarrow Cl_3COO^{\cdot} + Cl^- \rightarrow \tfrac{1}{2}Cl_3COOCCl_3 + \tfrac{1}{2}O_2$$

$$Cl_3COO^{\cdot} \xrightarrow{O_2^-} Cl_3COO^- + O_2$$

$$Cl_3COO^- \xrightarrow{3\,O_2^-,\ H_2O} HOC(O)O^- + 3Cl^- + \tfrac{5}{2}O_2 + \tfrac{1}{2}H_2O_2$$

Scheme 2

The initiation step for the O_2^-–CCl_4 reaction must be followed by rapid combination in the solvent cage of ${}^{\cdot}O_2^{\cdot}$ and $Cl_3C^{\cdot}$ to form the $Cl_3COO^{\cdot}$ radical. This radical is thought to initiate lipid peroxidation (Mason 1982), which would account for the hepatotoxicity of CCl_4 (Slater 1982).

The rates of reaction for O_2^- with $RCCl_3$ compounds are proportional to their reduction potentials, which is consistent with the s.e.t. mechanism (Roberts *et al.* 1983). A plot of $\lg k_1/[S]$ (table 1) against the reduction potentials of $RCCl_3$ compounds is approximately linear with a slope of -4.9 decade per volt. Such behaviour is consistent with a mechanism that occurs via simultaneous electron transfer and nuclear motion (Perrin 1974). This correlation indicates that in water, where the O_2/O_2^- redox potential is about 0.44 V more positive, the rate of the reaction for O_2^- with CCl_4 would be about 100–200 times slower than in aprotic solvents.

The stoichiometric data in table 1 for CCl_4 and $FCCl_3$ are consistent with a net chemical reaction that yields bicarbonate ion and four halide ions in the final aqueous workup of the reaction products. When O_2^- reacts with $RCCl_3$ compounds, the R—C bond is not cleaved (scheme 3). About 30 % of the product from the $PhCCl_3$–O_2^- reaction is perbenzoate ion. Hence, this may be the active intermediate responsible for the epoxidation of olefins by O_2^- in the presence of benzoyl chloride (Oae & Takata 1980).

$$PhCCl_3 + O_2^- \rightarrow [PhCl_2CCl^{\cdot -}O_2] \rightarrow PhCl_2COO^{\cdot} + Cl^-$$

$$PhCl_2COO^{\cdot} \xrightarrow{O_2^-} PhCl_2COO^- + O_2$$

$$PhCl_2COO^- \xrightarrow{2\,O_2^-} PhC(O)OO^- + 2Cl^- + \tfrac{3}{2}O_2$$

SCHEME 3

Superoxide ion reacts with vicinal dibromoalkanes to form aldehydes (see table 1). The mechanism proposed for these reactions (scheme 4) is a nucleophilic attack on carbon, followed by a one-electron reduction of the peroxy radical and nucleophilic displacement on the adjacent carbon to form a dioxetane that subsequently cleaves to form two moles of aldehyde (Calderwood & Sawyer 1984).

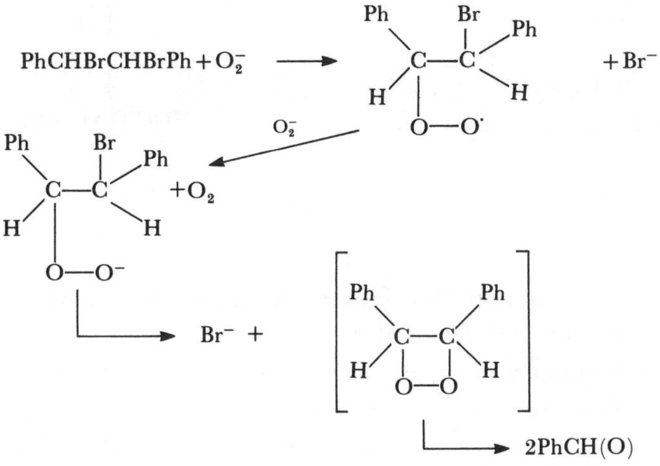

$$PhCHBrCHBrPh + O_2^- \longrightarrow PhCH(OO^{\cdot})CHBrPh + Br^-$$

$$PhCH(OO^{\cdot})CHBrPh \xrightarrow{O_2^-} PhCH(OO^-)CHBrPh + O_2$$

$$PhCH(OO^-)CHBrPh \longrightarrow Br^- + \left[\text{PhCH–CHPh dioxetane (O—O)}\right] \longrightarrow 2PhCH(O)$$

SCHEME 4

Both *p,p′*-DDT and Methoxychlor are rapidly deprotonated by OH^- in aprotic solvents with subsequent elimination of Cl^- to form the dehydrochlorination products, i.e. DDT forms DDE (scheme 5). The same products are formed in their reactions with O_2^-. Because the reaction rates that are measured by the rotated ring-disk electrode method are fairly rapid, the primary step must be a direct reaction with O_2^- and not with OH^- from the reaction of O_2^- with trace

p-ClPh(p-ClPh)C(H)—CCl_3 + O_2^- (DDT) → [p-ClPh(p-ClPh)C—C(Cl)(Cl)(Cl), H, $O^{-\cdot}$] → p-ClPh(p-ClPh)C=C(Cl)(Cl) (DDE) + $HO_2^{\cdot}$ + Cl^- → $\frac{1}{2}H_2O_2 + \frac{1}{2}O_2$

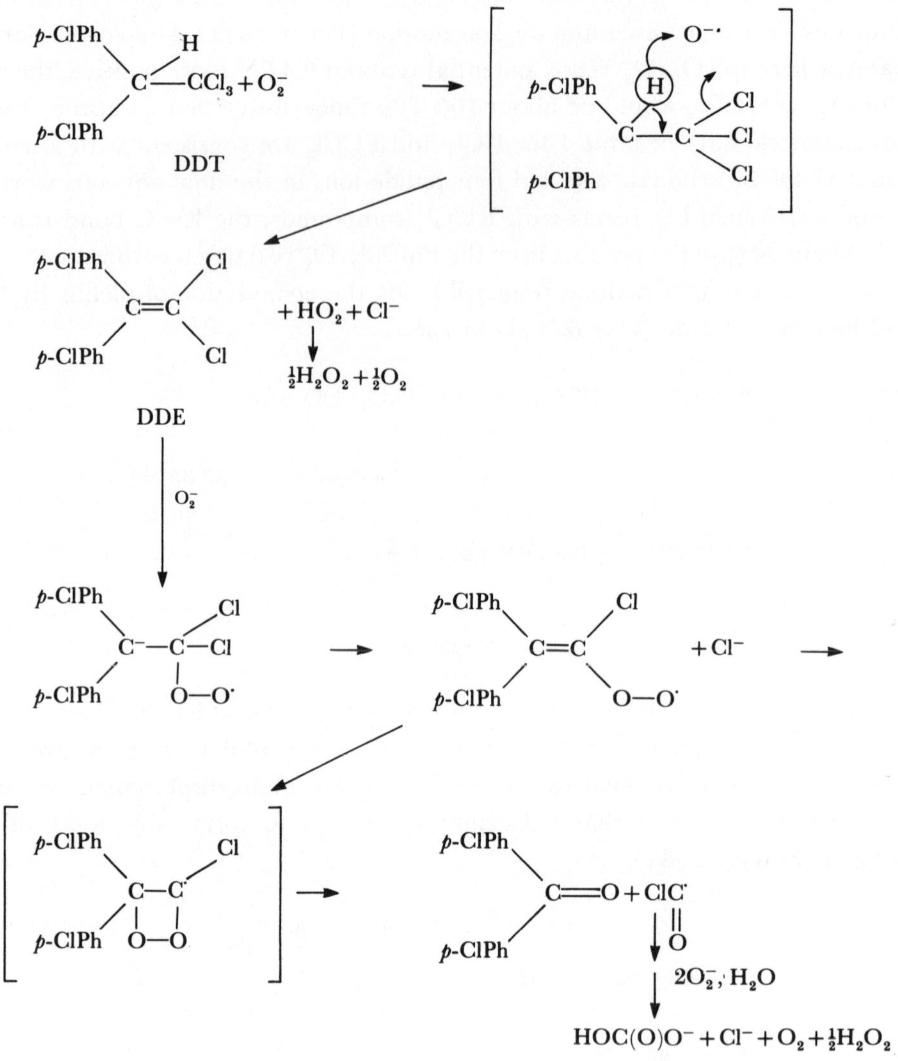

DDE —O_2^-→ p-ClPh(p-ClPh)C^-—C(Cl)(Cl)—O—$O^{\cdot}$ → p-ClPh(p-ClPh)C=C(Cl)—O—$O^{\cdot}$ + Cl^- → [p-ClPh(p-ClPh)C—$C^{\cdot}$(Cl), O—O] → p-ClPh(p-ClPh)C=O + $ClC^{\cdot}$(=O) —$2O_2^-$, H_2O→ $HOC(O)O^- + Cl^- + O_2 + \frac{1}{2}H_2O_2$

SCHEME 5

water in the solvent (see equation (4)). Hence the initial reaction with O_2^- is deprotonation followed by elimination of Cl^- to form DDE.

Although normal alkenes are not reactive with O_2^-, their chlorinated derivatives, including DDE, are readily oxidized (see table 1). Scheme 5 outlines a proposed mechanism for DDE that postulates nucleophilic addition to the activated olefin followed by ring closure of the peroxy radical and cleavage of the dioxetane-like radical to form products. After aqueous workup, the terminal olefinic carbon atom with a halogen is recovered as bicarbonate ion and halide ion. The peroxy radical (and chloracyl radical) intermediates should be effective initiators for the peroxidation of unsaturated lipids, as has been proposed for the $Cl_3COO^{\cdot}$ radical (Slater 1982).

(ii) *Nucleophilic addition to carbonyls*

Table 2 summarizes the available kinetic data for the reaction of O_2^- with esters, diketones and carbon dioxide (Gibian *et al.* 1979, Magno & Bontempelli 1976; Sawyer *et al.* 1983; Roberts *et al.* 1984). Esters react with superoxide ion to form diacyl peroxides or the carboxylate and the alcohol. Initial reaction is proposed to occur via a reversible addition–elimination reaction at the carbonyl carbon (scheme 6). This idea is supported by the products that are observed in the gas-phase reaction of O_2^- with phenyl acetate and phenyl benzoate, studied by Fourier-transform mass spectrometry (Johlman *et al.* 1983). In effect, there is a competition between loss of O_2^- and loss of the leaving group. Carbanions are poor leaving groups so that simple ketones without acidic α-hydrogen atoms are unreactive. The RC(O)OO˙ radical and the $RC(O)OO^-$ anion should be reactive intermediates for the initiation of autoxidations of allylic hydrogens and the epoxidation of olefins, respectively.

TABLE 2. PRODUCTS AND KINETICS FOR THE REACTION OF 1–5 mM O_2^- WITH CARBONYL COMPOUNDS AT 25 °C

substrate, S	solvent[a]	products/S	k_2 ($mol^{-1}\,s^{-1}$)
MeC(O)OEt	Py/0.1 M TEAP	—	0.01
MeC(O)OPh	Py/0.1 M TEAP	—	160.0
PhC(O)OPh	Py/0.1 M TEAP	—	5.0
PhCH(O)	Py/0.1 M TEAP	no reaction	—
PhC(O)C(O)Ph	DMF/0.1 M TEAP	$2PhC(O)O^-$	2000.0[b]
MeC(O)C(O)Me	DMF/0.1 M TEAP	$\frac{1}{2}H_2O_2$, enolate	4000.0[b]
MeC(O)C(O)OEt	DMF/0.1 M TEAP	$\frac{1}{2}H_2O_2$, enolate	4000.0[b]
CO_2	Me_2SO/0.1 M TEAP	$\frac{1}{2}\ {}^-OC(O)OC(O)OO^-$	1400.0[b]

[a] Py, pyridine; DMF, dimethylformamide; TEAP, tetraethylammonium perchlorate.
[b] Pseudo-first-order rate constants divided by substrate concentration, $k_1/[S]$, determined from measurements with a glassy carbon–glassy carbon ring-disk electrode that was rotated at 900 rev min^{-1}.

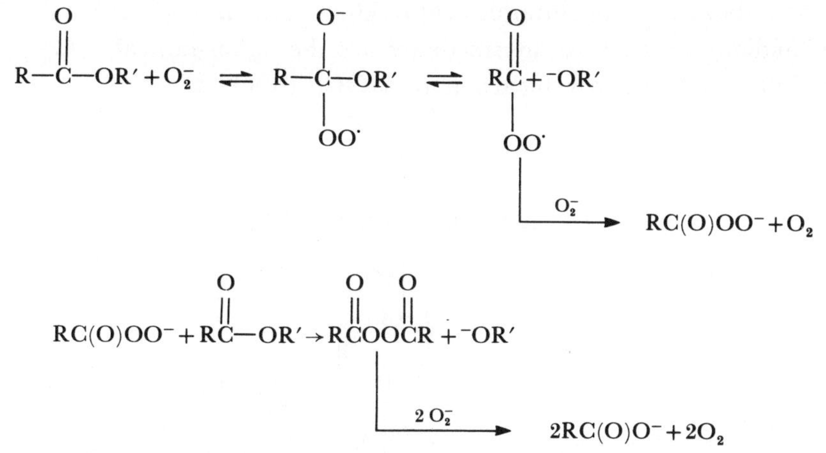

$$R\!-\!C(=O)\!-\!OR' + O_2^- \rightleftharpoons R\!-\!C(O^-)(OO^\cdot)\!-\!OR' \rightleftharpoons RC(=O)(OO^\cdot) + {}^-OR'$$

$$RC(=O)OO^\cdot \xrightarrow{O_2^-} RC(O)OO^- + O_2$$

$$RC(O)OO^- + RC(=O)\!-\!OR' \rightarrow RC(=O)OOC(=O)R + {}^-OR'$$

$$RC(=O)OOC(=O)R \xrightarrow{2\,O_2^-} 2RC(O)O^- + 2O_2$$

SCHEME 6

Simple diketones such as 2,3-butanedione are rapidly deprotonated by O_2^-, but the original diketone is recovered upon acidification (scheme 7). However, benzil (PhC(O)C(O)Ph) cannot enolize and is oxygenated by O_2^- to give two benzoate ions. Scheme 8 outlines a proposed mechanism that is initiated by nucleophilic attack. Frimer (1983) has discussed an alternative

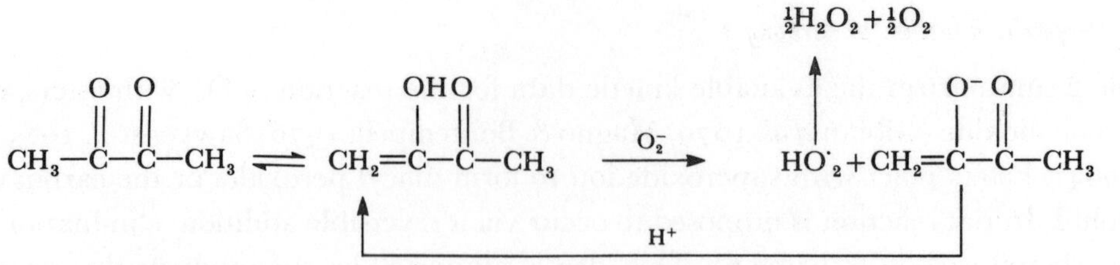

$$\mathrm{CH_3{-}C(=O){-}C(=O){-}CH_3 \rightleftharpoons CH_2{=}C(OH){-}C(=O){-}CH_3 \xrightarrow{O_2^-} HO_2^{\cdot} + CH_2{=}C(O^-){-}C(=O){-}CH_3}$$

$$\mathrm{HO_2^{\cdot} \rightarrow \tfrac{1}{2}H_2O_2 + \tfrac{1}{2}O_2}\qquad(\mathrm{H^+}\text{ returns to enol})$$

Scheme 7

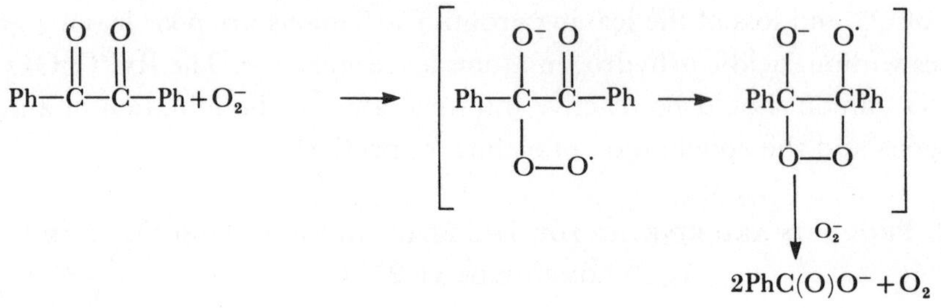

$$\mathrm{Ph{-}C(=O){-}C(=O){-}Ph + O_2^- \longrightarrow \left[Ph{-}C(O^-)(O{-}O^{\cdot}){-}C(=O){-}Ph \longrightarrow PhC(O^-)(O{-}O){-}C(O^{\cdot})Ph \right] \xrightarrow{O_2^-} 2PhC(O)O^- + O_2}$$

Scheme 8

pathway in which the initial step is electron transfer from O_2^- to the carbonyl, followed by coupling of the benzil radical with dioxygen to give the cyclic dioxetane-like intermediate.

Carbon dioxide reacts rapidly with O_2^- in aprotic solvents. The net stoichiometry of the reaction in acetonitrile is given by

$$2CO_2 + 2O_2^- \rightarrow C_2O_6^{2-} + O_2 \qquad (10)$$

and the proposed mechanism is outlined in scheme 9. This reaction is significant because it provides a route to an activated form of carbon dioxide that may be involved in the vitamin K-dependent carboxylation of glutamic acid residues (Esnouf *et al.* 1978). The results indicate that likely candidates for active intermediates are the anion radical, $CO_4^{-\cdot}$, or a hydrolysis product of $C_2O_6^{2-}$ such as peroxybicarbonate, $HOC(O)OO^-$.

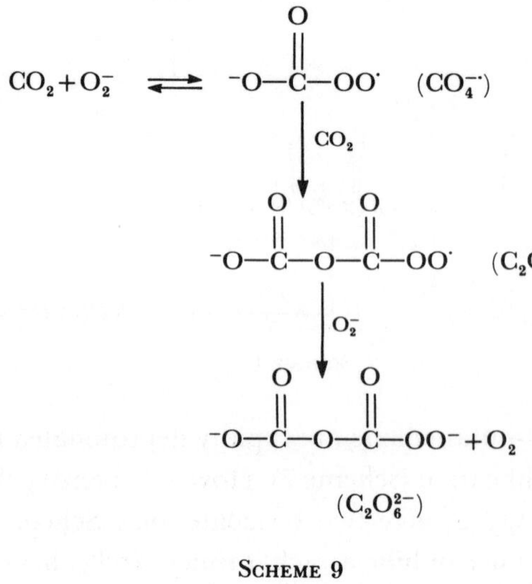

$$\mathrm{CO_2 + O_2^- \rightleftharpoons {}^-O{-}C(=O){-}OO^{\cdot}\quad (CO_4^{-\cdot})}$$

$$\mathrm{\xrightarrow{CO_2} {}^-O{-}C(=O){-}O{-}C(=O){-}OO^{\cdot}\quad (C_2O_6^{-\cdot})}$$

$$\mathrm{\xrightarrow{O_2^-} {}^-O{-}C(=O){-}O{-}C(=O){-}OO^- + O_2\quad (C_2O_6^{2-})}$$

Scheme 9

(iii) *Oxidation of compounds with hydrogen atoms on vicinal nitrogen or oxygen atoms*

Recent studies have demonstrated that 3,5-di-t-butylcatechol ($DTBCH_2$), ascorbic acid (H_2Asc), 1,2-disubstituted hydrazines, dihydrophenazine (H_2Phen), and dihydrolumiflavin (H_2Fl), are oxidized by O_2^- in aprotic media via a general mechanism (scheme 10) that involves the rapid sequential transfer to O_2^- of a proton and a hydrogen atom to form H_2O_2 and the anion radical of the dehydrogenated substrate (Sawyer *et al.* 1984; Calderwood *et al.* 1984; Sawyer *et al.* 1982*b*). Table 3 summarizes the stoichiometric and kinetic data for oxidation of these compounds by O_2^-.

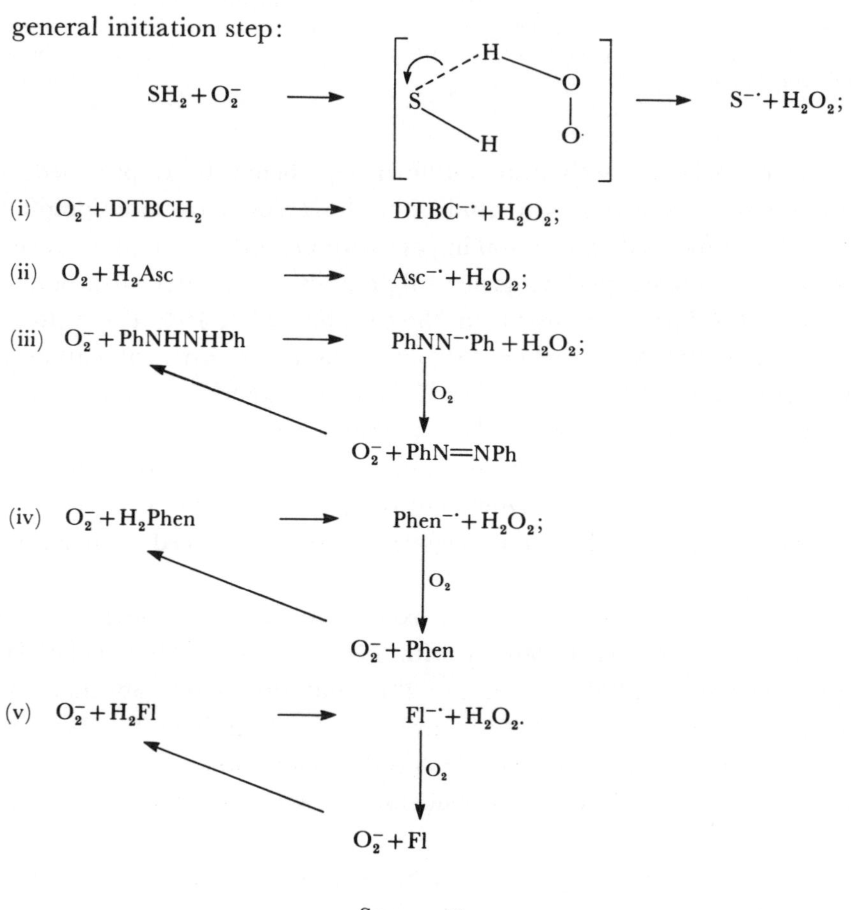

SCHEME 10

The azobenzene, phenazine and lumiflavin anion radicals are rapidly oxidized by dioxygen; hence O_2^- acts as an initiator for the autoxidation of these compounds (see scheme 10). For 1,2-diphenylhydrazine, turnover numbers in excess of 200 substrate molecules per O_2^- species have been observed. The 1,2-diphenylhydrazine autoxidation cycle can be initiated by OH^-, which indicates that O_2^- is formed in the OH^--initiated process. Superoxide ion also initiates the autoxidation of dihydrophenazine, which is a model for dihydroflavin. For example, the addition of 1 mM $(Me_4N)O_2$ in DMF to 10 mM H_2Phen in an O_2-saturated DMF solution results in the complete oxidation of the substrate (about 80% recovered as phenazine) and the production of 9–10 mM H_2O_2.

Table 3. Products and kinetics for the one-to-one combination of 2 mm $(Me_4N)O_2$ and 2 mm substrate in dimethylformamide (0.1 m tetraethylammonium perchlorate) at 25 °C

substrate, S[a]	anion radical/S[b]	H_2O_2/S	$k_1/[S]$ $(mol^{-1}\ s^{-1})$[a]
$DTBCH_2$	0.8	0.9	10^4
H_2Asc	0.8	0.9	1.8×10^4
H_2Phen	0.9	1.0	> 560
H_2Fl	0.8	0.9	> 340
PhNHNHPh	1.0	1.0	> 100

[a] $DTBCH_2$, 3,5-di-t-butylcatechol; H_2Asc, ascorbic acid; H_2Phen, dihydrophenazine; H_2Fl, dihydrolumiflavin.
[b] The u.v.–visible absorption spectra for the anion radical products were compared with those for the products from controlled potential electrolytic reduction of 3,5-di-t-butyl-*o*-benzoquinone, dehydroascorbic acid, phenazine, lumiflavin and azobenzene.

Support for the general mechanism outlined in scheme 10 is provided by gas-phase Fourier-transform–mass spectrometric studies of the anionic reaction products of several substrates with O_2^- (produced by electron impact with O_2; OH^- can be produced by electron impact with H_2O). In these experiments neutral products are not detected. Both O_2^- and OH^- react rapidly with 1,2-diphenylhydrazine in the gas phase ($P \approx 10^{-7}$ Torr†) to give the anion radical of azobenzene ($PhNN^{-\cdot}Ph$; $m/z = 182$) and the anion from deprotonation (PhN^-NHPh; $m/z = 183$), respectively. When O_2^- is ejected from the experiment, the peak at $m/z = 182$ disappears. In contrast to the exponential decay that is observed for the OH^- peak with time, the ion current for O_2^- decays to a steady-state concentration. Apparently, the $PhNN^{-\cdot}Ph$ product reacts with residual O_2 (which cannot be ejected from the F.t.–m.s. cell) to give O_2^- and azobenzene in a process that is analogous to the O_2^--induced autoxidation in aprotic solvents.

Analogous F.t.–m.s. studies with 1,2-dihydroxybenzenes also provide support for the general mechanism. Superoxide ion reacts rapidly with 3,5-di-t-butylcatechol ($DTBCH_2$) in the gas phase to give the anion ($DTBCH^-$; $m/z = 221$) and the anion radical of 3,5-di-t-butyl-*o*-benzoquinone ($DTBSQ^{-\cdot}$; $m/z = 220$) in an approximate ratio of 3:1. With hydroquinone [p-$Ph(OH)_2$] the dominant product (*ca.* 70%) is the anion radical ($SQ^{-\cdot}$; $m/z = 108$). When OH^- is the gas-phase reagent, the only product for $DTBCH_2$ (and for o-$Ph\ (OH)_2$) is the anion from deprotonation.

Parenthetically, an earlier study (Nanni & Sawyer 1980) formulated the reaction between O_2^- and dihydrophenazine (or dihydrolumiflavin) as

$$H_2Phen + O_2^- \rightarrow Phen + OH^{\cdot} + OH^-. \tag{11}$$

The experiments now appear to have been flawed by subsequent reaction of the anion radical of the dehydrogenated substrate with adventitious oxygen (or oxygen produced by base-catalysed decomposition of hydrogen peroxide). The more recent study (Calderwood *et al.* 1984) of these reactions confirms that they are not a source of hydroxyl radicals.

The fact that the anion radicals of the dehydrogenated substrates are produced in the gas phase as well as in aprotic solvents confirms that the reaction sequence deprotonation–

† 1 Torr = 101325/760 Pa.

hydrogen-atom abstraction to form H_2O_2 must either be a rapid sequence or a nearly concerted process. Because O_2^- is expected to abstract hydrogen atoms much less easily than $HO_2^{\cdot}$, the initial step is deprotonation of the substrate by O_2^- to form $HO_2^{\cdot}$; the latter (contained within the solvent cage or weakly bonded to the substrate in a 'sticky collision' in the gas phase) then abstracts a hydrogen atom from the substrate anion to form H_2O_2 and the anion radical of the dehydrogenated substrate.

Thus, the O_2^-- (or OH^--) induced autoxidations of 1,2-disubstituted hydrazines, dihydrophenazines, and dihydroflavins in aprotic media provide a simple pathway for rapid conversion of dioxygen to H_2O_2, and one that does not involve catalysis by metal ions or metalloproteins. This is exemplified by the net reaction for dihydrolumiflavin,

$$H_2Fl + O_2 \xrightarrow{O_2^-,\, OH^-} Fl + H_2O_2. \qquad (12)$$

(*b*) *Formation and reactivity of* $HO_2^{\cdot}$

In aprotic media, proton sources induce the rapid disproportionation of O_2^- to H_2O_2 and O_2 via formation of the perhydroxyl radical, $HO_2^{\cdot}$ (Chin *et al.* 1982),

$$O_2^- + HA \rightarrow A^- + HO_2^{\cdot} \xrightarrow{k} \tfrac{1}{2}H_2O_2 + \tfrac{1}{2}O_2. \qquad (13)$$

In aqueous media $HO_2^{\cdot}$ has a pK_a value of 4.9 and a disproportionation rate constant k of 10^6 mol^{-1} s^{-1}, but in dimethylformamide is estimated pK_a is 12 and k is greater than 10^7 mol^{-1} s^{-1}. Hence, formation of superoxide ion in an aprotic medium (biological membrane), that is at neutral pH, will result in $HO_2^{\cdot}$ as the dominant species.

Some years ago Howard & Ingold (1967) used radical-initiated autoxidation experiments in acetonitrile and chlorobenzene to demonstrate that $HO_2^{\cdot}$ abstracts hydrogen atoms from allylic hydrocarbons (1,4-cyclohexadiene). Because linoleic acid and arachidonic acid esters contain allylic groups and are important components of lipids, there is reason to believe that *in situ* generation of $HO_2^{\cdot}$ (via O_2^- plus H^+) can initiate lipid peroxidation and autoxidation. This has prompted some preliminary experiments on the reactivity of O_2^- with 1,4-cyclohexadiene in acidified dimethyl sulphoxide; the results are summarized in table 4. A reasonable reaction

Table 4. Oxidation of 1,4-cyclohexadiene (1,4-CHD) by $HO_2^{\cdot}$ (O_2^- plus HA) in dimethyl sulphoxide[a]

proton source (HA)	[HA]/mM	[O_2^-]/mM	[1,4-CHD]/mM	reaction efficiency (percentage)[b]	product distribution (percentage) 1,3-CHD	PhH
H_2O	100	5.3	10.6	25	74	26
H_2O	1000	3.3	6.6	90	92	8
$HClO_4$	1.6	3.2	6.4	100	79	21
$HClO_4$	3.2	3.2	6.4	92	69	31
$HClO_4$	6.4	3.2	6.4	31	54	46
$HClO_4$	8.4	8.4	17.0	40	75	25

[a] A 10 mM O_2^- solution (Me_2SO) was slowly added to an Me_2SO solution that contained 1,4-CHD and the proton source (HA). The indicated concentrations represent the initial values after mixing.

[b] 100% represents the reaction of one 1,4-CHD molecule per O_2^- added.

scheme involves the initial formation of $HO_2^{\bullet}$ with its subsequent disproportionation (second order in $HO_2^{\bullet}$) and attack of 1,4-cyclohexadiene (first order in $HO_2^{\bullet}$) (see scheme 11).

Other organic molecules with weak heteroatom—H bonds include Ph(Me)N—H, Ph(H)N—H and thiols (RS—H); these should be susceptible to $HO_2^{\bullet}$-initiated oxidations and autoxidations. Studies of the reactivity for $HO_2^{\bullet}$ with these substrates and with the esters of linoleic acid, arachidonic acid and other molecules with allylic groups are in progress.

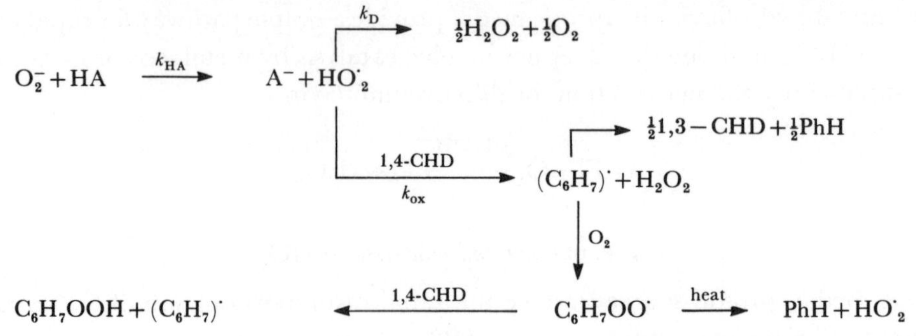

$$O_2^- + HA \xrightarrow{k_{HA}} A^- + HO_2^{\bullet}$$

$$A^- + HO_2^{\bullet} \xrightarrow{k_D} \tfrac{1}{2}H_2O_2 + \tfrac{1}{2}O_2$$

$$HO_2^{\bullet} \xrightarrow[k_{ox}]{1,4\text{-CHD}} (C_6H_7)^{\bullet} + H_2O_2$$

$$(C_6H_7)^{\bullet} \rightarrow \tfrac{1}{2}1,3-CHD + \tfrac{1}{2}PhH$$

$$(C_6H_7)^{\bullet} \xrightarrow{O_2} C_6H_7OO^{\bullet}$$

$$C_6H_7OO^{\bullet} \xrightarrow{1,4\text{-CHD}} C_6H_7OOH + (C_6H_7)^{\bullet}$$

$$C_6H_7OO^{\bullet} \xrightarrow{heat} PhH + HO_2^{\bullet}$$

SCHEME 11

(c) *Iron(II)-induced activation of* H_2O_2

Recent work (Sugimoto & Sawyer 1984) establishes that $Fe^{II}(MeCN)_4(ClO_4)_2$ in dry acetonitrile (MeCN) catalyses the rapid disproportionation of added 98% H_2O_2 to O_2 and H_2O, but all of the catalyst remains in the Fe^{II} oxidation state. Table 5 summarizes the results from the addition of dry H_2O_2 (98%, dissolved in MeCN) to solutions of various organic substrates in the presence of the Fe^{II} catalyst. Three classes of reaction occur on the basis of the substrate: (*a*) monoxygenations, (*b*) dehydrogenations and oxidations and (*c*) dioxygenations.

The products are in marked contrast to those observed for aqueous Fenton chemistry. However, the presence of 1% H_2O in the MeCN reaction system results in the oxidation of Fe^{II} and substrate products that are characteristic of the Fenton process. Fenton chemistry is generally believed to be induced by the $OH^{\bullet}$ radical that is produced from the reduction of H_2O_2 by Fe^{II} (Walling 1975):

$$Fe^{II} + H_2O_2 \rightarrow Fe^{III}(OH) + OH^{\bullet}. \tag{14}$$

The unique feature of the anhydrous system for the activation of H_2O_2 is that the acetonitrile matrix for the Fe^{II} catalyst causes the Fe^{III}/Fe^{II} redox potential to be greater than +1.8 V against n.H.e., compared to about +0.4 V against n.H.e. in water at pH 7. This large shift of redox potential precludes the reduction of H_2O_2 by Fe^{II}. As a result, the Fe^{II} catalyst remains in its reduced state for all of the reactions in dry acetonitrile.

At present, little is known about the structure of the activated Fe^{II}–H_2O_2 complexes. The disproportionation reaction for H_2O_2 and the three types of substrate reactions indicate that more than one kind of complex may be present, perhaps in dynamic equilibrium. Scheme 12 presents possible models for the oxidase–monoxygenase function and the disproportionase (catalase)–dioxygenase function.

TABLE 5. PRODUCTS FROM THE IRON(II)-INDUCED MONOXYGENATION, DEHYDROGENATION, AND DIOXYGENATION OF ORGANIC SUBSTRATES (RH) BY H_2O_2 IN DRY ACETONITRILE[a]

substrate	reaction efficiency (percentage)	products
monoxygenation		
blank (H_2O_2)	100	O_2, H_2O, Fe(II)
Ph_3P	100	Ph_3PO
Me_2SO	100	Me_2SO_2
Ph_2SO	100	Ph_2SO_2
EtOH	70	MeCH(O) (90%), MeC(O)OH (10%), O_2
$PhCH_2OH$	100	PhCH(O)
c-$C_6H_{11}OH$	47	$C_6H_{10}(O)$, O_2
MeCH(O)	20	MeC(O)OH, O_2
$Me_2C(O)$	NR	O_2
PhCH(O)	28	PhC(O)OH, O_2
dehydrogenation and oxidation		
cyclohexane	NR	O_2
1,4-*c*-C_6H_8	59	PhH, O_2
PhNHNHPh	100	PhN=NPh
H_2S	100	H_2SO_4
H_2O (56 mM)	100	Fe(III)
dioxygenation		
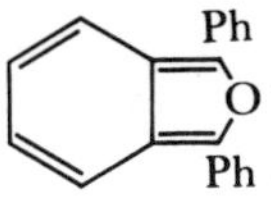 1,3-diphenylisobenzofuran (structure)	100	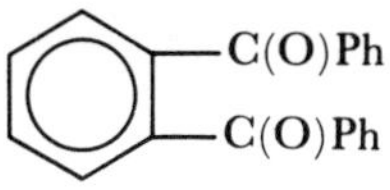 C_6H_4(C(O)Ph)$_2$ (structure: -C(O)Ph, -C(O)Ph)
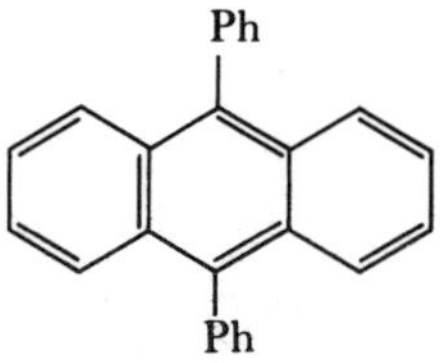 9,10-diphenylanthracene (structure: Ph, Ph)	69	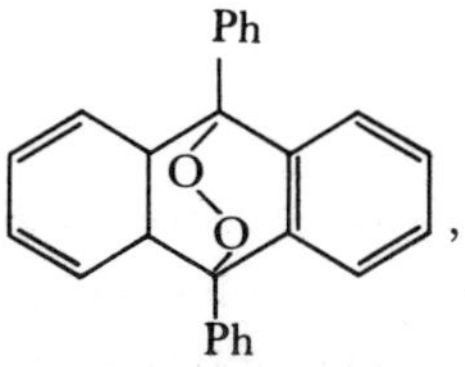 9,10-endoperoxide (structure: Ph, O–O, Ph), O_2
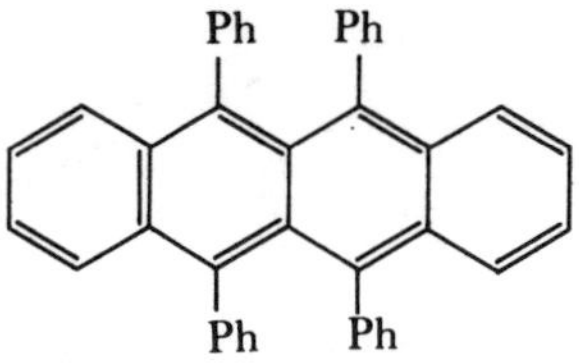 5,6,11,12-tetraphenylnaphthacene (structure: Ph Ph, Ph Ph)	83	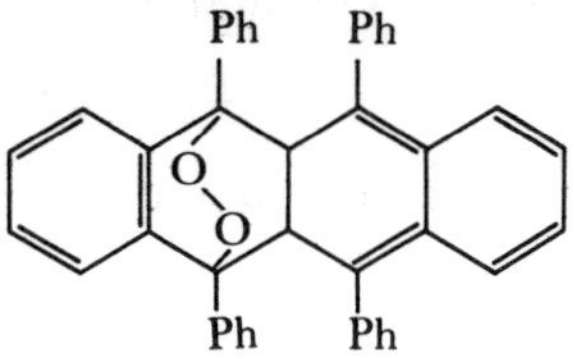 endoperoxide (structure: Ph Ph, O–O, Ph Ph), O_2
$Ph_2C{=}CPh_2$	22	$Ph_2C(O)$, O_2
PhC≡CPh	42	PhC(O)C(O)Ph, O_2
PhC≡CMe	26	PhC(O)C(O)Me, O_2
PhC≡CH	11	PhC(O)CH(O), O_2
c-PhCH=CHPh	52	PhCH(O) (98%), PhC≡CPh (2%), O_2
t-PhCH=CHPh	28	PhCH(O), O_2
PhCH=CHMe	32	PhCH(O) + MeCH(O) (85%), PhCHCHOMe (15%), O_2

[a] (Sugimoto & Sawyer 1984). Product solution [from the slow addition (*ca.* 5 min to give a final 2 mM concentration) of 1 M H_2O_2 (98% H_2O_2 in MeCN) to a solution of 1 mM $[Fe^{II}(MeCN)_4](ClO_4)_2$ plus 2 mM substrate] analysed by gas chromatography and assayed for residual Fe^{II} by MnO_4^- titration and by colorimetry with 1,10-*o*-phenanthroline.

[b] 100% represents one substrate oxygenation or dehydrogenation per H_2O_2 added. For dioxygenations, 100% represents one substrate converted per two H_2O_2 added.

[c] 100% represents one H_2S converted to H_2SO_4 per four H_2O_2 added.

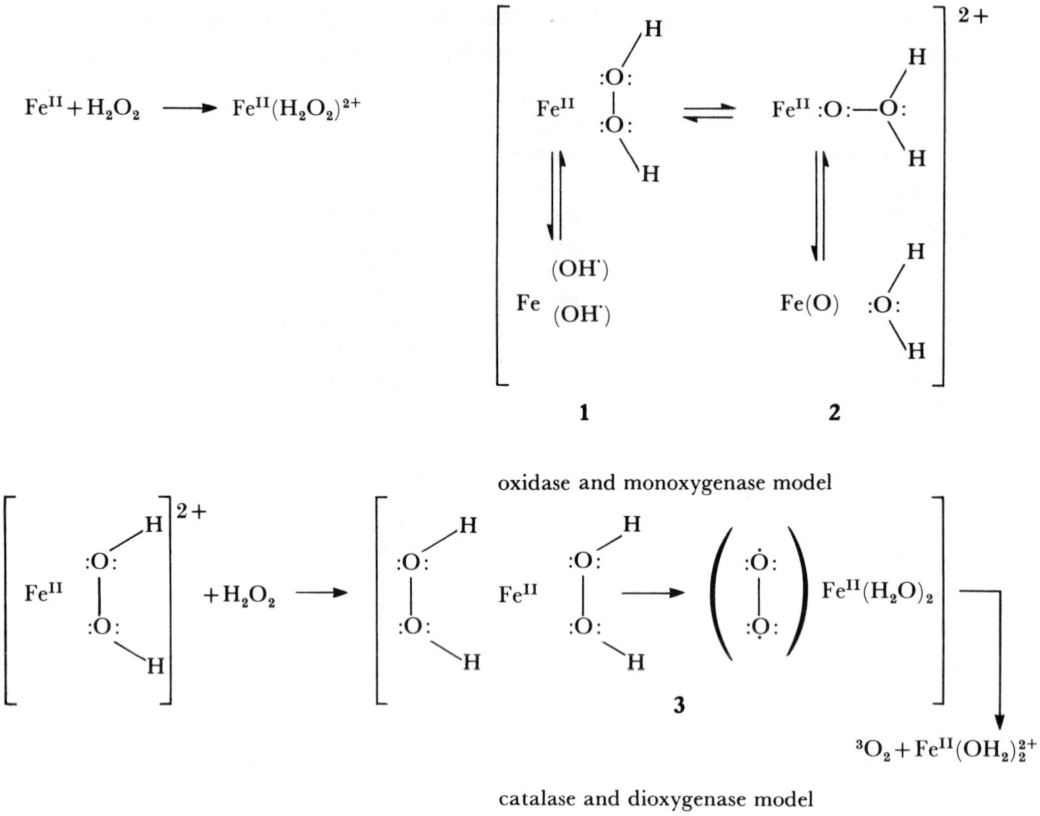

$Fe^{II} + H_2O_2 \longrightarrow Fe^{II}(H_2O_2)^{2+}$

oxidase and monoxygenase model

$^3O_2 + Fe^{II}(OH_2)_2^{2+}$

catalase and dioxygenase model

SCHEME 12

The elements of this chemistry can be coupled with the observed base-induced autoxygenation described in §2*c* to form a model chemical system for cytochrome P_{450} monoxygenase (Sawyer *et al.* 1984*b*). This is characterized by the net overall reaction

$$RH + DH_2 + O_2 \rightarrow ROH + D + H_2O, \tag{15}$$

where RH represents the substrate and DH_2 a two-electron reductant (donor), such as reduced flavin or ascorbic acid.

The model consists of (i) an O_2-activation segment that produces H_2O_2 from the base-initiated autoxidation of 1,2-diphenylhydrazine (a model for reduced flavin),

$$PhNHNHPh + O_2 \xrightarrow{O_2^-,\, OH^-} PhN{=}NPh + H_2O_2 \tag{16}$$

and (ii) a H_2O_2-activation component via the Fe^{II}–H_2O_2 complex in dry acetonitrile, shown as complex **2** in scheme 12. The combination of (i) and (ii) provides a catalytic system for the autoxygenation of organic substrates with reaction cycles that are similar to those for cytochrome P_{450} monoxygenases. Thus, when catalytic quantities of base (O_2^- or OH^-) are introduced into a dry acetonitrile solution that contains excess substrate (RH), ambient air (O_2), 1,2-diphenylhydrazine (PhNHNHPh) and Fe^{II}, the substrate is rapidly and efficiently monoxygenated (e.g. triphenylphosphine → triphenylphosphine oxide; benzyl alcohol → benzaldehyde; diphenylsulphoxide → diphenylsulphone) or dehydrogenated-(1,4-cyclohexadiene → benzene).

(*d*) $Fe^{III}Cl_3$-*induced activation of* H_2O_2

The observation (Sugimoto & Sawyer 1984) that iron(II) in ligand-free acetonitrile activates hydrogen peroxide to act as a monoxygenase and dehydrogenase (but not as an initiator of radical reactions via Fenton chemistry) has prompted the considerations of other iron salts. Thus, anhydrous ferric chloride ($Fe^{III}Cl_3$) in dry acetonitrile (MeCN) activates hydrogen peroxide to epoxidize alkenes, and to monoxygenate or dehydrogenate other organic substrates (Sugimoto & Sawyer 1985).

Table 6*a* summarizes the conversion efficiencies and product distributions for a series of alkene substrates subjected to the $Fe^{III}Cl_3$–H_2O_2/MeCN system. The extent of the $Fe^{III}Cl_3$-induced monoxygenations is enhanced by higher reaction temperatures and increased concentrations of the reactants (substrate, $Fe^{III}Cl_3$ and H_2O_2). For 1-hexene (representative of all of the alkenes) a substantial fraction of the product is the dimer of 1-hexene oxide, a disubstituted dioxane

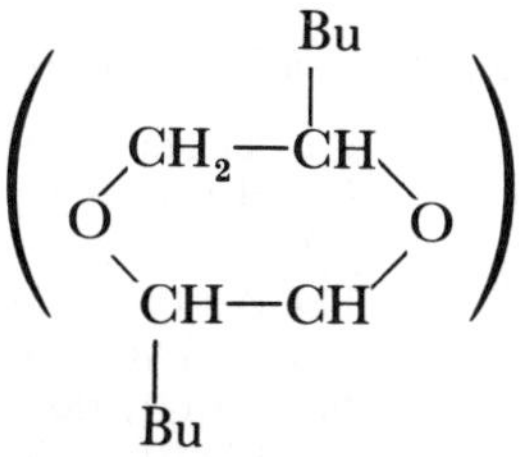

$$\left(\begin{array}{c} \text{Bu} \\ | \\ \text{O}\!\!\diagup\!\!\text{CH}_2\text{—CH}\!\diagdown\!\text{O} \\ \diagdown\text{CH—CH}\diagup \\ | \\ \text{Bu} \end{array}\right).$$

With other organic substrates (RH), $Fe^{III}Cl_3$ activates H_2O_2 for their monoxygenation and the reaction efficiencies and product distributions are summarized in table 6*b*. In the case of alcohols, ethers and cyclohexane, a substantial fraction of the product is the alkyl chloride, and with aldehydes [PhCH(O)] the acid chloride represents one-half of the product. In the absence of substrate the $Fe^{III}Cl_3$/MeCN system catalyses the rapid disproportionation of H_2O_2 to O_2 and H_2O.

Because $Fe^{III}Cl_3$ is an exceptionally strong Lewis acid and electrophilic centre, it activates H_2O_2 (which acts as a nucleophile) for the dehydrogenation of a second H_2O_2. On the basis of this disproportionation process, as well as the monoxygenation and dehydrogenation reactions of table 6, the activation of H_2O_2 by $Fe^{III}Cl_3$ probably involves the initial formation of at least two reactive forms of an $Fe^{III}Cl_3$ (HOOH) adduct that are in dynamic equilibrium,

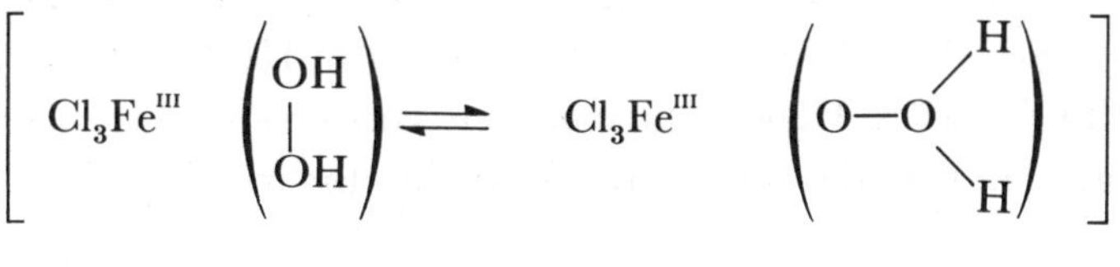

$$\left[\mathrm{Cl_3Fe^{III}}\left(\begin{array}{c}\mathrm{OH}\\|\\\mathrm{OH}\end{array}\right) \rightleftharpoons \mathrm{Cl_3Fe^{III}}\left(\mathrm{O{-}O}\begin{array}{l}\diagup\mathrm{H}\\ \diagdown\mathrm{H}\end{array}\right)\right].$$

The disproportionation of H_2O_2 occurs via a concerted transfer of the two hydrogen atoms from a second H_2O_2 to the $Fe^{III}Cl_3(H_2O_2)$ adduct. This dehydrogenation of H_2O_2 is a competitive process with the $Fe^{III}Cl_3$–substrate–H_2O_2 reactions. The controlled introduction of dilute H_2O_2 into the $Fe^{III}Cl_3$–substrate solution limits the concentration of H_2O_2 and ensures that the substrate–H_2O_2 reaction can be competitive with the second-order disproportionation process. The substrate reaction efficiencies in table 6 are proportional to the relative rates of reaction ($k_{RH}/k_{H_2O_2}$). The mode of activation of H_2O_2 by $Fe^{III}Cl_3$ is analogous to that of

Table 6. Products and conversion efficiencies for the ferric chloride ($Fe^{III}Cl_3$)-induced oxygenation–dehydrogenation of olefins and organic substrates (RH) by H_2O_2 in acetonitrile

substrate, RH	reaction efficiency (percentage)[a, b]	products[c]
	(a) olefins (−5 °C; 10 *min reaction times*)	
blank (H_2O_2)	100	O_2, H_2O
1-hexene	10	epoxide (1-hexene oxide) (71 %), dimer (dioxane) (10 %), others (19 %)
1-hexene (+5°)	23	epoxide (55 %), dimer (15 %), others (30 %)
1-octene	60	epoxide (53 %), dimer (10 %)
cyclohexene	25	epoxide (45 %), dimer (30 %)
$Me_2C{=}CMe_2$	40	$Me_2C\overset{O}{—}CMe_2$ (50 %), dimers and others (50 %)
	(b) other substrates (+5 °C, 20 *min reaction times*)	
cyclohexanol	52	cyclohexanone (88 %)
$PhCH_2OH$	63	PhCH(O) (51 %), $PhCH_2Cl$ (21 %), PhC(O)OH (14 %), PhC(O)Cl (14 %)
$PhCH_2OCMe_3$	56	PhCH(O) (72 %), $PhCH_2Cl$ (11 %), PhC(O)OH (3 %), PhC(O)Cl (14 %)
PhCH(O)	75	PhC(O)OH (55 %), PhC(O)Cl (45 %)
$PhCH_3$ (25 °C)	2	$PhCH_2OH$, PhCH(O), PhC(O)Cl, PhC(O)OH, cresols
cyclohexane	22	cyclohexylchloride (45 %), cyclohexanol (40 %), cyclohexanone (15 %)
Ph_2S	58	Ph_2SO (100 %)
Ph_2SO	60	Ph_2SO_2 (100 %)
Ph_3P	80	Ph_3PO (100 %)

[a] RH and $Fe^{III}Cl_3$ (1.0 mmol of each) combined in 10–20 ml dry MeCN, followed by the slow addition of 1 mmol H_2O_2 [1 M H_2O_2 (98 %) in MeCN].
[b] Percentage of substrate converted to products.
[c] After the indicated reaction time, the product solution was quenched with water, extracted with diethylether and analysed by capillary gas chromatography and g.c.–m.s.

$Fe^{II}(MeCN)_4^{2+}$; both are strong electrophiles in ligand-free dry MeCN and induce H_2O_2 to monoxygenate organic substrates.

The epoxidation of alkenes (table 6*a*) appears to involve an O-atom transfer from the end-on configuration of the $Fe^{III}Cl_3(HOOH)$ adduct. The electrophilicity of $Fe^{III}Cl_3$ promotes the initial activation of the alkene bond before the binding of H_2O_2. The resulting epoxides are rapidly dimerized to dioxanes. Hence, the complete conversion of an alkene to its epoxide is precluded; the more complete the conversion the higher the fraction of dioxane in the product mixture.

The results in table 6*b* indicate that the $Fe^{III}Cl_3(HOOH)$ adduct monoxygenates alkanes, alcohols and aldehydes. A mechanism that is consistent with this involves the homolytic scission of the HO–OH bond in the side-on configuration, induced by the bound substrate, and the subsequent abstraction by one $OH^{\cdot}$ of an H-atom from the α-carbon and addition of the second $HO^{\cdot}$ to the resulting carbon radical (equation (17)).

$$PhCH_2OH + Fe^{III}Cl_3(H_2O_2) \rightarrow [PhCH(OH)_2]Fe^{III}Cl_3(OH_2) \rightarrow PhCH(O) + Fe^{III}Cl_3(OH_2)_2. \quad (17)$$

An analogous process appears to occur for the oxygenation of benzaldehyde by the $Fe^{III}Cl_3$ (HOOH) adduct, but 50% of the product is the acid chloride.

This result indicates that the activated side-on complex has some hypochlorous acid (HOCl) character and can add a chlorine atom to the carbon radical that results from the H-atom abstraction by the $OH^{\cdot}$ group. This also occurs with alkanes, alcohols, and ethers (table 6*b*). Such chemistry is similar to the activation of chloride ion and H_2O_2 to HOCl by a haem protein, myeloperoxidase (Rosen & Klebanoff 1977; Held & Hurst 1978).

Phosphines, dialkylsulphides and sulphoxides are monoxygenated by the $Fe^{III}Cl_3(HOOH)$ adduct in a manner that appears to be analogous to that for the epoxidation of alkenes.

3. Model systems for biological O_2 and H_2O_2 activation

The base-catalysed autoxidations and the Lewis-acid-catalysed (iron(II) and $FeCl_3$) reactions of H_2O_2 with various substrates exhibit parallels to reactions that are catalysed by metallo-enzymes. They may, therefore, be useful models for various enzyme-catalysed reactions. For example, xanthine oxidase (XO) normally acts *in vivo* to catalyse the oxidation of reduced flavin,

$$H_2Fl + O_2 \xrightarrow{XO} H_2O_2 + Fl, \tag{18}$$

but is known to produce a flux of superoxide ions in the presence of xanthine and O_2 (Fridovich 1983). Thus there is a superficial resemblance between the xanthine oxidase-catalysed reaction and the superoxide ion-catalysed autoxidation of reduced flavin that is described in §2*b*(iii).

The Fe^{II}- and $FeCl_3$-catalysed disproportionation of H_2O_2 in dry acetonitrile (see §§2*c* and 2*d*) is analogous to that of catalase,

$$2H_2O_2 \xrightarrow{Fe^{II}} O_2 + 2H_2O \tag{19}$$

and in the presence of appropriate substrates these Lewis acids display a peroxidase-like activity,

$$H_2O_2 + RH \xrightarrow{Fe^{II}} H_2O + ROH \tag{20}$$

that is illustrated by the oxidation of benzaldehyde to benzoic acid.

Finally, the modelling of the cytochrome P_{450} monoxygenation reaction cycles, which is discussed in §2*c*, involves (i) an O_2-activation component whereby a donor molecule (DH_2; PhNHNHPh) is autoxidized to produce H_2O_2 and (ii) a peroxidase-like component whereby Fe^{II} activates H_2O_2 for the monoxygenation

$$DH_2 + O_2 \xrightarrow{O_2^-} H_2O_2 + D, \tag{21}$$

$$H_2O_2 + RH \xrightarrow{Fe^{II}} ROH + H_2O. \tag{22}$$

Although the reactions shown in table 5 parallel the 'oxene' monoxygenase chemistry that is catalysed by cytochrome P_{450}, the lack of significant reactivity with cyclohexene and norbornene indicates that $Fe^{II}(H_2O_2)^{2+}$ is an inadequate model for the reactive iron-oxygen centre of the enzyme (Sawyer *et al.* 1984). The $Fe^{III}Cl_3$–H_2O_2 system that is discussed in §2*d* is a much more effective monoxygenase and epoxidizing agent.

These models demonstrate that the chemistry of $Fe^{II}(MeCN)_4^{2+}$ and $Fe^{III}Cl_3$ is dramatically altered when water is removed to provide an aprotic and ligand-free environment and support the thesis that oxygen-activation processes in aprotic solvents are useful models for the chemistry of dioxygen in biological membranes and in the hydrophobic regions of metalloproteins.

This work was supported by the National Science Foundation under grant no. CHE-8212299.

References

Bielski, B. H. J. 1978 *Photochem. Photobiol.* **28**, 645–649.

Bielski, B. H. J. 1983 In *Oxy radicals and their scavenger systems* (ed. G. Cohen & R. A. Greenwald), vol. 1, pp. 1–7. New York: Elsevier.

Bull, C., McClune, G. J. & Fee, J. A. 1983 *J. Am. chem. Soc.* **105**, 5290–5300.

Calderwood, T. S., Johlman, C. L., Roberts, Jr, J. L., Wilkins, C. L. & Sawyer, D. T. 1984 *J. Am. chem. Soc.* **106**, 4683–4687.

Calderwood, T. S., Neuman, Jr., R. C. & Sawyer, D. T. 1983 *J. Am. chem. Soc.* **105**, 3337–3339.

Calderwood, T. S. & Sawyer, D. T. 1984 *J. Am. chem. Soc.* **106**, 7185–7186.

Chin, D.-H., Chiericato, G., Nanni, Jr., E. J. & Sawyer, D. T. 1982 *J. Am. chem. Soc.* **104**, 1296–1299.

Clark, J. H. 1980 *Chem. Rev.* **80**, 429–452.

Crooks, J. E. 1975 In *Proton-transfer reactions* (ed. E. Caldin & V. Gold). New York: Wiley.

Eberson, L. 1982 *Adv. phys. org. chem.* **18**, 79–185.

Esnouf, M. P., Green, M. R., Hill, H. A. O., Irvine, G. B. & Walter, S. J. 1978 *Biochem. J.* **174**, 345–348.

Foote, C. S. 1976 In *Free radicals in biology* (ed. W. A. Pryor), vol. 2, pp. 85–133. New York: Academic Press.

Fridovich, I. 1982 In *Superoxide dismutase* (ed. L. W. Oberley), vol. 1, pp. 1–9. Boca Raton, Florida: C.R.C. Press.

Frimer, A. A. 1983 In *The chemistry of functional groups, peroxides* (ed. S. Patai), pp. 429–461. New York: Wiley.

Gebicki, J. M. & Bielski, B. H. J. 1981 *J. Am. chem. Soc.* **103**, 7020–7022.

Gibian, M. J., Sawyer, D. T., Ungerman, T., Tangpoonpholvivat, R. & Morrison, M. M. 1979 *J. Am. chem. Soc.* **101**, 640–644.

Green, M. J. & Hill, H. A. O. 1984 In *Methods in enzymology* (ed. L. Packer, S. P. Colowick & N. O. Kaplan), vol. 105, pp. 3–22. New York: Academic Press.

Hamilton, G. A. 1974 In *Molecular mechanisms for oxygen activation* (ed. O. Hayaishi), pp. 405–451. New York: Academic Press.

Held, A. M. & Hurst, J. K. 1978 *Biochim. biophys. Res. Commun.* **81**, 878.

Howard, J. A. & Ingold, K. V. 1967 *Can. J. Chem.* **45**, 785–792.

Johlman, C. L., White, R. L., Sawyer, D. T. & Wilkins, C. L. 1983 *J. Am. chem. Soc.* **105**, 2091–2092.

Koppenol, W. H. 1983 In *Oxy radicals and their scavenger systems* (ed. G. Cohen & R. A. Greenwald), vol. 1, pp. 274–277. New York: Elsevier.

Krinsky, N. I. 1979 In *Singlet oxygen* (ed. H. H. Wasserman & R. W. Murray), pp. 597. New York: Academic Press.

Lim, M.-C. & Sawyer, D. T. 1982 *Inorg. Chem.* **21**, 2839–2841.

Magno, F. & Bontempelli, G. 1976 *J. electroanal. Chem.* **68**, 337–344.

Mason, R. P. 1982 In *Free radicals in biology* (ed. W. A. Pryor), vol. 6, pp. 161–222. New York: Academic Press.

McCandlish, E., Miksztal, A. R., Nappa, M., Sprenger, A. Q., Valentine, J. S., Strong, J. D. & Spiro, T. G. 1980 *J. Am. chem. Soc.* **102**, 4268–4271.

Merritt, M. V. & Johnson, R. A. 1977 *J. Am. chem. Soc.* **99**, 3713–3719.

Nanni, Jr., E. J. & Sawyer, D. T. 1980 *J. Am. chem. Soc.* **102**, 7593–7595.

Oae, S. & Takata, S. 1980 *Tetrahedron Lett.* **21**, 3689–3692.

Perrin, C. L. 1984 *J. phys. Chem.* **88**, 3611–3615.

Porter, N. A. 1984 In *Methods in enzymology* (ed. L. Packer, S. P. Colowick & N. O. Kaplan), vol. 105, pp. 273–282. New York: Academic Press.

Pryor, W. A. 1976 In *Free radicals in Biology* (ed. W. A. Pryor), vol. 1, pp. 1–49. New York: Academic Press.

Roberts, Jr., J. L., Calderwood, T. S. & Sawyer, D. T. 1983 *J. Am. chem. Soc.* **105**, 7691–7696.

Roberts, Jr., J. L., Calderwood, T. S. & Sawyer, D. T. 1984 *J. Am. chem. Soc.* **196**, 4667–4670.

Roberts, Jr., J. L., Morrison, M. M. & Sawyer, D. T. 1978 *J. Am. chem. Soc.* **100**, 329–330.

Roberts, Jr., J. L. & Sawyer, D. T. 1981 *J. Am. chem. Soc.* **103**, 712–714.

Roberts, Jr., J. L. & Sawyer, D. T. 1983 *Israel J. Chem.* **23**, 430–438.

Rosen, H. & Klebanoff, S. J. 1977 *J. biol. Chem.* **252**, 4803

Sawyer, D. T., Calderwood, T. S., Johlman, C. L. & Wilkins, C. L. 1985 *J. org. Chem.* **50**, 1409–1412.

Sawyer, D. T., Chiericato, Jr., G., Angelis, C. T., Nanni, Jr., E. J. & Tsuchiya, T. 1982*a* *Anal. Chem.* **54**, 1720–1724.
Sawyer, D. T., Chiericato, Jr, G. & Tsuchiya, T. 1982*b* *J. Am. chem. Soc.* **104**, 6273–6278.
Sawyer, D. T. & Gibian, M. J. 1979 *Tetrahedron* **35**, 1471–1481.
Sawyer, D. T., Gibian, M. J., Morrison, M. M. & Seo, E. T. 1978 *J. Am. chem. Soc.* **100**, 627–628.
Sawyer, D. T., Roberts, Jr., J. L., Tsuchiya, T. & Srivatsa, G. S. 1984 In *Oxygen radicals in chemistry and biology* (ed. W. Bors, M. Saran & D. Tait), pp. 25–33. Berlin: Walter de Gruyter & Co.
Sawyer, D. T., Stamp, J. J. & Menton, K. A. 1983 *J. org. Chem.* **48**, 3733–3736.
Sawyer, D. T., Sugimoto, H. & Calderwood, T. S. 1984*b* *Proc. natn. Acad. Sci. U.S.A.* **81**, 8025–8027.
Sawyer, D. T. & Valentine, J. S. 1981 *Acc. Chem. Res.* **14**, 393–400.
Simic, M. G. & Karel, M. (eds.) 1980 *Autoxidation in food and biological systems.* New York: Plenum Press.
Slater, T. F. 1982 In *Free radicals, lipid peroxidation and cancer* (ed. D. C. H. McBrien & T. F. Slater), pp. 243–274. New York: Academic Press.
Spiro, T. G. (ed.) 1980 *Metal ion activation of dioxygen.* New York: Wiley.
Sugimoto, H. & Sawyer, D. T. 1984 *J. Am. chem. Soc.* **106**, 4283–4285.
Sugimoto, H. & Sawyer, D. T. 1985 *J. org. Chem.* **50**, 1784–1786.
Taube, H. 1965 *J. gen. Physiol.* **49**, 29–50.
Valentine, J. S. 1979 In *Biochemical and clinical aspects of oxygen* (ed. W. S. Caughey), pp. 659–677. New York: Academic Press.
Walling, C. 1975 *Acc. Chem. Res.* **8**, 125–131.

Discussion

P. WARDMAN (*Gray Laboratory, Mount Vernon Hospital, Northwood, Middlesex HA*6 2*RN*). Professor Sawyer has stressed the increased oxidizing power of the superoxide radical compared with oxygen: this is indeed a most important point. However, this behaviour is not 'unique', at least in respect of physiological conditions. It seems the norm that oxidants capable of adding two electrons sequentially generally become more powerful oxidants when the first electron is added. Common examples include quinones, flavins, nicotinamides and dehydroascorbic acid as well as oxygen. (An alternative statement of this generalization is that the semiquinone formation constant, $[Q^{-\cdot}]^2/([Q]\,[QH_2])$, is less than unity in water at pH 7.) The only exceptions I can recall are those of the 2,2′- and 4,4′- bipyridinium dications (viologens; paraquat, diquat, etc.), where the reduction potential for adding the second electron is more negative than that for the first. This characteristic is responsible for some of their exceptional properties, of course.

Phil. Trans. R. Soc. Lond. B **311**, 505–516 (1985)
Printed in Great Britain

Radicals: their importance in synthetic chemistry and their relevance to biology

By Sir Derek Barton, F.R.S., and S. Z. Zard

Institut de Chimie des Substances Naturelles, C.N.R.S., 91190 *Gif-sur-Yvette, France*

Radical chain reactions are not often used in synthetic chemistry, although in fact such reactions can give good yields and show a selectivity that complements perfectly the ionic reactions that are in more general use. The design of radical chain reactions is discussed as well as a new method for obtaining carbon radicals in good yields under mild conditions.

Radical reactions have, of course, an enormous importance in the synthesis of polymers. However, we are concerned here with their use in the chemical synthesis of homogenous, low molecular mass molecules. In this area of scientific research, radical reactions are not often used because they are considered to be unselective and to give poor yields of products. It is the purpose of this article to show that well designed radical reactions can give high yields of single products and play an important role in organic synthesis. We will concentrate on reactions that give a high yield.

Radicals react with various functional groups at very different rates, which can vary over many powers of ten. By a judicious choice of reagents, solvents and other conditions one can devise a system capable of effecting a highly selective transformation. However, radical–radical interactions (such as coupling, disproportionation) are extremely fast processes and therefore much more difficult to control. If the sequence of radical reactions is conceived so as to constitute a linear chain process where the propagating steps are so fast that the concentration of radical species remains very small, radical–radical interactions can be largely avoided. Under such controlled conditions, clean high-yielding reactions become possible. Moreover, radical and radical chain reactions have several advantages over conventional ionic processes: neutral conditions; lower steric effects; lower polar effects; lower tendancy to unwanted elimination reactions and tolerance of many functional groups that have to be protected in ionic chemistry.

The following reactions will illustrate the scope and utility of well designed radical reactions.

Some years ago, we became interested in the problem of introducing the important 17α-hydroxy group in adrenocortical hormones such as **1**. An unexpected observation during

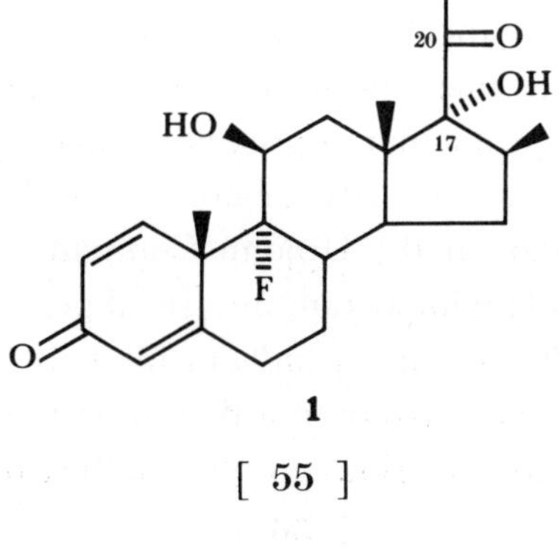

OR
20 O
OH
HO
17
F
O

1

the determination of the structure of limonin (Barton *et al.* 1961) suggested that the enolate of a suitable 20-ketopregnane derivative (**2**) would react with triplet oxygen with electron transfer to give a superoxide anion and a carbon radical (**3**). The latter would then react very rapidly with oxygen to produce a hydroperoxy radical (**4**), which by electron transfer from the enolate would give the hydroperoxide anion **5** and reform the carbon radical to carry the chain (scheme 1) (Barton *et al.* 1960, 1962).

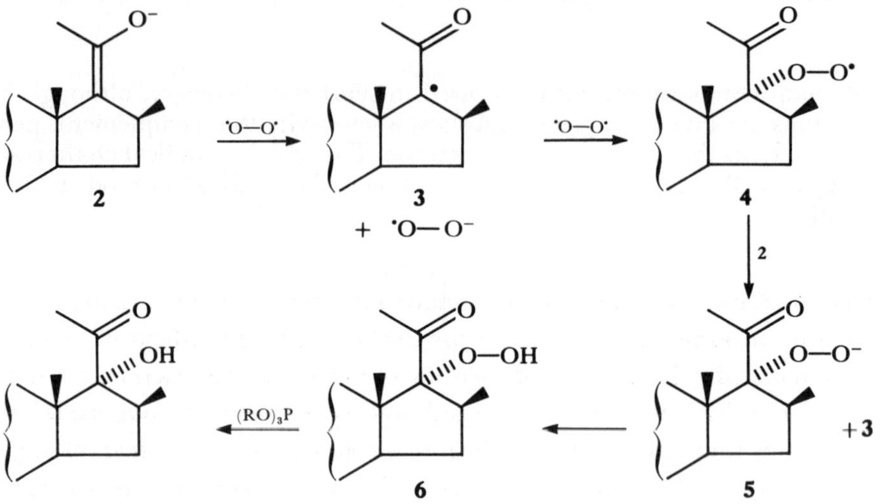

SCHEME 1

The 17α-hydroperoxide **6** could be easily reduced to the desired alcohol. However, the high reactivity of hydroperoxide **6** made the reaction difficult to reproduce, especially on a large scale, but operating in the presence of a phosphite to reduce the hydroperoxide as it is formed obviated this problem (Gardner *et al.* 1968). Thus with the use of simple and cheap reagents the desired transformation was achieved in nearly quantitative yield and has become an industrial process.

In the same field of corticosteroids, we were faced by the challenge to debrominate the hydrocortisone precursor **7** without concomitant elimination of the 11-β-hydroxy group. Conventional approaches completely failed, affording only 9(11)-olefin, so we had to devise a new method.

Consideration of the various mechanistic pathways for one-electron reduction by, for example, Cr^{II} led us to postulate that carbon radicals would be involved at least in the initial stages (scheme 2). The carbon radical **8**, produced on debromination, would presumably react to give an organochromium(III) species prone to further one-electron reduction resulting in elimination. We thought that a good hydrogen-atom donor would trap this initial carbon radical before it could give the organochromium complex.

Indeed, addition of a suitable mercaptan completely suppressed the elimination of the hydroxy group and resulted in a spectacular improvement in the yield (Barton *et al.* 1964, 1966). The efficiency of this method was recently demonstrated by Shephard & Van Rheehen (Shephard & Van Rheenen 1979) at the Upjohn Company, in a novel partial synthesis of hydrocortisone that is now part of an important industrial process. Yields exceeding 90 % were reported for the removal of the 9α-bromine using our method. This example clearly illustrates the superiority of a radical process in avoiding a β-elimination.

The process of β-elimination is also a serious problem in the carbohydrate and aminoglycoside

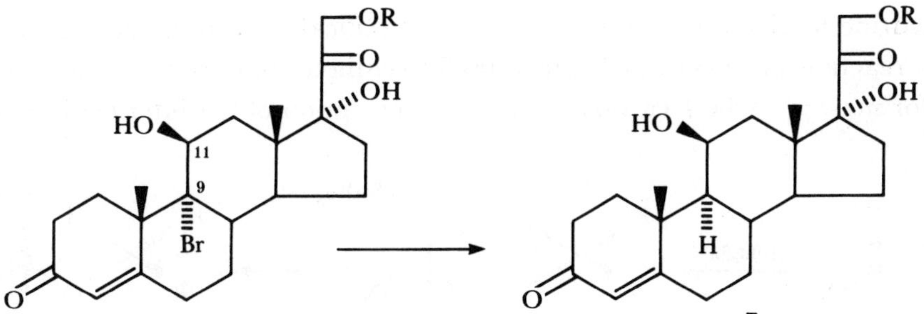

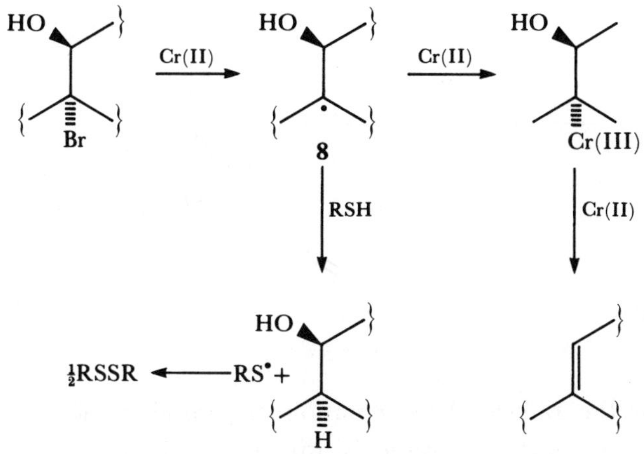

Scheme 2

fields. These important substances are usually heavily functionalized and selective deoxygenation or deamination present a formidable challenge.

We shall first consider the problem of deoxygenation. Many years ago, Van der Kerk (Van der Kerk *et al.* 1957) discovered accidentally the facile reduction of alkyl halides by stannanes. The mechanism involved is that of a typical radical chain reaction. The stannyl radicals, generated photochemically, thermally or by a chemical initiator (such as azoisobutyronitrile, AIBN), abstract the halogen to give a carbon radical. This carbon radical abstracts a hydrogen

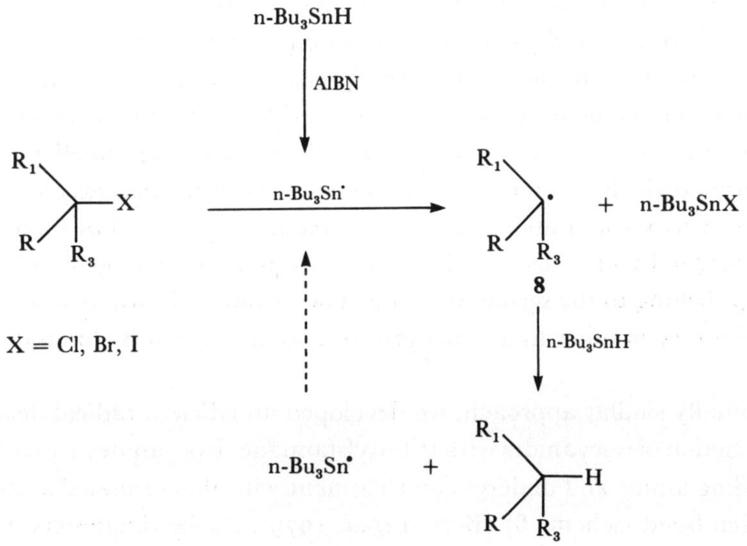

Scheme 3

atom from the hydride to give the alkane and a stannyl radical, thus propagating the chain (scheme 3). This reaction has developed into a useful synthetic tool, not only for reducing halides but also for making carbon–carbon bonds by interception of the intermediate carbon radical **8**.

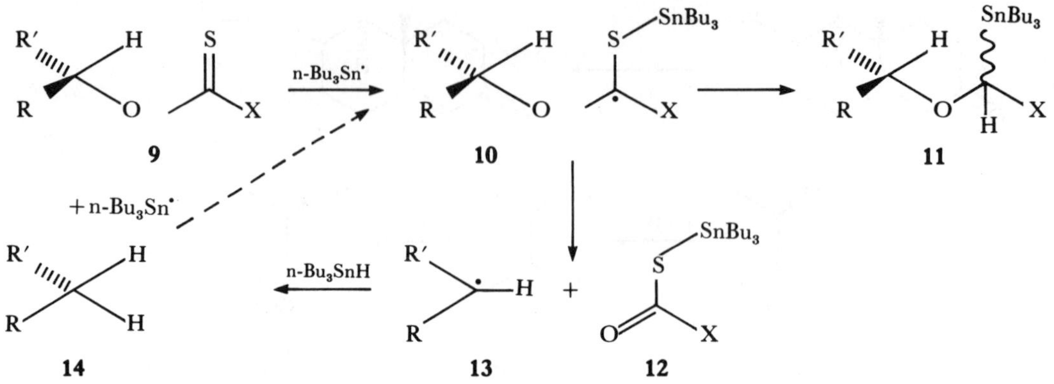

15 X = Ph
16 X = imidazole
17 X = SMe

SCHEME 4

Our conception for the radical deoxygenation of secondary alcohols is summarized in scheme 4. In the presence of stannyl radicals, thionoester derivatives of the alcohol such as **9** would react to give a carbon radical (**10**). This radical can, of course, react with the stannane to give **11**. If, however, the temperature is sufficiently high, fragmentation of **10** may take place to form a carbonyl derivative (**12**) and another radical (**13**). This can be reduced by the stannane to give the desired alkane (**14**). The chain is propagated by the stannyl radical that is also produced. The driving force for the process is the conversion of a thiocarbonyl into a carbonyl and the increase of entropy produced by the fragmentation.

After some experimentation, we found that thionobenzoates (**15**), thionoimidazolides (**16**) and especially xanthates (**17**) were suitable substrates (Barton & McCombie 1975). For secondary alcohols, moderate temperatures (80–110 °C) are sufficient to cause the fragmentation of the intermediate radical (**10**) and yields are generally excellent. For primary alcohols, higher temperatures are required and yields are moderate to good (Barton *et al.* 1981). With tertiary alcohols, problems were encountered in the preparation of the various thionesters. These derivatives underwent Chugaev elimination too readily to survive the reaction conditions. We found, however, that thionoformates were thermally stable enough to allow the deoxygenation to take place in good yields, but these derivatives were relatively inaccessible (Barton *et al.* 1982).

Despite these various limitations, this deoxygenation reaction has found widespread use, especially (as originally intended) in the carbohydrate and aminoglycoside field where most hydroxy groups belong to the secondary type. The example shown in scheme 5 illustrates the exceptional tolerance of other functional groups and absence of β-elimination (Hayashi *et al.* 1978).

In a conceptually similar approach, we developed an efficient radical deamination method based on the reaction of isocyanides with tributylstannane. Isocyanides are easily accessible from the corresponding amine and undergo, on treatment with the stannane, a smooth fission of the carbon–nitrogen bond (scheme 6) (Barton *et al.* 1979). As for thionesters, the carbon radical formed is reduced to the alkane. Although this reaction was independently discovered by

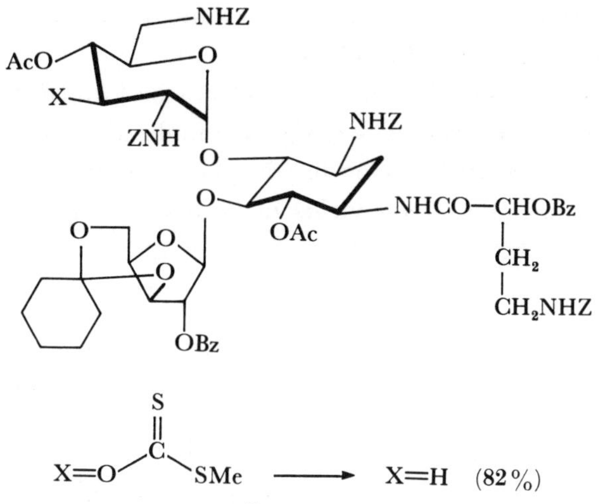

AcO, X, ZNH, NHZ, O, NHCO—CHOBz, CH_2, CH_2NHZ, OAc, OBz

$$X{=}O{-}C({=}S){-}SMe \longrightarrow X{=}H \quad (82\,\%)$$

Scheme 5

Saegusa, the yields reported were only moderate and the synthetic potential somewhat underestimated (Saegusa *et al.* 1968). Isothiocyanates and isoselenocyanates also undergo a similar reduction. This deamination reaction was successfully tested on a variety of substrates, including a dipeptide, which suggested possible applications in the peptide field. Even the 6-amino group in penicillins could be removed without harm to the sensitive β-lactam moiety (Ivor John *et al.* 1979).

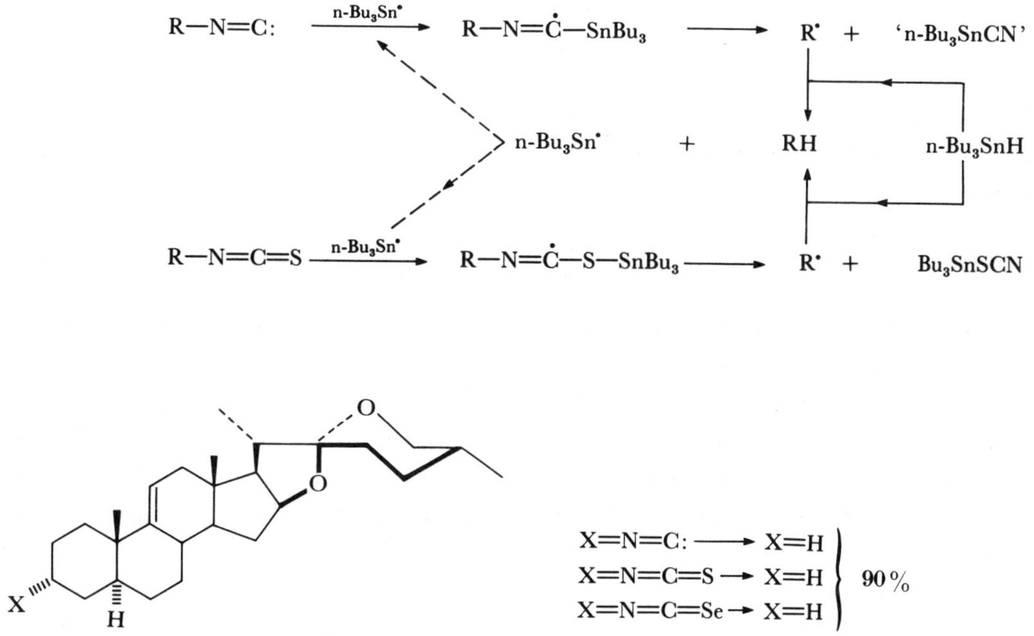

$$R{-}N{=}C: \xrightarrow{n\text{-}Bu_3Sn^\bullet} R{-}N{=}\dot{C}{-}SnBu_3 \longrightarrow R^\bullet + \text{'}n\text{-}Bu_3SnCN\text{'}$$

$$n\text{-}Bu_3Sn^\bullet \quad + \quad RH \quad n\text{-}Bu_3SnH$$

$$R{-}N{=}C{=}S \xrightarrow{n\text{-}Bu_3Sn^\bullet} R{-}N{=}\dot{C}{-}S{-}SnBu_3 \longrightarrow R^\bullet + Bu_3SnSCN$$

$$\left.\begin{array}{l} X{=}N{=}C: \longrightarrow X{=}H \\ X{=}N{=}C{=}S \rightarrow X{=}H \\ X{=}N{=}C{=}Se \rightarrow X{=}H \end{array}\right\} 90\,\%$$

Scheme 6

As expected, the order of reactivity is tertiary > secondary > primary isocyanide. Consequently, selective deamination is possible by simple adjustment of the reaction temperature. We have thus been able to prepare a number of deaminated neamine derivatives, useful in structure–activity determinations (Barton *et al.* 1980*a*).

Another important problem that attracted our attention is the decarboxylation of carboxylic acids by radicals. Existing methods lack mildness and generality and hence are ill-suited for complex and often fragile natural products.

We were aware that a carboxylic radical would not be formed by β-elimination from an alicyclic or aliphatic radical. We therefore designed a molecule where the double bond formed is incorporated in an aromatic array, as for dihydrophenanthrene derivatives (scheme 7). With this aromatization driving force, good yields of noralkanes are obtained by β-elimination (Barton *et al.* 1980*b*).

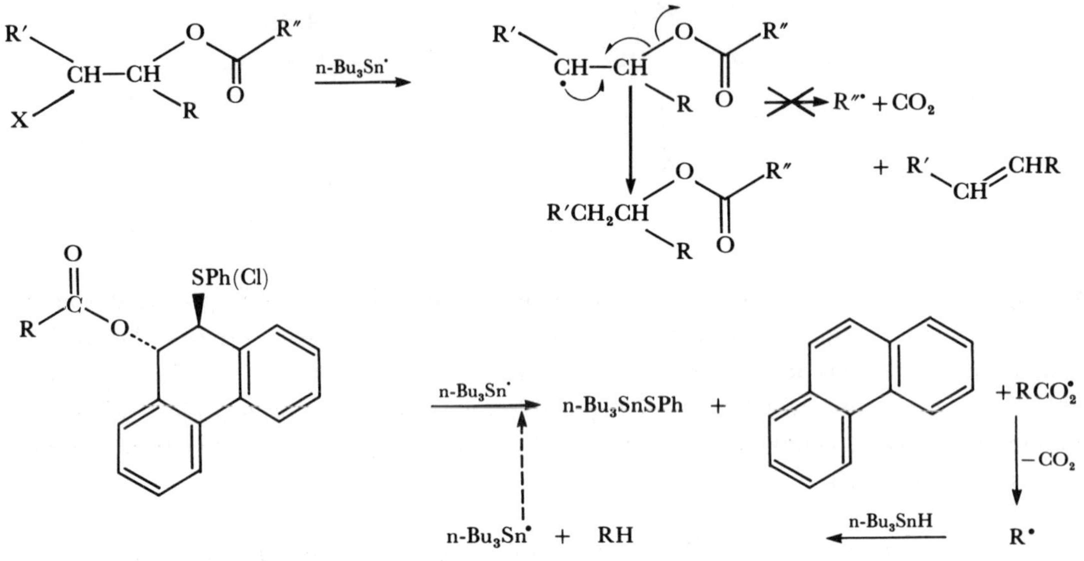

SCHEME 7

The difficulty in preparing the various ester intermediates, especially when hindered, was a drawback to this method. We conceived that the esters (mixed anhydrides) of thiohydroxamic acids of the type

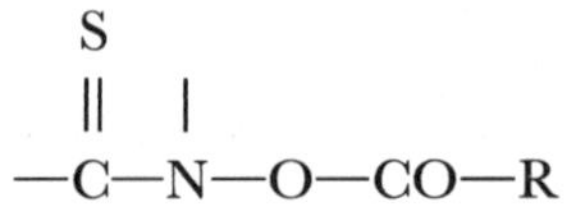

$$\begin{array}{c} \text{S} \quad\;\; \\ \| \quad \mid \quad\quad\quad \\ -\text{C}-\text{N}-\text{O}-\text{CO}-\text{R} \end{array}$$

should show a facile radical fragmentation based on the same considerations that had been taken into account in the design of the deoxygenation reaction described earlier. Thus these esters have a relatively weak (N—O) β-bond that, in principle, makes fragmentation easy. Drawing on our previous experience, we incorporated the C=N double bond to be formed by the fragmentation into an aromatic system for additional driving force. Specifically, we examined esters (**18**), derived from *N*-hydroxy thiopyridone (**19*a***). This compound exists in equilibrium with its tautomer, 2-thiopyridine-*N*-oxide (**19*b***), but the derived esters are entirely in the form shown (**18**).

These derivatives (**18**) are easily accessible from acids or from acid chlorides (scheme 8). A non-negligible advantage is the commercial availability of **19** and of its sodium salt.

The reaction with tributylstannane occurred smoothly to give, in high yield, the corresponding noralkane from a variety of aliphatic and alicyclic acids (scheme 8) (Barton *et al.* 1983*a*).

For primary acids, however, we were surprised to find that reactions were faster and gave a higher yield in benzene at 80 °C than in toluene at 110 °C. Careful examination of the reaction mixture in a particular case showed the presence of another compound, identified as the sulphide (**20**). This is slowly reduced by the stannane to the same noralkane (scheme 9).

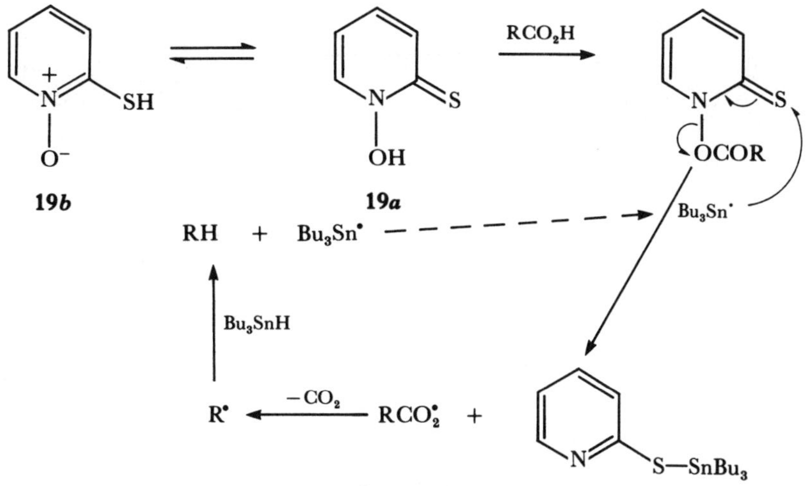

19b 19a RCO$_2$H OCOR $Bu_3Sn^{\bullet}$ RH + $Bu_3Sn^{\bullet}$ Bu_3SnH $R^{\bullet}$ $-CO_2$ $RCO_2^{\bullet}$ + S—SnBu$_3$

SCHEME 8

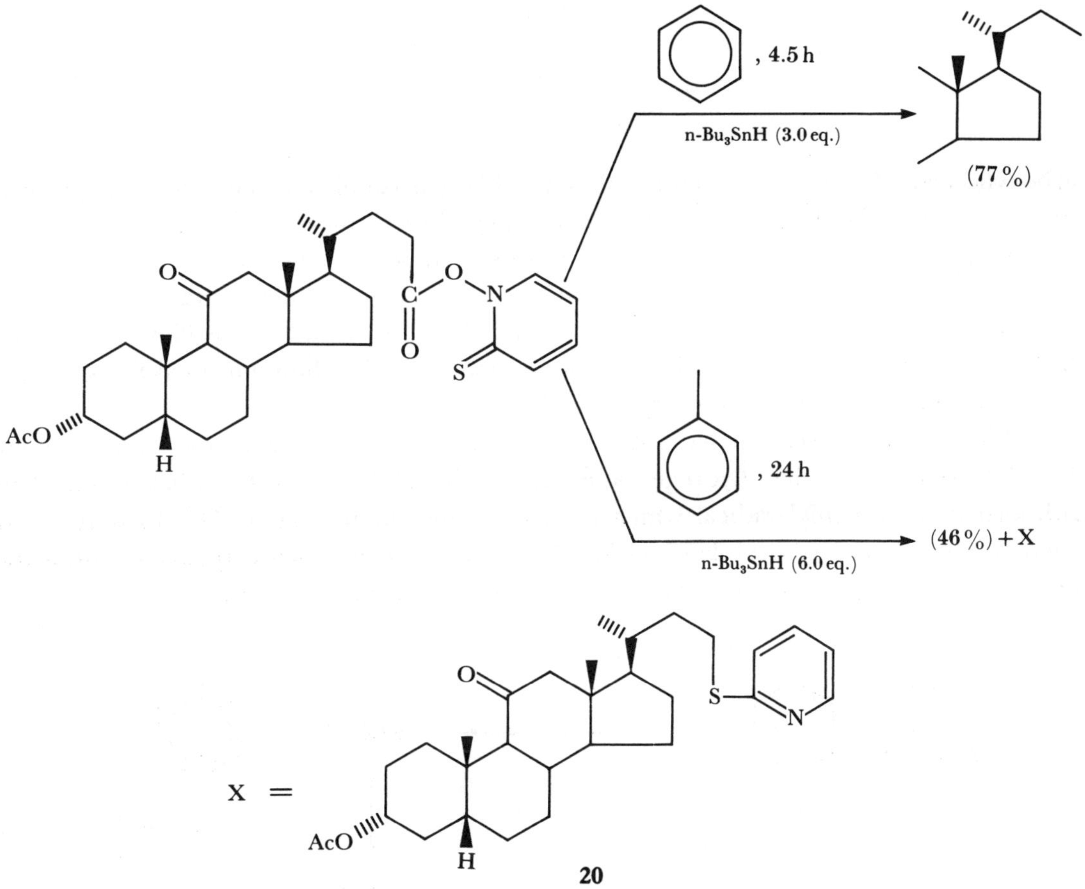

, 4.5 h n-Bu$_3$SnH (3.0 eq.) (77%)

AcO H , 24 h n-Bu$_3$SnH (6.0 eq.) (46%) + X

X = AcO H 20

SCHEME 9

Apparently, at the higher temperature of refluxing toluene the formation of sulphide can compete with the 'normal' reduction. Indeed, in the absence of the reducing agent, the sulphide is the exclusive product. We found this novel decarboxylative rearrangement to be a general high-yielding reaction proceeding by the simple radical chain mechanism depicted in scheme 10. We had in essence discovered a new and mild way of generating carbon radicals from

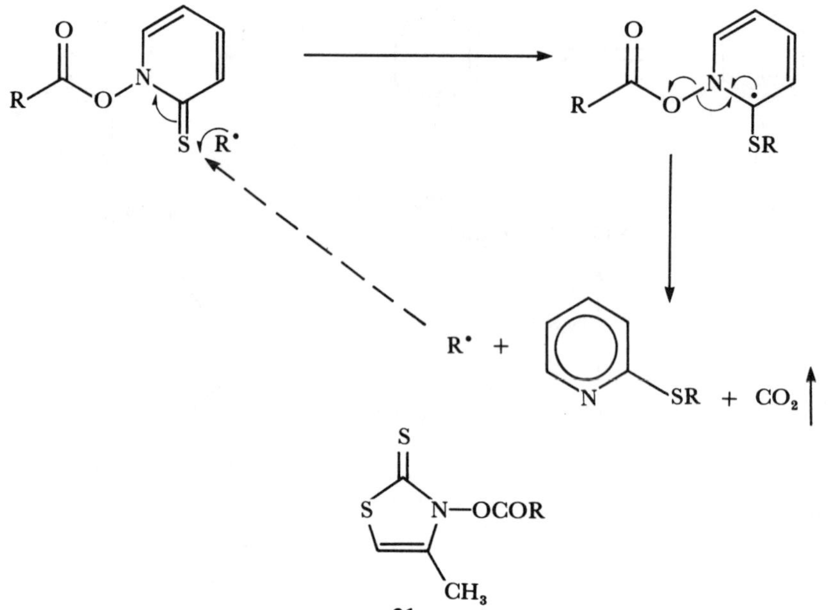

SCHEME 10

carboxylic acids. The reaction occurs with other thiohydroxamic esters having an appropriate structure, such as **21** (Barton & Kretzschmar 1983). Furthermore, the decarboxylation may be promoted thermally or photochemically at room temperature or lower.

Having at hand a convenient source of carbon radicals, we considered ways of intercepting them by various reagents, thus diverting the reaction from its normal course. However, to retain all the advantages of radical reactions we had to maintain the chain mechanism by using an appropriate propagating (chain-carrying) radical.

To illustrate this conception, let us consider performing the decarboxylation in the presence of a thiol. Being an excellent hydrogen-atom donor, the thiol reduces the radical to the alkane with formation of a thiyl radical which propagates the chain (scheme 11). In practice, the reaction proceeds as predicted. High yields of noralkane are obtained without the purification problems usually encountered in stannane reductions (Barton *et al.* 1983*a*).

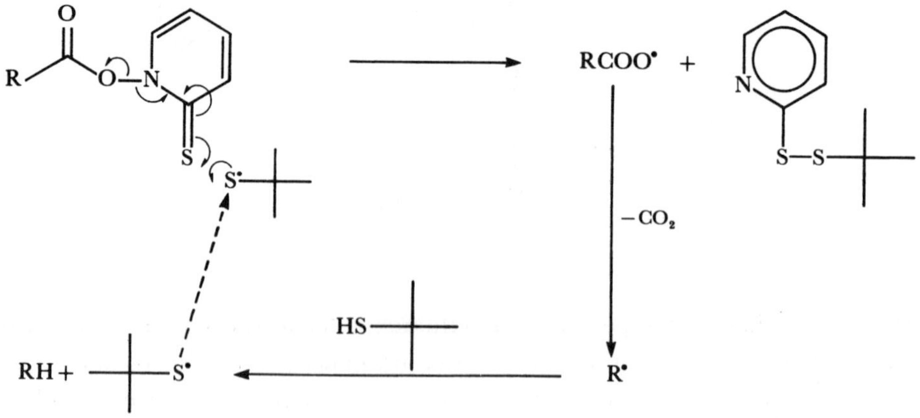

SCHEME 11

As a consequence of similar mechanistic considerations, we have succeeded in preparing chlorides, bromides and iodides in excellent yield simply by operating in the presence of carbon tetrachloride, bromotrichloromethane and iodoform respectively (scheme 12) (Barton *et al.* 1983*b*). In terms of mildness of conditions, generality and yields this method is far superior

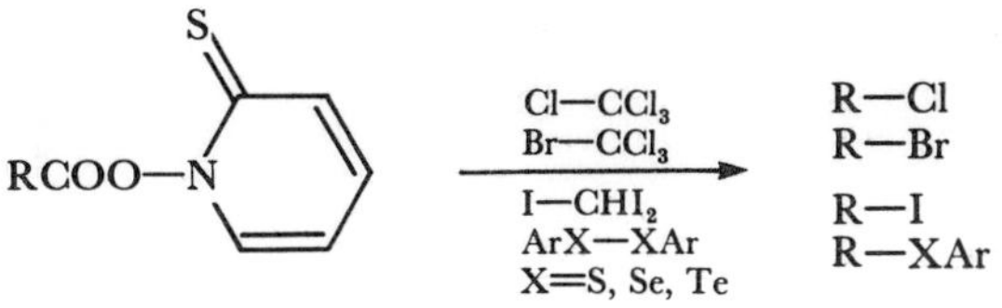

$$RCOO{-}N(\text{pyridine-2-thione}) \xrightarrow[\substack{I{-}CHI_2 \\ ArX{-}XAr \\ X{=}S,\ Se,\ Te}]{\substack{Cl{-}CCl_3 \\ Br{-}CCl_3}} \begin{matrix} R{-}Cl \\ R{-}Br \\ R{-}I \\ R{-}XAr \end{matrix}$$

SCHEME 12

to the classical Hunsdiecker reaction and its variants, which are practically all based on the use of toxic or expensive heavy metal salts. As a demonstration of the mildness of the conditions, we can cite the successful decarboxylative bromination of the heavily functionalized acid **22** with the use of our method (S. Ikegami 1984, personal communication). All other classical methods failed in this case.

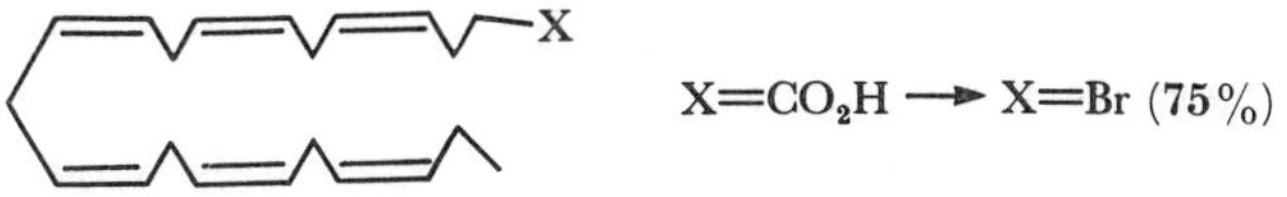

$X{=}CO_2H \longrightarrow X{=}Br$ (75%)

22

Reductive chalcogenation can similarly be achieved, leading to sulphides, selenides and even tellurides. The carbon radical in this case is trapped with a diaryl disulphide, diselenide or a ditelluride respectively (scheme 12) (Barton *et al.* 1984*a*).

We have also found it possible to intercept the carbon radical with oxygen. The reaction has to be performed in the presence of a thiol to reduce the intermediate hydroperoxy radical to the hydroperoxide and to furnish the thiyl (propagating) radical. The hydroperoxide can be reduced further, to the corresponding alcohol, by a phosphite, or transformed into an aldehyde or a ketone depending on whether the starting acid is primary or secondary (scheme 13) (Barton *et al.* 1984*b*).

We next examined a transformation of crucial importance to organic synthesis, the formation of carbon–carbon bonds. We were encouraged in this respect by the work of Giese (Giese 1983) and others on the addition of carbon radicals to electron-deficient double bonds. We found that similar addition could be achieved with our system if the olefin was sufficiently activated for the addition to compete successfully against the background decarboxylative rearrangement (scheme 14) (Barton *et al.* 1984*c*). Yields were variable, being highest for reactive, not easily polymerized, olefins. An interesting reagent is the acrylic ester derivative **23**. The sulphur moiety, by acting as a leaving group, prevents polymerization and efficiently propagates the chain. The products (**24**) have considerable synthetic potential (Barton & Crich 1984*a*).

Somewhat inevitably, our work on the decarboxylation of acids provided us with the key to the so-far inadequately solved problem of radical deoxygenation of tertiary alcohols.

If a carboxyl radical derived from the hemi-ester of oxalic acid such as **25** can be produced it will, by the loss of two molecules of carbon dioxide, give a tertiary carbon radical which can

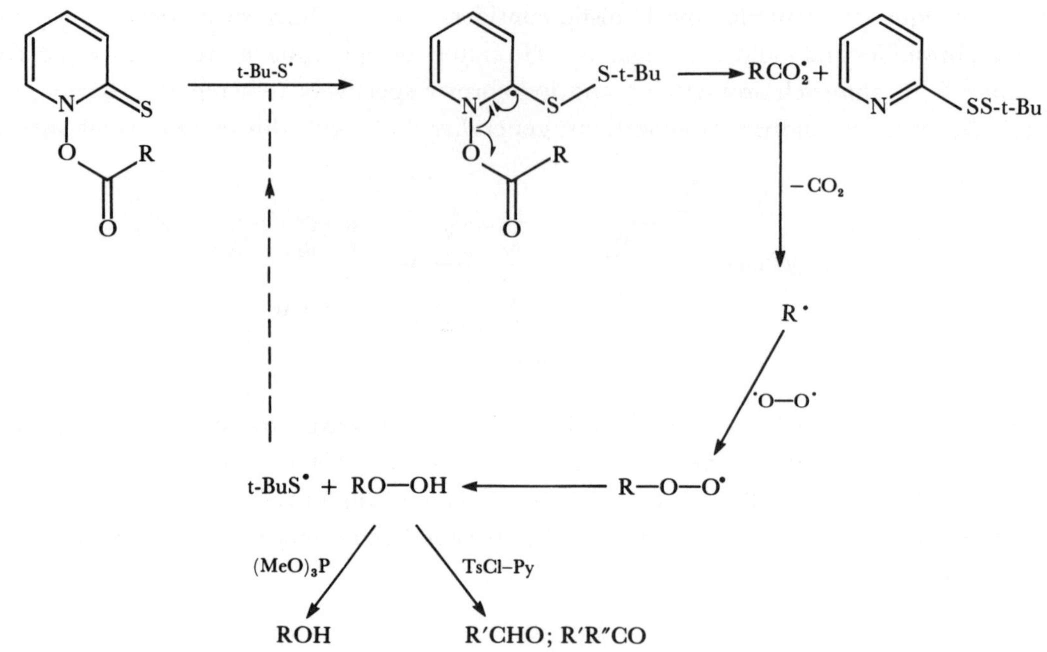

SCHEME 13

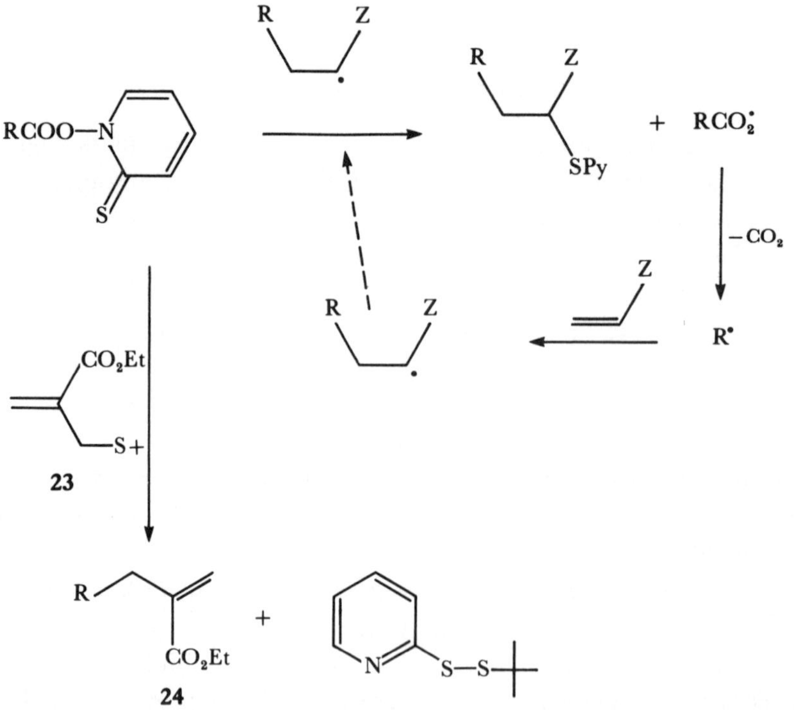

SCHEME 14

be reduced as before by a thiol (scheme 15). This conception was readily applied in practice and a variety of tertiary alcohols were deoxygenated in good yield (Barton *et al.* 1984*b*).

Interception of the tertiary radical with the acrylate reagent **23** results in the formation of a quaternary centre directly from an alcohol (scheme 15) (D.H.R. Barton & D. Crich 1984, unpublished results). The synthetic potential of this transformation is considerable given the difficulties usually encountered in creating quaternary centres.

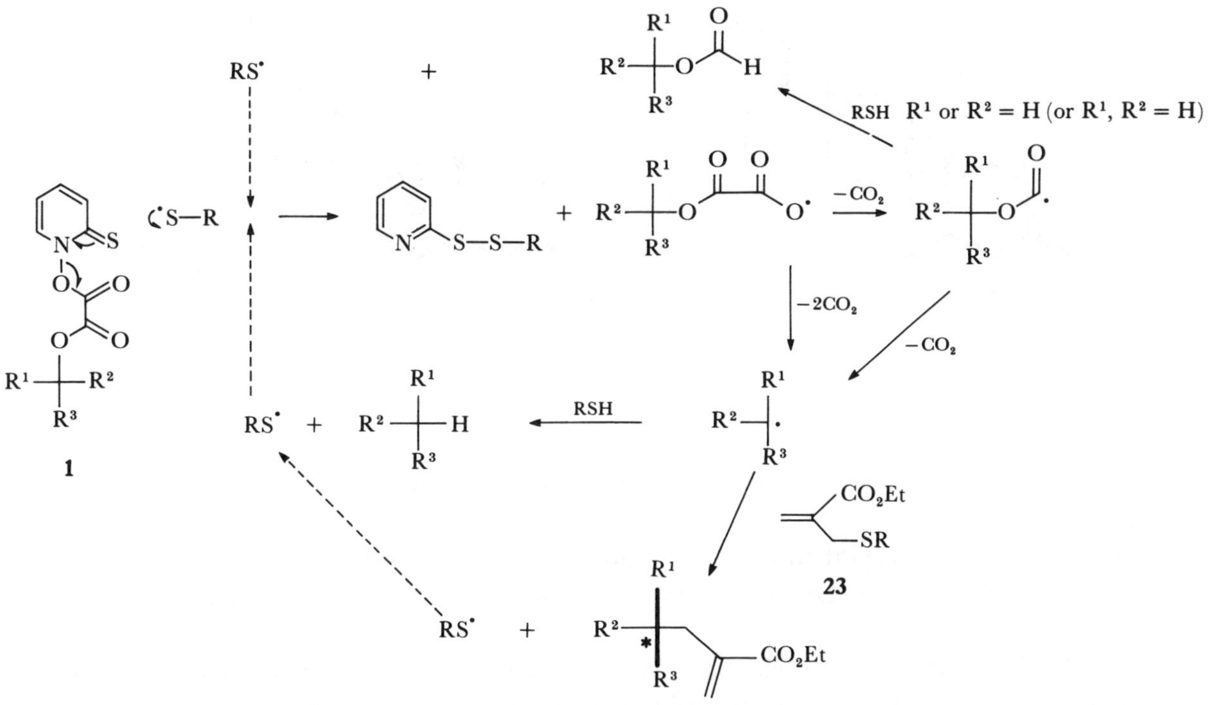

* quaternary centre

SCHEME 15

This type of radical chemistry is well suited for the manipulation of amino acids and peptides. By using a mixed anhydride coupling procedure, *N*-hydroxy-2-thiopyridone esters of *N*-protected amino acids or peptides can be synthesized and decarboxylated in very high yield (scheme 16) by using a thiol as hydrogen-atom transfer reagent. Alcoholic, phenolic and even

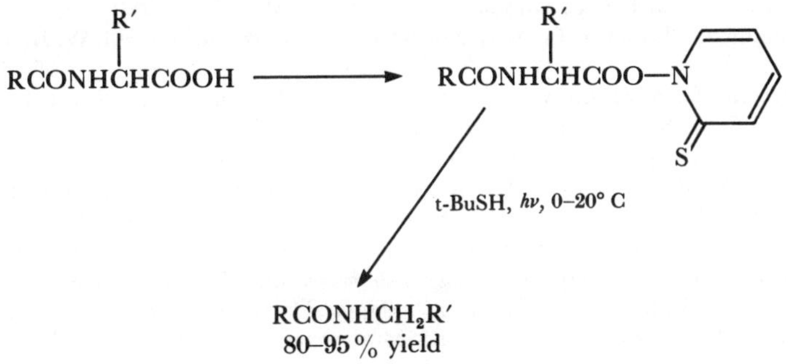

RCONHCH(R′)COOH ⟶ RCONHCH(R′)COO—N(2-thiopyridone)

t-BuSH, *hν*, 0–20° C

$RCONHCH_2R'$
80–95% yield

SCHEME 16

indolic groups do not need protection in this reaction. Equally important is the manipulation of side-chain carboxyl groups, made possible when the α-carboxyl is appropriately protected (Barton *et al.* 1984*d*).

The variety of synthetic possibilities uncovered by the decarboxylation reaction is remarkable and is probably only an indication of what may be in store. Radical reactions can achieve amazingly selective transformations, given proper control of the various parameters. An additional example to the reactions discussed so far is the ingenious device conceived by Breslow (Breslow *et al.* 1977) for remote functionalization of molecules. This has been elegantly applied

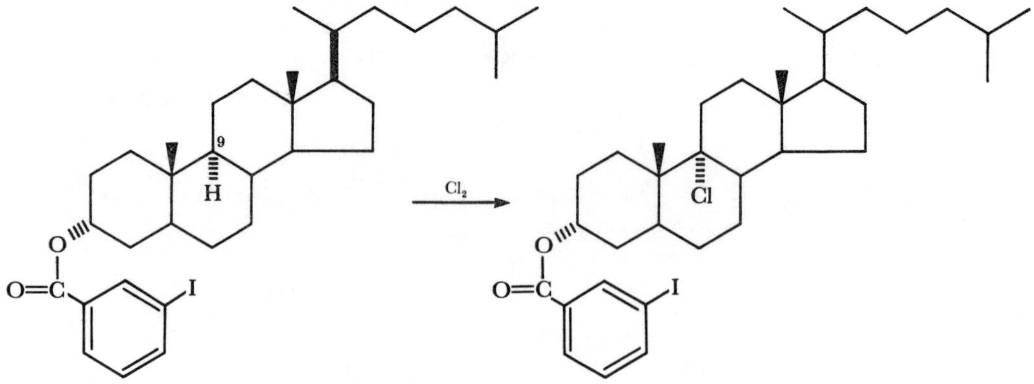

SCHEME 17

in the selective intramolecular radical chlorination of the important 9α-positions of steroids (scheme 17).

This short review of synthetically useful radical chemistry demonstrates that high-yielding and selective reactions can be effected if the propagation steps are properly designed.

We thank all our colleagues who have participated in this work and whose names are cited in the References. Dr David Crich played the pioneering role in the early work. We thank also Roussel–Uclaf for their generous support.

REFERENCES

Barton, D. H. R., Bailey, E. J. & Elks, J. 1960 *Proc. chem. Soc.* pp. 214–215.
Barton, D. H. R., Bailey, E. J., Elks, J. & Templeton, J. F. 1962 *J. chem. Soc.* pp. 1578–1591.
Barton, D. H. R. & Basu, N. K. 1964 *Tetrahedron Lett.* pp. 3151–3153.
Barton, D. H. R., Basu, N. K., Hesse, R. H., Morehouse, F. S. & Pechet, M. M. 1966 *J. Am. chem. Soc.* **88**, 3016–3021.
Barton, D. H. R., Bridon, D. & Zard, S. Z. 1984*a* *Tetrahedron Lett.* **25**, 5777–5780.
Barton, D. H. R., Bringmann, G., Lamotte, G., Hay-Motherwell, R. S. & Motherwell, W. B. 1979 *Tetrahedron Lett.* pp. 2291–2294.
Barton, D. H. R., Bringmann, G. & Motherwell, W. B. 1980*a* *J. chem. Soc. Perkin Trans.* I, pp. 2665–2669.
Barton, D. H. R. & Crich, D. 1984*a* *Tetrahedron Lett.* **25**, 2787–2790
Barton, D. H. R. & Crich, D. 1984*b* *J. chem. Soc. chem. Commun.* 744–775.
Barton, D. H. R., Crich, D. & Motherwell, W. B. 1983*a* *J. chem. Soc. chem. Commun.* pp. 939–941.
Barton, D. H. R., Crich, D. & Motherwell, W. B. 1983*b* *Tetrahedron Lett.* **24**, 4979–4982.
Barton, D. H. R., Crich, D. & Motherwell, W. B. 1984*b* *J. chem. Soc. chem. Commun.* 242–244.
Barton, D. H. R., Crich, D. & Kretzschmar, G., 1984*c* *Tetrahedron Lett.* **25**, 1055–1058.
Barton, D. H. R., Dowlatshahi, H., Motherwell, W. B. & Villemin, D. 1980*b* *J. chem. Soc. chem. Commun.* pp. 32–733.
Barton, D. H. R., Hartwig, W., Hay-Motherwell, R. S., Motherwell, W. B. & Stange, A. 1982 *Tetrahedron Lett.* **23**, 2019–2022.
Barton, D. H. R. & Kretzschmar, G. 1983 *Tetrahedron Lett.* **24**, 5889–5892.
Barton, D. H. R. & McCombie, S. 1975 *J. chem. Soc. Perkin Trans.* I, pp. 1574–1585.
Barton, D. H. R., Motherwell, W. B. & Stange, A. 1981 *Synthesis*, pp. 743–745.
Barton, D. H. R., Pradhan, S. K., Sternhell, S. & Templeton, J. F. 1961 *J. chem. Soc.* pp. 255–275.
Gardner, J. N., Carlon, F. E. & Gnoj, O. 1968 *J. org. Chem.* **33**, 3294–3297.
Giese, B. 1983 *Angew. Chem. int. Edn Engl.* **22**, 753.
Hayashi, T., Iwakoa, T., Takeda, N. & Ohki, E. 1978 *Chem. pharm. Bull., Tokyo* **26**, 1786–1797.
Ivor John, D., Thomas, E. J. & Tyrell, N. D. 1979 *J. chem. Soc. chem. Commun.*, pp. 345–347.
Van der Kerk, G. J. M., Noltes, J. G. & Luitjen, J. G. A. 1957 *J. appl. Chem.* **7**, 356–365.
Saegusa, T., Kobayashi, S., Ito, T. & Yasuda, N. 1968 *J. Am. chem. Soc.* **90**, 4182.
Shephard, K. P. & Van Rheenen, V. 1979 *J. org. Chem.* **44**, 1582–1584.

Phil. Trans. R. Soc. Lond. B **311**, 517–529 (1985)
Printed in Great Britain

Chemical models and radiation damage

By M. Fitchett, B. C. Gilbert and M. Jeff
Department of Chemistry, University of York, Heslington, York YO1 5DD, U.K.

E.s.r. spectroscopy has been used in conjunction with an aqueous flow system to investigate both the metal-catalysed decomposition of hydrogen peroxide to $OH^{\bullet}$ and the subsequent reactions of this radical with a variety of biomolecules. Particular emphasis is placed on the effects of pH and ligand on the Fe^{II}–H_2O_2 reaction and on the sites of attack by $OH^{\bullet}$ in its reaction with pyranose and furanose sugars, sugar phosphates, nucleosides and nucleotides. Attention is focused on subsequent reactions (for example, of radicals formed by attack in the ribofuranose moiety of adenosine) which may be involved in radiation damage.

Introduction

Free radicals are increasingly implicated in the initiation and progression of diseases and the toxic action of drugs, chemicals, and radiation (Willson 1977). Oxygen-centred radicals, in particular, are thought to be involved in the irreversible damage of vitally important biomolecules, as in the degeneration of synovial membranes (McCord 1974), and in many radiation-induced reactions (see, for example, Adams & Wardman 1978). Radiation damage to genetic material is thought to arise both via direct interaction of the ionizing radiation with DNA, giving radical cations and radical anions localized on the polynucleotide bases (Boon *et al.* 1984), and a secondary process in which the polynucleotide components react with $OH^{\bullet}$, $H^{\bullet}$ and e^-_{aq} (formed by the interaction of the radiation with water) (see, for example, von Sonntag & Schulte-Frohlinde 1978).

The radiation chemistry of aqueous solutions of polynucleotide components has been studied extensively, mainly by product studies following attack by $OH^{\bullet}$ (generated from pulse radiolysis) on model compounds (von Sonntag *et al.* 1981). E.s.r. spectroscopy, which allows the direct detection of free radicals, has previously been used in studies of the reaction of $OH^{\bullet}$ with pyrimidines (Nicolau *et al.* 1969) and purines (Schmidt & Borg 1976) as well as some models for sugar phosphates (Steenken *et al.* 1974; Samuni & Neta 1973). It has subsequently been suggested that radical formation from the *sugar* moiety of DNA is responsible for strand breakage (Behrens *et al.* 1982; Bothe *et al.* 1984), and pulse radiolysis experiments indicate that these radicals may be derived via secondary reactions of base-derived radicals both in the absence (Deeble & von Sonntag 1984) and presence of oxygen (Schulte-Frohlinde & Bothe 1984).

The Ti^{III}–H_2O_2 redox couple can be used in conjunction with e.s.r. spectroscopy (see, for example, Norman 1979) to study the complex mixture of radicals formed from the reaction of $OH^{\bullet}$ with a variety of carbohydrates (Gilbert *et al.* 1981), and subsequent acid- and base-catalysed transformations of these first-formed radicals can be characterized. This relatively direct approach to the study of radical reactions in biomimetic and radiomimetic systems has now been extended to an investigation of the corresponding Fenton system

(Fe^{II}–H_2O_2) and to the reaction of $OH^{\cdot}$ with compounds chosen to provide information on the damage inflicted by radiation on nucleic acids and their components. Particular emphasis has been placed on the selectivity of attack and on reactions which lead to fragmentation.

E.S.R. STUDIES OF THE Ti^{III}–H_2O_2 AND Fe^{II}–H_2O_2 COUPLES

(*a*) Ti^{III}–H_2O_2

The experiments described here are based on previous extensive use of the Ti^{III}–H_2O_2 couple as a source of $OH^{\cdot}$ (reaction (1); see, for example, Norman 1979). We have used a continous-flow system in which three streams (containing, respectively, Ti^{III} (typically 5×10^{-3} mol dm^{-3}), hydrogen peroxide (4×10^{-2} mol dm^{-3}), and the organic substrate (typically in the range 0.1–1 mol dm^{-3})) are mixed a short time (20–50 ms) before their passage through the cavity of an e.s.r. spectrometer. The radicals observed are produced by a reaction of the first-formed hydroxyl radical with the organic substrate; the relatively simple reaction scheme shown in reactions (1)–(3) is applicable under circumstances where [RH] is sufficiently high to prevent further reaction of $OH^{\cdot}$ with Ti^{III} or H_2O_2 (i.e. to 'scavenge' $OH^{\cdot}$) and in which the radicals decay by bimolecular termination (typically at a diffusion-controlled rate, $2k_3 \approx 10^9$ dm^3 mol^{-1} s^{-1}).

$$Ti^{III} + H_2O_2 \rightarrow Ti^{IV} + OH^{\cdot} + HO^-, \tag{1}$$

$$OH^{\cdot} + RH \rightarrow R^{\cdot} + H_2O, \tag{2}$$

$$2R^{\cdot} \rightarrow \text{molecular products.} \tag{3}$$

It has been shown that under these conditions a pseudo-steady-state is achieved in the cavity (Czapski 1971; Samuni *et al.* 1972; Gilbert *et al.* 1973) and it follows that $k_1[Ti^{III}]_t[H_2O_2]_t = 2k_3[R^{\cdot}]^2$ (where the subscript t refers to concentrations in the cavity at time t (in seconds after mixing). Because $[H_2O_2] \gg [Ti^{III}]$, the concentration of the latter falls exponentially, and

$$[R^{\cdot}]^2 = k_1[Ti^{III}]_0[H_2O_2]_0 \exp(-k_1[H_2O_2]_0 t)/2k_3 \tag{4}$$

so that, for given t and $[H_2O_2]_0$, $[R^{\cdot}]$ depends on $[Ti^{III}]^{\frac{1}{2}}$ (and, of course, on k_1 and k_3). It can also be shown that the concentration of $R^{\cdot}$ as a function of $[H_2O_2]_0$ for given t should have a maximum given by (5), from which it follows that if t is known then k_1 can be obtained. For several simple radicals in the Ti^{III}–H_2O_2 system the behaviour expected on the basis of (4) and (5) is observed: these equations form the basis for determining k_1 and k_3 and also rate constants for other processes which compete with bimolecular termination.

$$[H_2O_2]_0 = 1/k_1 t. \tag{5}$$

When ethanol is used as a substrate with the Ti^{III}–H_2O_2 couple (with $[H_2O_2]_0 \approx 10^{-2}$ mol dm^{-3}) at *ca.* pH 2, signals from both $^{\cdot}CHMeOH$ (α) and $^{\cdot}CH_2CH_2OH$ (β) are observed, with relative concentrations (as determined by computer simulation of spectra) of *ca.* 10:1. This ratio reflects the relative rates of attack of $OH^{\cdot}$ at the two positions (and highlights the electrophilic character of $OH^{\cdot}$), though the ratio decreases as $[H_2O_2]$ is increased (for example, to *ca.* 4.5:1 for $[H_2O_2]_0 = 0.07$ mol dm^{-3}); this is evidently the result of the reaction of the oxygen-conjugated α-radical with H_2O_2 (reaction (6)) (Gilbert *et al.* 1974),

$$\dot{C}HMeOH + H_2O_2 \rightarrow {}^{+}CHMeOH + OH^{\cdot} + OH^{-}. \tag{6}$$

To confirm this and to obtain more detailed kinetic information we have used a kinetic simulation program (kindly provided by Professor D. J. Waddington) to compute steady-state values of $[R^{\cdot}]$, $[H_2O_2]$, etc., as a function of rate constants, concentrations of reagent, and time after mixing. (When some competing secondary reactions are involved (such as reaction (6)) the appropriate steady-state equations can be derived and solved, but simulation allows more complex systems to be treated (see later).) In a set of preliminary experiments we oxidized Me_3COH to $^{\cdot}CH_2CMe_2OH$, a radical for which only simple bimolecular decay would be expected. We observed both the predicted dependence of $[^{\cdot}CH_2CMe_2OH]$ on $[Ti^{III}]^{\frac{1}{2}}$ (with $[H_2O_2]$ and t constant) and the maximum in $[^{\cdot}CH_2CMe_2OH]$ as $[H_2O_2]$ alone was varied (see equation (5)): measurement of the mixing-time t (by using a spectrophotometric method) leads to a value for k_1 of 2250 dm^3 mol^{-1} at 24 °C.

Use of this value for k_1 in the kinetic simulation program, together with reported rate constants for the $OH^{\cdot}$ reaction with ethanol (at C_α and C_β) and for radical termination (Gilbert *et al.* 1982) leads (when a value $k_6 = 1.6 \times 10^5$ dm^3 mol^{-1} s^{-1} is incorporated) to exact agreement of the observed and calculated values of the α:β ratio and its variation with $[H_2O_2]$.

$$(HO_2CCH_2)_2NCH_2CH_2N(CH_2CO_2H)_2$$

1

$$(HO_2CCH_2)_2NCH_2CH_2N(CH_2CO_2H)CH_2CH_2N(CH_2CO_2H)_2$$

2

While the effects of pH and ligands on k_1 remain to be fully explored, we have found that use of ethylenediaminetetraacetic acid (EDTA, **1**) as a ligand makes relatively little difference to k_1 (though Ti^{IV}–EDTA itself appears to be a relatively good oxidant for $^{\cdot}CHMeOH$); on the other hand the ligand diethylenetriaminepentaacetic acid (DTPA, **2**) retards the initiation reaction dramatically.

(*b*) Fe^{II}–H_2O_2

The Fenton system is reported to give e.s.r. signals from ethanol (but with $[\beta] > [\alpha]$) under certain circumstances (Shiga 1965). The use of phosphate buffer (at pH 7) as previously employed is found to be unnecessary; greater radical concentrations with a range of substrates are obtained when EDTA, for example, is used to complex Fe^{II}, particularly at the upper end of the pH range 1.5–8 (with the use of typical conditions as follows: $[Fe^{II}] = [EDTA] = 2.5 \times 10^{-3}$ mol dm^{-3} and $[H_2O_2] = 2.5 \times 10^{-2}$ mol dm^{-3}). $[H_2O_2]$ has to be significantly increased (to *ca.* 0.1 mol dm^{-3}) for the successful detection of radicals in experiments (at pH 1.5–2) in which EDTA is omitted.

Kinetic analysis of experiments with $^{\cdot}CH_2CMe_2OH$ (from t-butyl alcohol) formed in the Fe^{II}–EDTA–H_2O_2 system (based on the parallel reactions and equations for Ti^{III}) leads to a value of $k_7 = 9 \times 10^3$ dm^3 mol^{-1} s^{-1} (at pH 7). In contrast, at pH 1.85 in the absence of EDTA the initiation reaction has a value $k_7 \approx 200$ dm^3 mol^{-1} s^{-1}. We conclude that complexing, and a change in pH, can have a dramatic effect on the initiation rate, and preliminary investigations indicate that use of DTPA between pH 2 and 7 leads to an increase in rate similar to that observed for EDTA.

$$Fe^{II} + H_2O_2 \rightarrow Fe^{III} + OH^{\cdot} + OH^{-}. \quad (7)$$

As with the previous report of Fenton-system oxidation of ethanol (Shiga 1965), we find that under all conditions $[^{\cdot}CH_2CH_2OH] > [^{\cdot}CHMeOH]$ (typically *ca.* 5:1), in marked contrast with the result for Ti^{III}–H_2O_2 (similar results are obtained for related radicals from other substrates, such as propan-2-ol). By the use of the kinetic simulation program we find that this behaviour is reproduced if rapid oxidation of the α radical by Fe^{III} is incorporated (see, for example, Walling 1975). In experiments with Fe^{II}, EDTA, H_2O_2 and ethanol at pH 7, for example, both the $\alpha:\beta$ ratio and the absolute radical concentration of each is reproduced with a value of $k_8 = 10^9\ dm^3\ mol^{-1}\ s^{-1}$.

$$CH_3\dot{C}HOH + Fe^{III}\text{–EDTA} \rightarrow CH_3CHO + H^+ + Fe^{II}\text{–EDTA}. \quad (8)$$

While the role of pH and ligand remains to be investigated further, it is clear that, at least under the conditions described above, the Fenton system is an excellent source of the hydroxyl radical; it is also confirmed that oxygen-conjugated radicals are rapidly oxidized by Fe^{III} under these conditions, to perpetuate a chain process.

Reactions of OH$^{\cdot}$ with biomolecules

The extent of damage sustained, and by what part of the molecule, is governed both by the initial site of attack (and hence the relative reactivities of different moieties) and the subsequent fate of first-formed radicals. We have been particularly interested in those processes which compete with the commonly encountered radical 'repair' mechanisms (for example, by hydrogen-abstraction from a thiol) or dimerization and which, via fragmentation or other reactions, serve to 'fix' the initial damage. These processes include reaction with, for example, metal ions, hydrogen peroxide (see above) or oxygen, but also a variety of fragmentation reactions, many of which are related to the much simpler and well established decomposition of $^{\cdot}CH(OH)CH_2OH$ to $^{\cdot}CH_2CHO$ (see, for example, Gilbert *et al.* 1972); this is subject to both acid and base catalysis, and at least in the former case is believed to involve the formation of an incipient radical cation (reaction (9)). This reaction appears to be encouraged by the mesomeric effect of oxygen at the α-radical centre and the possession of a good β-leaving group (H_2O in this case); we have investigated (for radicals from a variety of biomolecules) variations of this process which involve other leaving groups (such as phosphate and nucleotide bases), and in which steric and electronic features are revealed, and other reactions involving long-range electron transfer in the incipient radical cations.

$$^{\cdot}CH(OH)\text{—}CH_2(OH) \xrightarrow[-H_2O]{H^+} [\dot{C}H(OH)\text{—}\overset{+}{C}H_2] \xrightarrow{-H^+} O{=}CH\text{—}\dot{C}H_2 \quad (9)$$

(a) *Sugars and sugar phosphates*

As with the α-hydrogen in ethanol, the —CH(OH)— fragments in sugars are reactive targets for OH$^{\cdot}$ attack, and, at least for pyranose sugars, there appears to be little selectivity between the different positions. For example, when ice-cooled solutions of β-D-fructose (at a concentration after mixing of *ca.* 0.05 mol dm^{-3}) are reacted at *ca.* pH 4 with the Ti^{III}–H_2O_2 couples (generally preferred to Fe^{II}–H_2O_2 so that radical oxidation by Fe^{III} is avoided), attack at OH˙ gives all five possible radicals from the *pyranose* form (present almost exclusively under these conditions, and in the $^2C_5$ conformation). Details of just one of these are given here (see structure **3**). As the pH is lowered, changes occur which indicate that the first-formed radicals are susceptible to acid-catalysed rearrangement (cf. reaction(9)) and it is significant that the conversion of **3** into **4** occurs much faster than the corresponding reaction of the other hydroxyalkyl radicals: the reaction evidently has a stereoelectronic requirement and is facilitated when, as in **3**, the (axial) leaving group eclipses the orbital containing the unpaired electron at the radical centre.

Solutions of fructose that have been allowed to stand for a few minutes before flowing show new features (see figure 1) which are consistent with the oxidation of a significant proportion

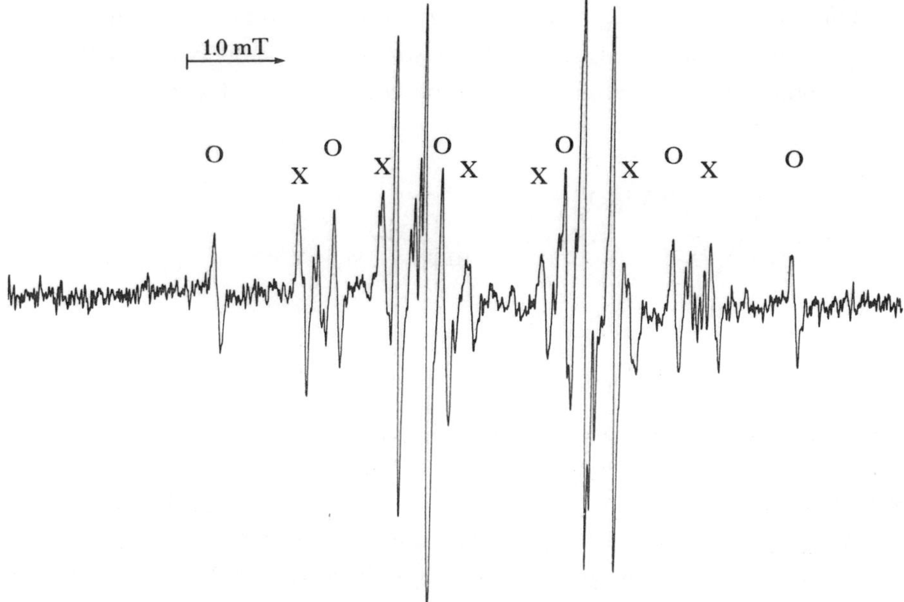

FIGURE 1. E.s.r. spectrum obtained from β-D-fructose (after mutarotation) and OH˙ at pH 4, showing (o) signals assigned to the C-5 (pyranose) radical, and (×) the C-5 radical (**5**) from the minor furanose structure.

of β-D-fructo*furanose*, formed by mutarotation (the predominant forms of fructose in aqueous solution at room temperature are β-D-fructopyranose (*ca.* 75%) and β-D-fructofuranose (*ca.* 20%); Angyal & Bethell (1976)). The detection of a significant concentration of the furanose C-5 radical **5**, rather than other radicals from this ring, is consistent with a preference

$\xrightarrow{H^+}$

3: $a(\beta\text{-H}) = \begin{cases} 3.360\ \text{mT} \\ 0.785 \end{cases}$ $g = 2.0031$

4: $a(\alpha\text{-H}) = 1.763$ $a(\beta\text{-H}) = \begin{cases} 4.010\ \text{mT} \\ 3.520 \end{cases}$ $a(\gamma\text{-H}) = 0.370$ $g = 2.0042$

5: $a(\beta\text{-H}) = \begin{cases} 2.520\ \text{mT} \\ 0.887\ (2) \end{cases}$ $g = 2.0031$

for attack by OH˙ at the hydrogen on the carbon adjacent to the ring-oxygen in the five-membered ring (see a comparable result for sucrose, in which pyranose and furanose rings are linked together (Gilbert *et al.* 1983)). This is interpreted in terms of the stabilization offered in the transition state for hydrogen abstraction by the optimum overlap between the unpaired electron and the lone pair of electrons on oxygen (Malatesta & Ingold 1981). This may be particularly relevant for the ribofuranose rings in nucleic acid components (see later).

The effect of incorporation of a phosphate group has been investigated through the study of both pyranose and furanose substrates. For example, reaction of OH˙ with α-D-glucose-1-phosphate at pH 4 gives complex spectra corresponding closely to those from α-D-glucose under similar conditions (Gilbert *et al.* 1981), except that signals from **7** replace those anticipated for the C-2-derived radical **6**. The detection of **7** rather than **6** indicates that the loss of the β-phosphate group, reaction (10), is much faster than the corresponding loss of β-OH or β-alkoxy groups (in, say, glucose or dextran (Gilbert *et al.* 1984)), for which acid catalysis is required. The rate constant k_{10} for the *overall* process, obtained from pulse radiolysis experiments performed in collaboration with Professor R. L. Willson, is 1.2×10^4 dm^3 mol^{-1} s^{-1} (the loss of phosphate and a proton may not be concerted, and a radical cation may be involved).

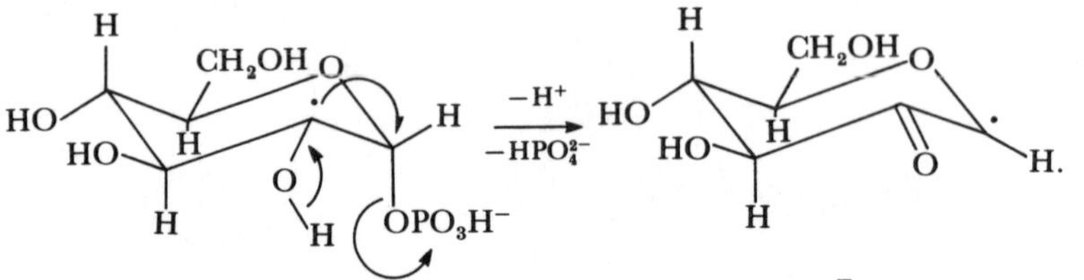

$$\mathbf{6} \xrightarrow[-HPO_4^{2-}]{-H^+} \mathbf{7} \qquad (10)$$

Related fragmentation pathways are revealed for fructose-6-phosphate and fructose-1,6-diphosphate, in both of which the sugar is furanose. For fructose-1,6-diphosphate, for example, reaction of OH˙ leads to e.s.r. signals which indicate that there is preferential attack at C-5 (as with the formation of **5** from fructose itself) and that there is also a significant extent of reaction at the C-1 hydroxymethyl group. Though **8** is itself detected, the additional observation of the signal attributed to **10** is interpreted (scheme 1) in terms of attack at C-5 followed by rapid loss of phosphate and subsequent hydration of the resulting radical cation (loss of phosphate, with $k \approx 10^4$ s^{-1}, is evidently somewhat faster than for some related acyclic

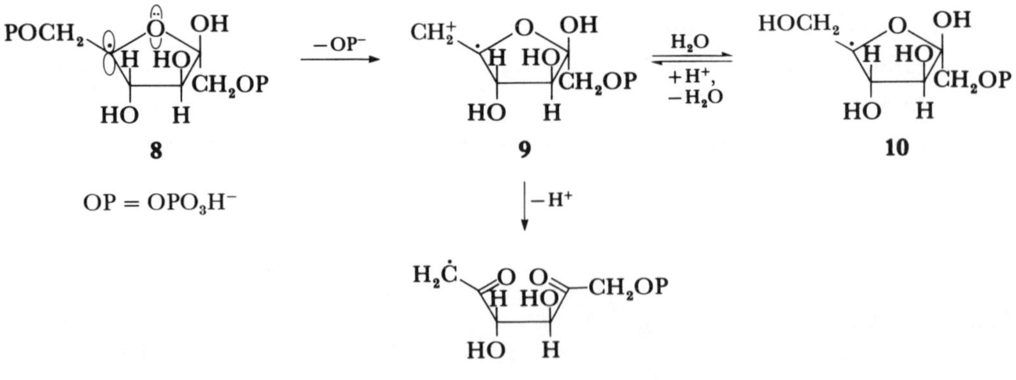

$$\mathbf{8} \xrightarrow{-OP^-} \mathbf{9} \underset{+H^+,\ -H_2O}{\overset{H_2O}{\rightleftharpoons}} \mathbf{10}; \qquad \mathbf{9} \xrightarrow{-H^+} \mathbf{11}$$

$OP = OPO_3H^-$

Scheme 1

radicals; this is *ca.* 10^3 s^{-1} for $^{\cdot}CH(OMe)CH_2OPO_3H^-$ (Behrens *et al.* 1978)). As the pH is lowered (to *ca.* 1.5), signals from **8** and **10** are themselves replaced by those attributed to a ring-opened carbonyl-conjugated radical **11**, which evidently derives from **9**; the role of acid is presumably to accelerate the regeneration of **9** via catalysis of the loss of β-OH in **10**. (Such acid-catalysed rearrangements are revealed by e.s.r. when the rates of the appropriate reactions become faster than radical–radical termination rates (which in the systems described here limit radical lifetimes to several milliseconds). When radical concentrations are lower and lifetimes longer (as under physiological conditions), these fragmentation processes can play an important role at higher pH values.) The ring-opening must involve the loss of the proton on the C_2-hydroxyl group (possibly in concerted fashion, see **12**), because the corresponding radical from sucrose, **13**, fails to undergo cleavage even at pH 0.25.

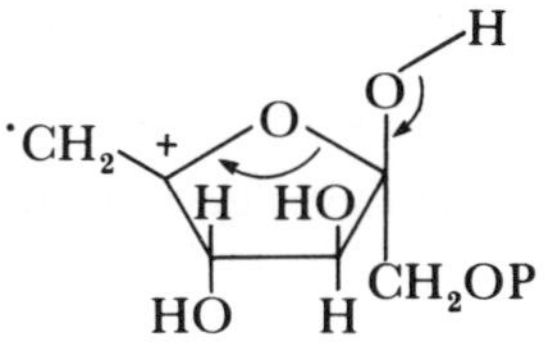

12 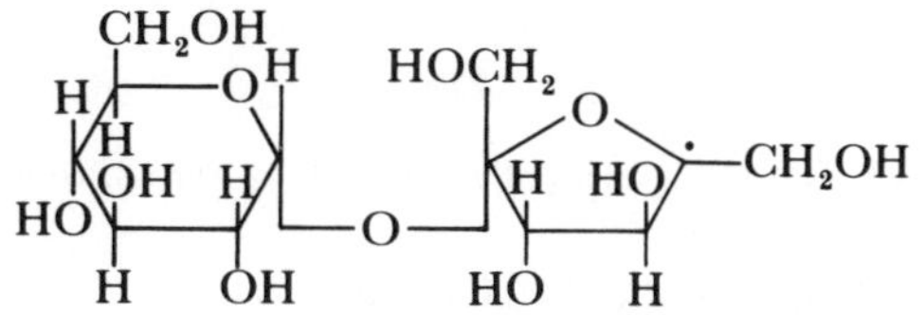 **13**

Another rearrangement revealed for a radical derived from fructose-1,6-diphosphate is reaction (11); the detection of the *cis* and *trans* isomers of the α-hydroxyketo-substituted radical (**15** and **16**) is attributed to breakage of the β C—O bond in the C_1-derived radical (**14**), with, presumably, encouragement from the remaining (β) OH group in stabilizing the intermediate radical cation.

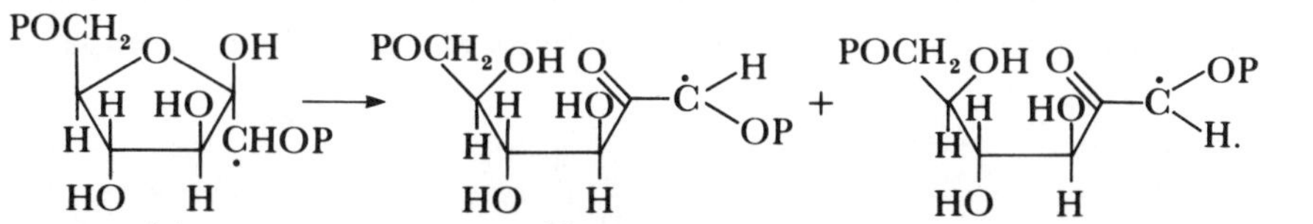

14 → **15** + **16** (11)

Ribose-5-phosphate reacts with $OH^{\cdot}$ at *ca.* pH 4 in a slightly less selective fashion, giving rise to signals from **17**, **18**, **19** (R = H), **20** and **24** (see table 1). Radical **19** (R = H) evidently arises (scheme 1) via loss of phosphate and rehydration. As the pH is lowered, radical **19** is replaced by the ring-opened radical **22** (pH < 2) (*cf.* **11**) and acid-catalysed dehydration accounts for the conversion of **20** into **23** (pH < 1.5). The observation of **24** under all conditions implies that the dehydration of its precursor, **21**, is extremely rapid even at *ca.* pH 4; the rapidity of this fragmentation may reflect strain in the ring, or (as with reaction of **8**, scheme 1) the overlap between the β C—O bond and the orbital of the unpaired electron, or both.

The results described here for furanose phosphates provide crucial direct evidence to support the suggestion (Behrens *et al.* 1978, 1982) that attack by $OH^{\cdot}$ in the ribofuranose rings of DNA, and subsequent loss of phosphate are important components of a mechanism for radiation-induced strand breakage.

TABLE 1. RADICALS FORMED FROM THE REACTION BETWEEN OH˙ AND RIBOSE-5-PHOSPHATE AT pH 4 AND BELOW

17 **18** **19**† **20**

21‡ ·CH_2C(=O)CH(OH)CH(OH)CHO **22**§ **23**‖ **24**

† Radical **19** (R=H) is evidently derived by rehydration of the radical cation from **19** ($R{=}PO_3H^-$); see scheme 1.

‡ This radical, the precursor of **24**, is not detected.

§ Replaces **19** (R=H) below pH 2.

‖ Replaces **20** below pH 1.5.

(*b*) *Pyrimidines, purines and nucleosides*

Reaction of OH˙ (from the Ti^{III}–H_2O_2 couple) with the pyrimidine nucleosides uridine **25** and thymidine gives signals attributable to the radicals formed by addition to the C=C bond in the base (for example, **26** for uridine; see figure 2). This pattern of behaviour resembles the reactions of the pyrimidines themselves (Nicolau *et al.* 1979), though steric effects are evidently more important here, and reflects the greater reactivity of alkenic bonds compared with sugar moieties.

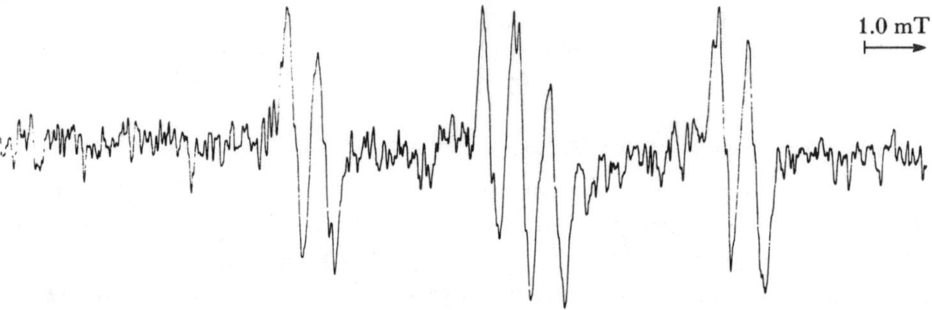

FIGURE 2. E.s.r. spectrum of the radical **26**, obtained from uridine and OH˙ at *ca.* pH 4.

These findings contrast markedly with results for the purine nucleosides adenosine (**27**, R = H) and inosine (**28**), and their respective monophosphates (for example, **27**, $R = HPO_3^-$)). Although it has previously been suggested that the (weak) signals from purine nucleosides may be assigned to OH˙ adducts at the purine bases themselves (Schmidt & Borg 1976), our spectra have significantly higher signal:noise ratios and allow a different interpretation. We have been unable to detect any significant signals in experiments in which the reactions of purines themselves were studied (which may reflect, at least in part, some deactivation of the aromatic ring towards attack by the electrophilic hydroxyl radical). However, inosine and adenosine give identical spectra, which closely resemble those obtained

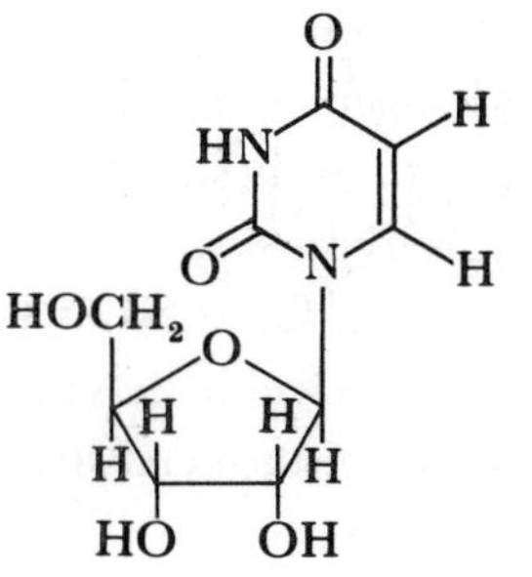

25 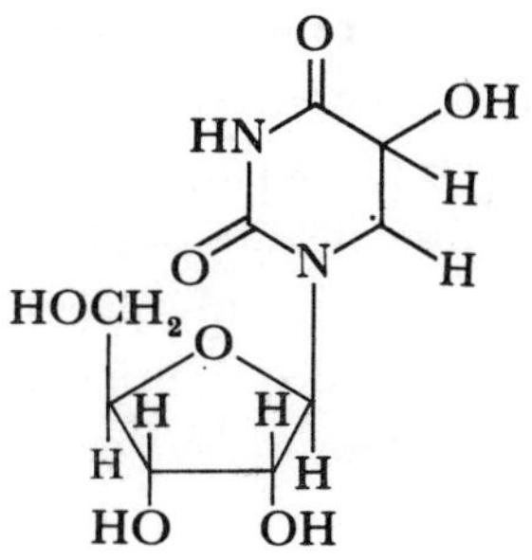 **26**

from ribose-5-phosphate and which hence characterize attack in the ribofuranose ring. (The contrasting reactivities observed for purine and pyrimidine nucleosides reflect the difference in rate constants for the addition of OH˙ to the bases (9×10^8 dm^3 mol^{-1} s^{-1} for adenine, 5×10^9 dm^3 mol^{-1} s^{-1} for uracil, both at pH 2; Anbar (1977).) This indicates a substantial proportion of attack on the sugar under these conditions (though product studies (Scholes *et al.* 1960) suggest that this route represents only *ca.* 10–20% of all products from reaction of OH˙ with polynucleotides).

The most striking feature of the e.s.r. results for adenosine, inosine, and their 5′-monophosphates is the detection at all pH values in the range 2–8 of a signal from a radical that is both carbonyl and oxygen conjugated (cf. **24** from ribose-5-phosphate). This is assigned in each case to the radical derived by rapid fragmentation ($k > ca.\ 10^4$ s^{-1}) of the radical formed via attack at C-2′ (see, for example, **29** formed from adenosine; reaction (12)).

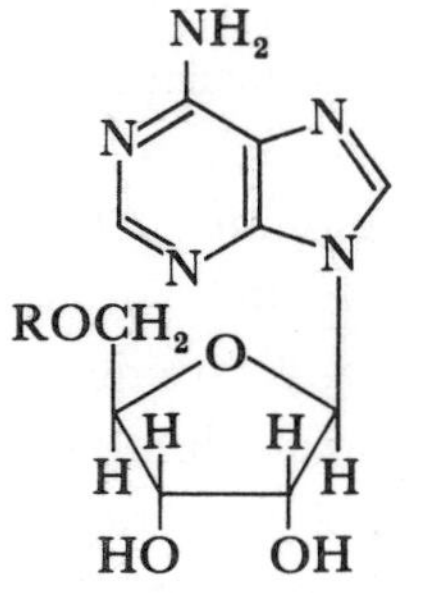

27 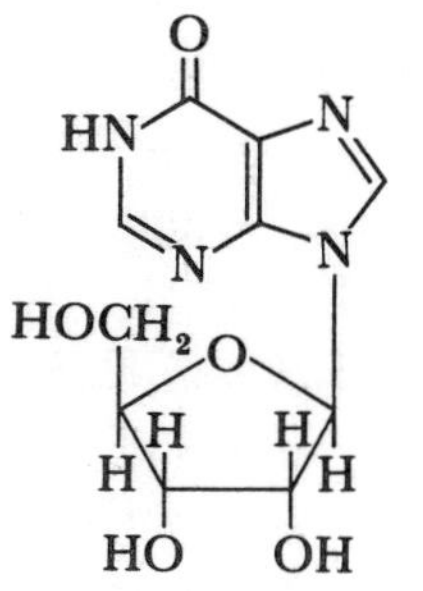 **28**

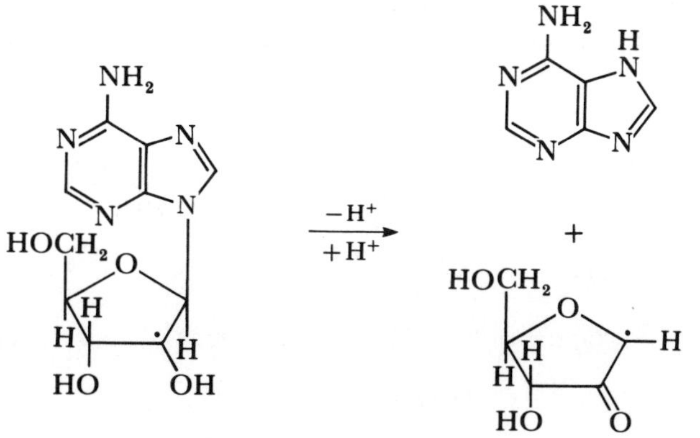

$$\xrightarrow[+H^+]{-H^+} \qquad (12)$$

29

Signals are also observed from radicals formed by initial attack at positions 2′–4′ in the sugar ring, and it is particularly significant that for both adenosine and adenosine-5′-phosphate the radical **30** is clearly seen. This confirms not only that there is significant attack at C-4′ but also that in the latter substrate, loss of phosphate (and rehydration of a radical cation) has also occurred. Our results thus not only provide a mechanism for the loss of base in radiolysed nucleotides and polynucleotides (Ullrich & Hagen 1971; Deeble & von Sonntag 1984), but also provide strong support for the suggestions that radiation-induced strand breakage in DNA and RNA involves attack at C_4' (either directly by $OH^{\cdot}$ (Behrens *et al.* 1982) or via secondary reaction of a peroxy radical derived from initial base attack (Schulte-Frohlinde & Bothe 1984)), followed by loss of the C-5′ (or C-3′) phosphate group.

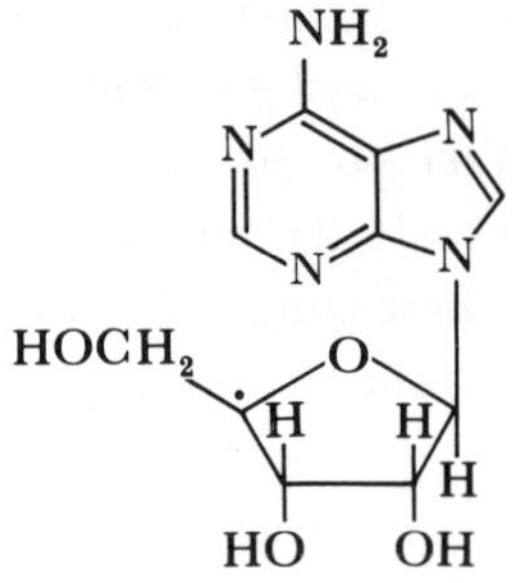

30

(c) *Aromatic and sulphur-containing amino acids*

Finally, we draw attention to other ways in which initial $OH^{\cdot}$ attack can be followed by novel, irreversible fragmentation reactions. The substrates concerned are amino acids possessing either an aromatic ring or a sulphur substituent, both of which provide a reactive 'target' for $OH^{\cdot}$.

Radicals that are detected from methionine, $^{\cdot}CH_2SCH_2CH_2CH(CO_2^-)NH_3^+$, $^{\cdot}CH(SMe)CH_2CH(CO_2^-)NH_3^+$ (and their protonated counterparts) and, perhaps more surprisingly, $^{\cdot}CH(NH_2)CH_2CH_2SMe$ (above *ca.* pH 2) are derived by subsequent reaction of an adduct (formed by attack of $OH^{\cdot}$ at sulphur) and a sulphur-centred radical cation, **31**. The decarboxylation step implicated in the formation of these radicals evidently involves overall one-electron transfer either from CO_2^- (Davies *et al.* 1983) or NH_2 (Asmus *et al.* 1985). A similar and perhaps parallel example is provided by the observation that whereas $OH^{\cdot}$ reacts readily with phenylalanine to give cyclohexadienyl adducts (such as **32**) above pH 4, spin-trapping experiments with t-BuNO show that in the pH range 2–4 the decarboxylated radical

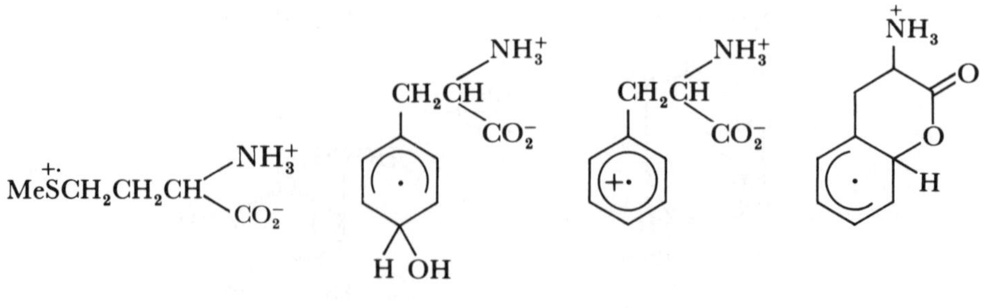

$\overset{+\cdot}{MeS}CH_2CH_2CH(NH_3^+)CO_2^-$

31 **32** **33** **34**

$^{\bullet}CH(NH_2)CH_2C_6H_5$ is formed (the corresponding adduct with t-BuNO has a = 1.45 mT (1N), 0.250 mT (1N), 0.105 mT (1H), 0.0375 mT (2H) (see, for example, Davies *et al.* 1983). As with the similar behaviour of simpler aromatic carboxylic acids ($Ph(CH_2)_nCO_2H$, where n = 2–4 (Davies *et al.* 1984)), this is believed to involve the acid-catalysed formation of an aromatic radical cation **33** (*cf.* **31**) and effective overall one-electron transfer in which a discrete but short-lived cyclic intermediate, **34**, may be formed.

Support for this work from the SERC and Interox Chemicals Ltd (for a CASE award) and the Association for International Cancer Research is gratefully acknowledged.

References

Anbar, M. (ed.) 1977 Selected specific rates of reaction of transients from water in aqueous solution. *National Standards Reference Data Series*, no. 3. U.S. National Bureau of Standards.

Adams, G. E. & Wardman, P. 1978 Free radicals in biology: the pulse radiolysis approach. In *Free radicals in biology*, vol. 3 (ed. W. Pryor), pp. 53–95. New York, London: Academic Press.

Angyal, S. J. & Bethell, G. S. 1972 Conformational analysis in carbohydrate chemistry. III. The ^{13}C n.m.r. spectra of the hexuloses. *Aust. J. Chem.* **29**, 1249–1265.

Asmus, K.-D., Göbl, M., Hiller, K.-O., Mahling, S. & Mönig, J. 1985 S···N and S···O three-electron bonded radicals and radical-cations in aqueous solutions. *J. chem. Soc. Perkin Trans.* II, 641–646.

Behrens, G., Koltzenburg, G., Ritter, A. & Schulte-Frohlinde, D. 1978 The influence of protonation or alkylation of the phosphate group on the e.s.r. spectra and on the rate of phosphate elimination from 2-methoxyethyl phosphate 2-yl radicals. *Int. J. radiat. Biol.* **33**, 163–171.

Behrens, G., Koltzenburg, G. & Schulte-Frohlinde, D. 1982 Model reactions for the degradation of DNA-4′ radicals in aqueous solution. Fast hydrolysis of α-alkoxyalkyl radicals with a leaving group in the β-position followed by radical rearrangement and elimination reactions. *Z. Naturf.* **37** c, 1205–1227.

Boon, P. J., Curtis, P. M., Symons, M. C. R. & Wren, B. W. 1984 Effects of ionizing radiation on DNA and related systems. Part I. The role of oxygen. *J. chem. Soc. Perkin Trans.* II, 1393–1399.

Bothe, E., Qureshi, E. A. & Schulte-Frohlinde, D. 1984 Rate of OH radical induced strand break formation in single stranded DNA under anoxic conditions. An investigation in aqueous solutions using conductivity methods. *Z. Naturf.* **38** c, 1030–1042.

Czapski, G. 1971 E.s.r. studies of short-lived radicals generated by fast flow techniques in aqueous solution. *J. phys. Chem.* **75**, 2957–2967.

Davies, M. J., Gilbert, B. C., McCleland, C. W., Thomas, C. B. & Young, J. 1984 E.s.r. evidence for the multiplicity of side-chain oxidation pathways in the acid-catalysed decomposition of substituted hydroxycyclohexadienyl radicals. *J. chem. Soc. chem. Commun.*, 966–967.

Davies, M. J., Gilbert, B. C. & Norman, R. O. C. 1983 E.s.r. studies. Part 64. The hydroxyl radical-induced decarboxylation of methionine and some related compounds. *J. chem. Soc. Perkin Trans.* II, 731–738.

Deeble, D. J. & von Sonntag, C. 1984 γ-Radiolysis of poly(U) in aqueous solution. The role of primary sugar and base radicals in the release of undamaged uracil. *Int. J. radiat. Biol.* **46**, 247–260.

Gilbert, B. C., King, D. M. & Thomas, C. B. 1981 Radical reactions of carbohydrates. Part 2. An e.s.r. study of the oxidation of D-glucose and related compounds with the hydroxyl radical. *J. chem. Soc. Perkin Trans.* II, 1186–1199.

Gilbert, B. C., King, D. M. & Thomas, C. B. 1983 Radical reactions of carbohydrates. Part 4. E.s.r. studies of radical-induced oxidation of some aldopentoses, sucrose, and compounds containing furanose rings. *J. chem. Soc. Perkin Trans.* II, 675–683.

Gilbert, B. C., King, D. M. & Thomas, C. B. 1984 The oxidation of some polysaccharides by the hydroxyl radical: an e.s.r. investigation. *Carbohydr. Res.* **125**, 217–235.

Gilbert, B. C., Larkin, J. P. & Norman, R. O. C. 1972 E.s.r. studies. Part 33. Evidence for heterolytic and homolytic transformations of radicals from 1,2-diols and related compounds. *J. chem. Soc. Perkin Trans.* II, 794–802.

Gilbert, B. C., Norman, R. O. C. & Sealy, R. C. 1973 E.s.r. studies. Part 39. A kinetic investigation of the role of radical-reduction in metal-ion-hydrogen peroxide systems. *J. chem. Soc. Perkin Trans.* II, 2174–2180.

Gilbert, B. C., Norman, R. O. C. & Sealy, R. C. 1974 E.s.r. studies. Part 40. Kinetic investigation of the oxidation of carbon radicals by hydrogen peroxide. *J. chem. Soc. Perkin Trans.* II, 824–830.

Gilbert, B. C., Norman, R. O. C., Williams, P. S. & Winter, J. N. 1982 E.s.r. studies. Part 63. Selective radical oxidation by titanium(IV) complexes. *J. chem. Soc. Perkin Trans.* II, 1439–1445.

McCord, J. 1974 Free radicals and inflammation: protection of synovial fluid by superoxide dismutase. *Science, Wash.* **185**, 529–530.

Malatesta, V. & Ingold, K. U. 1981 Kinetic applications of e.p.r. spectroscopy. Part 36. Stereoelectronic effects in hydrogen abstraction from ethers. *J. Am. chem. Soc.* **103**, 609–614.

Nicolau, C., McMillan, M. & Norman, R. O. C. 1969 An e.s.r. study of short-lived radicals derived from pyrimidines in aqueous solution. *Biochim. biophys. Acta* **174**, 413–422.

Norman, R. O. C. 1979 Applications of e.s.r. spectroscopy to kinetics and mechanism in organic chemistry. *Chem. Soc. Rev.* **8**, 1–27.

Samuni, A., Meisel, D. & Czapski, G. 1972 The kinetics of the oxidation of chromium(II), titanium(III) and vanadium(IV) by hydrogen peroxide and hydroxyl radicals. *J. chem. Soc. Dalton Trans.*, 1273–1277.

Samuni, A. & Neta, P. 1973 Hydroxyl radical reaction with phosphate esters and mechanisms of phosphate cleavage. *J. phys. Chem.* **77**, 2425–2429.

Schmidt, J. J. & Borg, D. C. 1976 Free radicals from purine nucleosides after hydroxyl radical attack. *Radiat. Res.* **65**, 220–237.

Scholes, G., Ward, J. F. & Weiss, S. J. 1960 Mechanisms of the radiation-induced degradation of nucleic acids. *J. molec. Biol.* **2**, 379–391.

Schulte-Frohlinde, D. & Bothe, E. 1984 Identification of the major pathway of stand break formation in poly-U induced by OH radicals in the presence of oxygen. *Z. Naturfor.* **39** c, 315–319.

Shiga, T. 1965 An e.p.r. study of alcohol oxidation by Fenton's reagent. *J. phys. Chem.* **69**, 3805–3814.

von Sonntag, C. & Schulte-Frohlinde, D. 1978 Radiation-induced degradation of the sugar in model compounds and of DNA. In *Effects of ionizing radiation on DNA* (ed. J. Huttermann, W. Kohnlein, R. Teoule & A. J. Bertinchamps), pp. 204–225. Berlin: Springer-Verlag.

von Sonntag, C., Hagen, U., Schön-Bopp, A. & Schulte-Frohlinde, D. 1981 Radiation induced strand breakage in DNA: chemical and enzymatic analysis of end groups and mechanistic aspects. *Adv. radiat. Biol.* **9**, 109–142.

Steenken, S., Behrens, G. & Schulte-Frohlinde, D. 1974 Radiation chemistry of DNA model compounds. IV. Phosphate ester cleavage in radicals derived from glycerol phosphates. *Int. J. radiat. Biol.* **25**, 205–210.

Ullrich, M. & Hagen, U. 1971 Base liberation and concomitant reactions in irradiated DNA solution. *Int. J. radiat. Biol.* **19**, 509–517.

Walling, C. 1975 Fenton's reagent revisited. *Acct. chem. Res.* **8**, 125–131.

Willson, R. L. 1977 Chemically reactive species in biology. Free radicals and electron-transfer in biology and medicine. *Chemy Ind., Lond.*, 183–193.

Discussion

B. J. Parsons (*Research Division, North East Wales Institute, Clwyd*). I would like to comment that it is not always necessary to deal with a mixture of free radicals. We have shown, for example (B. J. Parsons *et al.* 1978) that $Br^{2-\cdot}_{3}$ reacts selectively at C-1 on 2-deoxy-D-ribose. Such selectivity would make analysis of Dr Gilbert's e.s.r. experiments more straightforward.

Reference

Parsons, B. J., Schulte-Frohlinde, D. & von Sonntag, C. 1978 *Z. Naturf.* **33** b, 666.

B. C. Gilbert. We have attempted to study the reaction of carbohydrates with, for example, the radicals $Cl_2^{-\cdot}$ and $Br_2^{-\cdot}$, which would be expected to be more selective than $OH^{\cdot}$. We suspect that it is their comparative lack of reactivity that precludes the detection of e.s.r. signals from radicals derived from their reaction.

Catherine Rice-Evans (*Department of Biochemistry, Royal Free Hospital School of Medicine*). Dr Gilbert has described the attack of hydroxyl radicals on purine and pyrimidine nucleotides at pH 4. Could he explain the relevance of his observations to radiation-induced hydroxyl radical attack on nucleic acids at physiological pH values in vivo?

B. C. Gilbert. In our experiments, signals are generally unchanged in the pH range 4–7 (except for a slight decrease in intensity). The choice of *ca.* pH 4 (in contrast to pH $\lesssim 2$ for the investigation of the acid-catalysed processes described) reflects the optimum conditions for $OH^{\cdot}$ generation (as well as the need to avoid interference from rapid base-catalysed reactions above *ca.* pH 7).

R. L. Willson (*Department of Biochemistry, Brunel University*). I feel that attack must occur, with uridine and adenosine, at both the base and the sugar. Dr G. Scholes many years ago showed this elegantly by product analysis of irradiated solutions. Is it that Dr Gilbert is just not observing a particular radical because it is oxidized or reduced by other species present in the flowing system? Pulse radiolysis studies show that the different radicals from uracil and adenine and have a wide range of redox properties.

B. C. Gilbert. It is clear from the results that Professor Willson mentions that there must be a significant amount of attack by ˙OH at the base in purine nucleosides: it seems likely that the appropriate adducts have e.s.r. spectra possessing many weak lines (so that detection and recognition is hampered), rather than that they are destroyed by further reactions. On the other hand, the e.s.r. results clearly indicate both the importance of attack at identifiable positions in the sugar ring and the reactivity of the pyrimidine ring (cf. figure 2).

Phil. Trans. R. Soc. Lond. B **311**, 531–544 (1985)
Printed in Great Britain

Concerning the intermediacy of organic radicals in vitamin B_{12}-dependent enzymic reactions

By Ruth M. Dixon, B. T. Golding, S. Mwesigye-Kibende and D. N. Ramakrishna Rao

Department of Organic Chemistry, Bedson Building, The University, Newcastle upon Tyne NE1 7RU, U.K.

The vitamin B_{12} coenzyme adenosylcobalamin assists the enzymic catalysis of molecular rearrangements of the type

$$\underset{\text{substrate}}{-\overset{\overset{\displaystyle H}{|}}{\underset{|}{C}}-\overset{\overset{\displaystyle X}{|}}{\underset{|}{C}}-} \longrightarrow \underset{\text{intermediate or product}}{-\overset{\overset{\displaystyle X}{|}}{\underset{|}{C}}-\overset{\overset{\displaystyle H}{|}}{\underset{|}{C}}-}\ ,$$

in which the migrating group X can be OH, NH_2 or a suitable substituted carbon atom such as $C(=CH_2)CO_2H$. This paper discusses evidence for the participation of organic radicals as intermediates in these reactions. Theoretical and model studies supporting the intermediacy of radicals in the reactions catalysed by the enzymes diol dehydratase and α-methyleneglutarate mutase are described. For the model studies, alkyl radicals, alkylcobaloximes (alkyl represents, for example, ethoxycarbonyl substituted, but-3-enyl and cyclopropylmethyl) and also dihydroxyalkylcobalamins have been investigated. The $Co–C_\alpha–C_\beta$ angle of 125° in adenosylcobalamin is shown to be an 'especial' angle by analysis of the crystal structures of *R*- and *S*-2,3-dihydroxypropylcobalamin.

1. Introduction

The vitamin B_{12} coenzyme adenosylcobalamin (AdoCbl, **1*a***; see Appendix for structures) assists in the enzymic catalysis of molecular rearrangements of the following kind

$$\underset{\text{substrate}}{a-\overset{\overset{\displaystyle H}{|}}{\underset{\underset{\displaystyle b}{|}}{C}}-\overset{\overset{\displaystyle X}{|}}{\underset{\underset{\displaystyle c}{|}}{C}}-d} \longrightarrow \underset{\text{intermediate or product}}{a-\overset{\overset{\displaystyle X}{|}}{\underset{\underset{\displaystyle b}{|}}{C}}-\overset{\overset{\displaystyle H}{|}}{\underset{\underset{\displaystyle c}{|}}{C}}-d}\ . \qquad (1)$$

Examples of such systems are diol dehydratase when X = OH, a = OH, b = c = H, d = Me; and α-methyleneglutarate mutase when $X = C(=CH_2)CO_2H$, $a = CO_2H$, b = c = d = H. For recent reviews of these reactions see Finke *et al.* (1984), Golding (1982) and Zagalak (1982).

Scheme 1 presents a pathway for diol dehydratase operating on propane-1,2-diol (Finlay *et al.* 1972), that is applicable in principle to all of the AdoCbl-dependent enzymic reactions. In an initiation step, homolysis of the Co–C σ-bond of AdoCbl gives cob(II)alamin (Cbl(II), **1*b***) and the adenosyl radical (Ado˙), a primary organic radical. This propagates the reaction by abstracting a particular hydrogen atom from a substrate molecule (for example, H_R from C-1 of *R*-propane-1,2-diol) to give deoxyadenosine (AdoH) and a substrate-derived radical S˙, which is converted into a product-related radical P˙. Product PH is derived by the removal of a hydrogen atom from the methyl group of deoxyadenosine by radical P˙, thus regenerating Ado˙. The reaction is terminated by the recombination of Cbl(II) with Ado˙.

To solve the 'B_{12} mystery' with scheme 1 requires that a number of questions be answered.

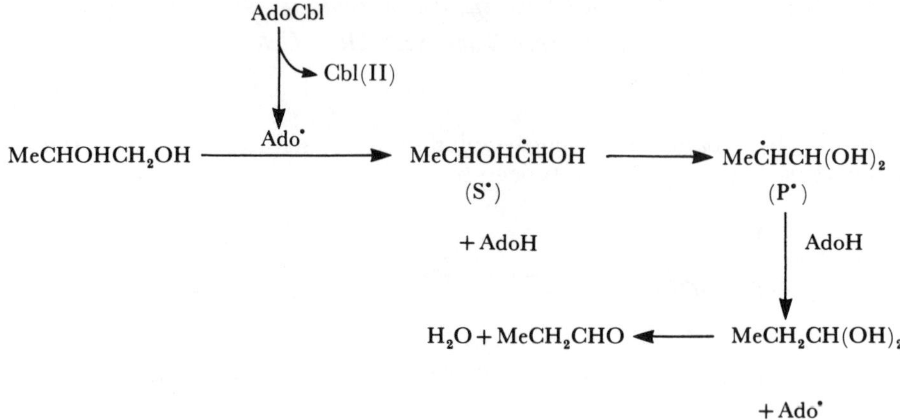

SCHEME 1. Diol dehydratase pathway via radical intermediates (see also Abeles *et al.* 1972).

(*a*) *What is the driving force for the cleavage of the* Co–C *bond of* AdoCbl?

The way in which strain induced within the enzyme–coenzyme–substrate complex might bring about cleavage of the Co–C bond has been discussed (Pratt 1982; Halpern 1982; Summers 1984). Of importance in this context is the meaning of the 125° angle found for Co–C_α–C_β of AdoCbl, and this is addressed later.

(*b*) *Are organocorrinoids obligatory intermediates for the conversion of* S˙ *into* P˙?

Critics of the alternative postulate of radicals in AdoCbl-dependent enzymic reactions are apt to refer to 'free-radical mechanisms' and to cite the difficulty of achieving 1,2-shifts in organic radicals. However, it has been made abundantly clear (Golding 1979, 1982; Finke *et al.* 1984) that protein-bound radicals are envisaged (for example, for diol dehydratase (see scheme 2)) and that rearrangements of these radicals could occur by one of the following routes.

(i) Via a cyclopropylmethyl radical (reaction catalysed by α-methyleneglutarate mutase, see equation (1)). This possibility is suggested by the rapidity of the interconversion of the parent radicals (cyclopropylmethyl and but-3-enyl) (Effio *et al.* 1980).

Model studies in support of this proposition are presented in §3*b*.

(ii) Via protonation of the migrating group ($OH \rightarrow {}^+OH_2$ in the diol dehydratase reaction (see equation (1)); $NH_2 \rightarrow {}^+NH_3$ in the ethanolamine ammonia lyase reaction and other aminomutases; CoSCoA→C(=O⁺H)SCOA in the methylmalonyl CoA mutase reaction). Recent studies have shown a 1,2-migration of protonated carboxyl (i.e. $CO_2^+H_2$) in a gas-phase reaction (Weisle *et al.* 1985). Migration of protonated OH is discussed in §3*a* with model studies for diol dehydratase.

(iii) Via dissociation–recombination (for glutamate mutase).

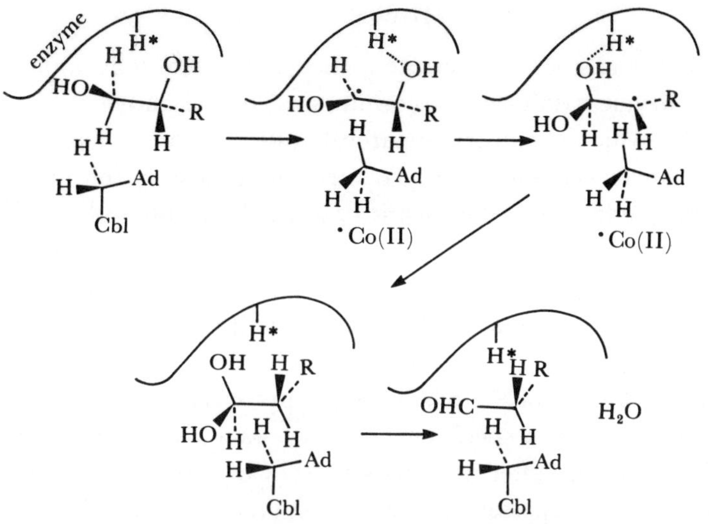

SCHEME 2. Hypothetical pathways (1–4) for AdoCbl-dependent enzymatic reactions. Pathways 1 and 2 involve electron transfer (e.g. $S^{\cdot} + Cbl(II) \rightleftharpoons S^{-} + Cbl(III)$); pathway 3 proceeds via protein-bound organic radicals; pathway 4 goes via σ-bonded organocorrinoids.

(c) *Does scheme 1 account for the stereochemical features of the reactions?*

In a recent comment (Rooney 1984) on the so-called 'free-radical' mechanism for ethanolamine ammonia lyase, it was stated that the 'highly stereospecific' nature of this reaction is 'difficult to explain on the basis of a ...radical mechanism'. As first recognized by Ogston and clearly reviewed by Alworth (1972) many years ago, the discrimination between enantiotopic atoms or groups by chiral reagents (such as enzymes) is an inevitable consequence of the symmetry properties of prochiral molecules (those with lack of rotational symmetry). All of the AdoCbl-dependent enzymic reactions exhibit sharp discrimination whenever there is a choice in respect of enantiotopic atoms or groups (Retey 1982). While these results cannot define a unique mechanism, any mechanism suggested must be in accord with them. The bound radical mechanism for diol dehydratase (scheme 2, R = Me) is in perfect accord with the experimental evidence from studies with labelled propane-1,2-diols (as stated by Golding (1979) and fully explained by Finke *et al.* (1984). The product-related radical of scheme 2 (R = Me) receives a hydrogen atom at a prochiral centre and delivery is therefore to a particular face. When either *R*- or *S*-[1-^{2}H,1-^{3}H]ethane-1,2-diol was used as substrate in the diol dehydratase reaction, racemic [2-^{2}H,2-^{3}H]ethanal was obtained (Retey 1982). With ethane-1,2-diol as substrate, the faces of the product-related radical (scheme 2, R = H) are equivalent, provided that rotation about the C-1–C-2 bond is faster than transfer of a hydrogen atom from deoxyadenosine. Substitutions of the hydrogen atoms at the radical centre, one by deuterium and the other by tritium, make the faces of this centre enantiotopic. In practice it is highly unlikely that the enzyme can discriminate sufficiently between deuterium and tritium to prevent rotation about C-1–C-2 occurring at a rate similar to that with the corresponding unlabelled substrate, and so racemic [2-^{2}H, 2-^{3}H]ethanal results. The evidence from isotopic labelling of ethane-1,2-diol is therefore fully consistent with the bound-radical hypothesis for diol dehydratase (for detailed accounts see Finke *et al.* (1984) and Retey (1982)).

2. Model studies

The purpose of a model study is to create a chemical system that faithfully reproduces the main features of an enzyme-catalysed transformation. A model study is most valuable with those enzymic reactions for which chemistry apparently offers no precedent and are not presently amenable to study at an enzymic level. To attempt to answer the questions posed in the introduction we have studied both the chemistry of alkyl radicals and alkylcobalt compounds. The alkyl species are structurally related to the substrate-derived and product-related entities of scheme 1. Our approach is based on the recognition that cob(II)alamin could be a 'conductor' or 'spectator' in scheme 1. By 'conductor' it is meant that for S˙ to be converted into P˙ it is necessary for Cbl(II) either to join with S˙ to give an organcorrinoid (S–Cbl), or to effect oxidation or reduction of S˙ (to S^+ or S^-, respectively; cf. scheme 3 (Abeles & Dolphin 1976)). One of these modified S species has the necessary reactivity to be converted into P. By 'spectator' it is meant that after homolysis of AdoCbl, Cbl(II) plays no further part until it terminates the reaction by repossessing the adenosyl radical. If this is the role of cobalt, then AdoCbl evolved as a masked primary organic radical and no more.

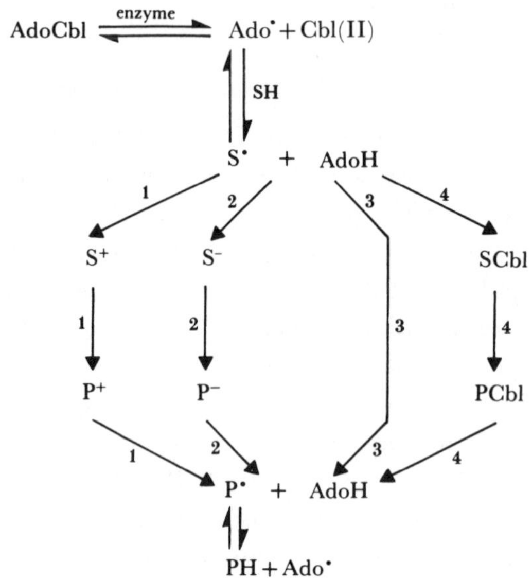

Scheme 3. Pathway for AdoCbl-dependent diol dehydratase via protein-bound organic radicals (H* is an acidic group of the proteins, R = H or Me).

Early attempts were made to explain the action of AdoCbl by means of known organometallic chemistry (Whitlock 1963). It was important to be aware of the properties of alkyl groups bound to cobalt. Although this knowledge is best acquired with the 'real thing' (i.e. alkylcobalamin) a valuable innovation was the development of model compounds for alkylcobalamins, the alkylcobaloximes, which are easy to prepare and characterize (Schrauzer 1968). Studies of the model compounds provide information about the reactivities of alkyl groups σ-bonded to cobalt in a corrinoid-like environment. (For a review of similarities and differences between cobalamins and cobaloximes see Elliott *et al.* 1981.)) It was shown that remarkable activation of simple ester groupings could be achieved in 2-acetoxyalkylcobaloximes (Golding *et al.* 1970).

$$\mathrm{AcOCH_2CH_2Co(dmgH)_2py} \xrightarrow[\text{several hours } 25\,°\mathrm{C}]{\mathrm{MeOH}} \mathrm{MeOCH_2CH_2Co\,(dmgH)_2\,py}. \tag{2}$$

(dmgH = monoanion of dimethylglyoxime, py = pyridine).

This $B_{AL}1$ cleavage of an ester is facilitated by σ–π hyperconjugation (electron donation from the Co–C σ–bond to the developing vacant p-orbital at C-2), leading to an intermediate ethene–Co^{III} complex that is captured by methanol. This finding strengthened the proposal of π-complexes as intermediates in certain AdoCbl-dependent enzymic reactions (Babior 1970; Silverman *et al.* 1972). However, the possibility that the enzymatic reactions might proceed via organic radicals was also considered (Eggerer *et al.* 1960; Cockle *et al.* 1972). This was taken more seriously when it was noticed that non-enzymatic chemistry of 1,2-diols and vicinal amino-alcohols (see equations (3) and (4); Gilbert *et al.* 1972; Bansal *et al.* 1973), known to occur via free radicals, bore a suggestive similarity to the enzymic reactions catalysed by AdoCbl-dependent diol dehydratase and ethanolamine ammonia lyase (Golding & Radom 1973, 1976).

$$\mathrm{HOCH_2CH_2OH} \xrightarrow{\mathrm{OH^{\cdot}}} \mathrm{HOCH_2\dot{C}HOH + H_2O} \xrightarrow{\mathrm{H^+}} \mathrm{H_2O + {}^{\cdot}CH_2CHO}, \tag{3}$$

$$\mathrm{HOCH_2CH_2OH} \xrightarrow{\mathrm{OH^{\cdot}}} \mathrm{HOCH_2\dot{C}HOH + H_2O} \xrightarrow{\mathrm{OH^-}} \mathrm{HOCH_2\dot{C}HO^-} \longrightarrow \mathrm{OH^- + {}^{\cdot}CH_2CHO}.$$

3. Conversion of S˙ into P˙

(*a*) *Modelling diol dehydratase*

Once we had become aware of the type of non-enzymic chemistry exemplified by equations (3) and (4), we set about the design of an 'ideal model system' for diol dehydratase that would illustrate two main points.

(i) The regioselective attack by a primary organic radical, derived by homolysis of a Co–C σ-bond, at C-1 of a 1,2-diol giving a 1,2-dihydroxyalkyl radical (this would mimic the highly regioselective attack by the adenosyl radical on propane-1,2-diol in the diol dehydratase reaction).

(ii) The conversion of the 1,2-dihydroxyalkyl radical into an aldehyde, preferably *via* a 2,2-dihydroxyalkyl radical, formed by a 1,2-oxygen shift (the occurrence of 1,2-oxygen shifts in the conversion of propane-1,2-diols into propanal, and ethane-1,2-diol into ethanal, have been proved by oxygen-labelling studies with diol dehydratase (Retey 1982)).

We showed initially that irradiation of methyl(pyridine)cobaloxime in aqueous ethane-1,2-diol at pH 3 gave a low yield of ethanal (Golding *et al.* 1975). This was interpreted by postulating attack of methyl radicals (from photohomolysis of the methyl–cobalt bond) on ethane-1,2-diol to give 1,2-dihydroxyethyl radicals. These underwent acid-catalysed conversion into the formylmethyl radical or 2,2-dihydroxyethyl radical, which then abstracted a hydrogen atom from dimethylglyoxime to give ethanal (hydrate). More efficient models were obtained by making intramolecular hydrogen transfer from vicinal diol to radical (Golding *et al.* 1980):

$$\mathrm{HOCH_2CHOH(CH_2)_3Co(dmgH)_2py} \xrightarrow[\mathrm{pH3}]{h\nu} \text{pentanal (10\%) + other products;} \quad (5)$$

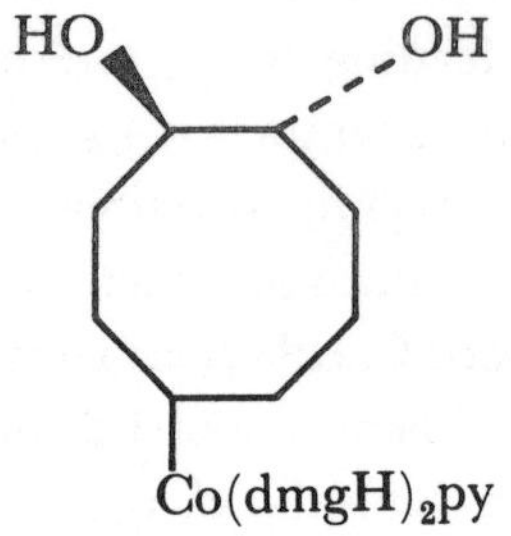

$$\text{[cyclo-octane-1,2-diol bearing Co(dmgH)}_2\text{py]} \xrightarrow[\mathrm{pH3}]{h\nu} \text{cyclo-octanone (30\%) + other products.} \quad (6)$$

HO OH

Co(dmgH)$_2$py

Equation (5) reproduces one important feature of the diol dehydratase reaction (see (i)), and at least the overall conversion 1,2-diol → aldehyde (see (ii)); but is the mechanism the same as that of diol dehydratase?

Ab initio molecular orbital calculations showed that conversion of, for example, the 1,2-dihydroxyethyl radical into the 2,2-dihydroxyethyl radical would be facilitated by protonation of the migrating OH, enabling the reaction to occur via a bridged transition state or intermediate (Golding & Radom 1973, 1976):

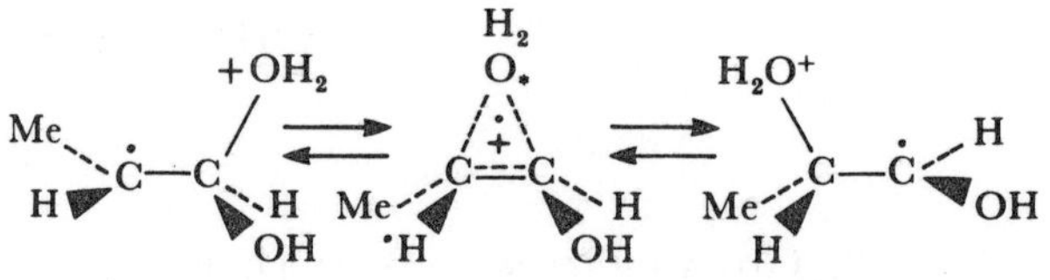

$$\text{Me,H}\dot{\mathrm{C}}\text{—C(}\overset{+}{\mathrm{O}}\mathrm{H_2}\text{)(H)(OH)} \rightleftharpoons \text{[bridged } \mathrm{H_2O}^{+}\text{: C}\text{=}\text{C}\text{]} \rightleftharpoons \mathrm{H_2O^+}\text{—C(Me)(H)—}\dot{\mathrm{C}}\text{(H)(OH)} \quad (7)$$

This mechanism is also applicable to the non-enzymic reactions of equations (3), (5) and (6). Recently, mass spectrometric studies have detected the protonated 2-hydroxyethyl radical $\cdot CH_2CH_2O^{+}H_2$, while further calculations have shown that it and its bridged form are significantly more stable than the well known ethanol radical cation (Bouma *et al.* 1983). It was recognized (Golding & Radom 1976) that the model reactions (equations (3), (5) and (6)) and diol dehydratase could proceed via an alternative acid-catalysed fragmentation, as proposed for (3) by Gilbert *et al.* (1972). However, this mechanism requires that, at least for diol dehydratase, it is necessary for water to recombine with the intermediate 2-oxoalkyl radical to give a 2,2-dihydroxyalkyl radical. It was also recognized (Golding & Radom 1976) that the fragmentation of the conjugate base of dihydroxyalkyl radicals postulated for reactions observed at higher pH (equation (4)) suggested an alternative dissociation–recombination pathway for diol dehydratase (see also discussion on this enzyme by Finke *et al.* 1984)).

Recently, we have studied a series of dihydroxyalkylcobalamins in the search for an ideal model system for diol dehydratase. When either *R*- or *S*-2,3-dihydroxypropylcobalamin (**1*c*** and **1*d***, respectively) were photolysed or thermolysed at pH 7–8 in aqueous solution (D_2O) the main products were [H]-3-hydroxypropanone and prop-2-en-1-ol. The yields of these products depended on the dioxygen content of the system and the mode of cleavage of the Co–C bond (table 1). The formation of these products can be rationalized by scheme 4. Thus,

TABLE 1. PRODUCTS FROM THERMAL AND PHOTOLYTIC REACTIONS OF DIHYDROXYALKYL-COBALAMINS (RCbl)[a]

R[b]	products	Δ (anaerobic)	$h\nu$ (anaerobic)[c]	Δ (O_2)	$h\nu$ (O_2)
R-2,3-dihydroxypropyl	DCH_2COCH_2OH	20	65	70	45
	$CH_2 = CHCH_2OH$	75	35	15	0
S-2,3-dihydroxypropyl	DCH_2COCH_2OH	5	65	70	70
	$CH_2 = CHCH_2OH$	50	35	30	0
4,5-dihydroxypentyl	$CH_2 = CHCH_2CHOHCH_2OH$	70	60	100	100
	t-$MeCH = CHCHOHCH_2OH$	30	40	0	0
5,6-dihydroxyhexyl	$CH_2 = CHCH_2CH_2CHOHCH_2OH$	0	20	100	100
	t-$MeCH = CHCH_2CHOHCH_2OH$	20	80	0	0

[a] All reactions were performed (normally in duplicate) in n.m.r. tubes in unbuffered D_2O(p[^{2}H]7–8). Thermolyses were done in darkness at 90 °C for *ca.* 8 h. Photolyses were done at 20 °C, irradiating with a 100 W lamp at a distance of 20 cm. Products were identified by ^{1}H n.m.r. spectroscopy (after purging the tubes with oxygen), gas chromatography, and isolation in some cases (carbonyl compounds as 2,4-dinitrophenylhydrazones). Yields are absolute and are accurate to $\pm 10\,\%$.

[b] The 4,5-dihydroxypentyl- and 5,6-dihydroxyhexylcobalamin were mixtures of diastereoisomers.

[c] One experiment was done with 4,5-dihydroxypentylcobalamin at p[^{2}H] 3.4 in D_2O–CD_3CO_2D and did not yield detectable pentanal (less than 1 %).

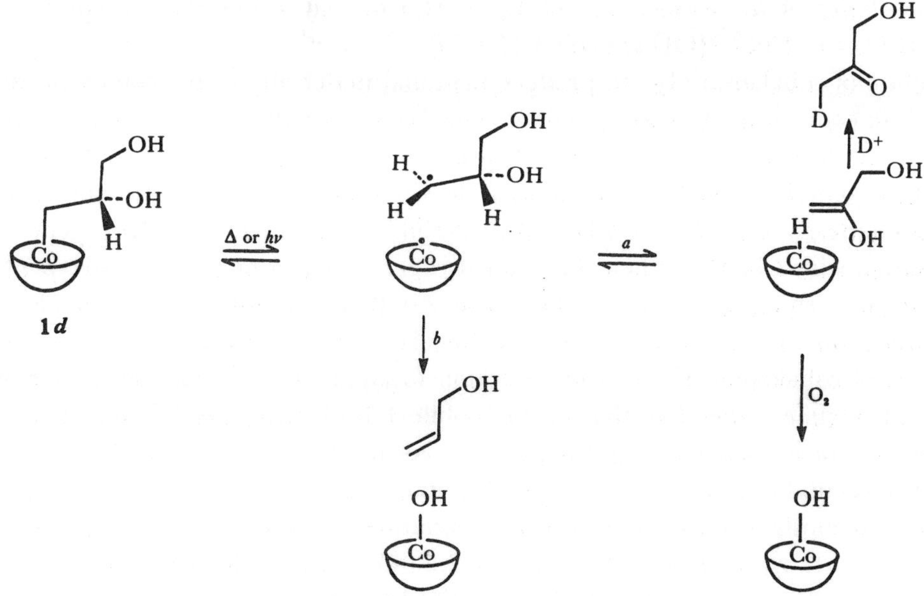

SCHEME 4. Pathways for thermal (Δ) or photochemical ($h\nu$) decomposition of *S*-2,3-dihydroxypropylcobalamin.

following homolysis of the Co–C bond, the derived radical pair could undergo elimination of the hydrogen β to the radical, giving hydridocobalamin (H–Cbl, **1*e***), and the enol of hydroxypropanone, prop-2-ene-1,2-diol (pathway *a*, scheme 4). Alternatively, the β-hydroxyl group is eliminated giving prop-2-en-1-ol and hydroxocobalamin, OH–Cbl (**1*f***) (pathway *b*, scheme 4). Under anaerobic conditions, the formation of H–Cbl and prop-2-ene-1,2-diol is reversible and relatively more prop-2-en-1-ol is formed by pathway *b*, which is assumed to be irreversible. In the presence of oxygen, H–Cbl is scavenged (probably via its conjugate base cob(I)alamin) to give OH–Cbl irreversibly, and the prop-2-ene-1,2-diol tautomerizes to

[1-^{2}H]3-hydroxypropanone. The different ratios of products obtained under thermal or photolytic conditions may be ascribed to one or both of the following reasons:

(i) temperature effects on the rates of pathways *a* and *b* (table 1);

(ii) the mechanism of cleavage may differ (for example, thermal radical pair formation (as in scheme 4) or photochemical concerted β-elimination from an excited state of R–Cbl).

The anaerobic thermolyses and photolyses of the dihydroxyalkylcobalamins **1g** and **1*h*** gave mixtures of dihydroxyalkenes (table 1), but no aldehyde (for example, pentanal from **1g**) or ketone (such as hexan-2-one from **1*h***) under either neutral or acidic conditions. Aerobically, the terminal alkene was the sole organic product. The behaviour described can be rationalised in a manner similar to the explanation for *R*- and *S*-2,3-dihydroxypropylcobalamin. In the presence of oxygen, H–Cbl is scavenged leaving 4,5-dihydroxypent-1-ene [from **1g**] or 5,6-dihydroxyhex-1-ene (from **1*h***) as the exclusive organic product. In the absence of oxygen, β-elimination to H–Cbl and terminal alkene is followed by the recombination of these species to afford, for example, either **1g** or the secondary alkylcobalamin **1*i***. The latter dissociates to H–Cbl and an internal alkene (mainly *t*-4,5-dihydroxypent-2-ene). The 4,5-dihydroxypent-1-ene obtained from **1g** by anaerobic thermolysis is partly deuterium labelled, whereas the alkene obtained in aerobic thermolysis is unlabelled. These observations support the postulated reversibility of the dissociation of **1g** to H–Cbl and 4,5-dihydroxypent-1-ene (N.B. H–Cbl + $D_2O \rightleftharpoons$ D–Cbl + HOD via cob(I)alamin).

The failure of cobalamin (**1g**) to produce pentanal under any of the conditions described (table 1) is in contrast to the corresponding cobaloxime (equation (5); see also Golding *et al.* 1980). For the formation of pentanal a 1,5-shift of a hydrogen atom is obligatory in the 4,5-dihydroxypentyl radical from homolysis of the Co–C bond of **1g**. The 4,5-dihydroxypentyl radical can undergo, *inter alia* a 1,5-H shift or β-elimination (to 4,5-dihydroxypent-1-ene) or recombination with Cbl(II). When this radical is generated from 4,5-dihydroxypentyl(pyridine)cobaloxime (equation (5)), two of these processes (β-elimination involving cobaloxime(II), or recombination of a radical with cobaloxime(II)) are less likely because of the rapid degradation of cobaloxime(II) in acidic solution to aquated Co^{II}. The behaviour observed with the cobaloxime is therefore that of the mobile 4,5-dihydroxypentyl radical (i.e. radical surrounded by solvent molecules and not influenced by cobalt ions), which rearranges to the 1,2-dihydroxypentyl radical, the precursor of pentanal. To observe analogous chemistry with cobalamins obviously requires that both β-elimination and recombination are blocked. To achieve this we are synthesizing 4,5-dihydroxy-2,2-dimethylpentylcobalamin.

Our studies to date of dihydroxyalkylcobalamins have not provided a better model for diol dehydratase than that developed with cobaloximes (Golding *et al.* 1980). However, we believe that they support the premise that the function of deoxyadenosine in the mechanism for diol dehydratase of scheme 2 is to prevent the interaction of substrate-derived radical with Cbl(II), which would lead to non-productive elimination of either H or OH β to the radical centre.

(*b*) *Modelling α-methyleneglutarate mutase*

Following the model studies of Dowd *et al.* (1975) and Chemaly & Pratt (1976) concerning the possible intermediacy of dicarboxy-substituted but-3-enyl- and cyclopropylmethylcobalamins in the rearrangement catalysed by α-methyleneglutarate mutase (equation (1)), we undertook the synthesis and study of but-3-enyl- and cyclopropylmethylcobaloximes. We have reported thermally induced and trifluoroacetic acid-catalysed interconversions of these compounds, which were rationalized by pathways in which the organic moiety never left cobalt (the 'conductor' role) (Atkins *et al.* 1980). To improve the correspondence of the model with the 'real thing' we synthesized the ethoxycarbonyl-substituted cobaloximes **2a**–**2d**. In contrast to the methyl-substituted series, the ethoxycarbonyl-substituted compounds were much less readily interconverted (either thermally or by acidic catalysis), perhaps because intermediate η^3 homoallyl species are destabilized by replacing methyl by ethoxycarbonyl (Golding & Mwesigye-Kibende 1983).

Attempts to synthesize the cobaloxime **2c** by reacting cobaloxime(I) with the *cis* iodo compound **3c** were unsuccessful, the predominant product of this reaction being the but-3-enylcobaloxime **2a**. With the *trans* iodo compound **3d**, cobaloxime(I) gave a mixture of cobaloximes **2d** and **2a**, the latter usually predominating. For the reaction of alkyl halides with cobaloxime(I), two competing mechanisms have been established: S_N2 displacement, and a stepwise pathway via an intermediate organic radical produced by electron transfer from one reactant to another (Okabe & Tada 1982). Which pathway is preferred depends on the nature of the alkyl halide, the stepwise route being favoured with hindered alkyl iodides. The results described for iodides **3c** and **3d** can be rationalized by an electron-transfer pathway leading to cyclopropylmethyl radicals (**4c** from **3c**, **4d** from **3d**). These can either be trapped by cobaloxime(II), or ring-open primarily to radical **4a**, which reacts with cobaloxime(II) to yield cobaloxime **2a**. For steric reasons the *cis* radical **4c** may open more rapidly than the *trans* radical **4d**. Thus, the *cis* cobaloxime **2c** is not observed, in contrast to the *trans* isomer **2d**. Spectroscopic characterization of the radicals **4a** and **4b**, and preliminary information on the reactivity of radicals **4a** and **4d**, has been obtained. After γ-radiolysis the halides **3a** and **3b** gave the corresponding radicals at 77 K (**4a** and **4b** respectively), characterized by e.s.r. spectroscopy (see figure 1). However, the cyclopropylmethyl iodides **3c** and **3d** were converted into butenyl radical **4a** by γ-radiolysis at 77 K. Irradiations (visible light) of cobaloximes **2a** and **2b** in deuteriochloroform, followed by exposure to air, gave the peroxyalkylcobaloxime **2e** (Alcock

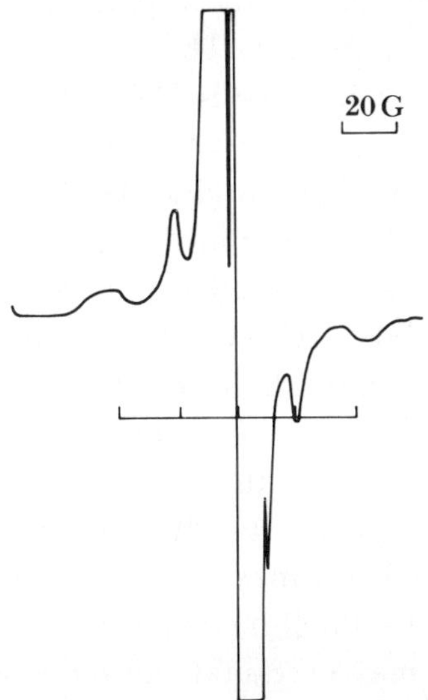

FIGURE 1. First-derivative X-band e.s.r. spectrum for a solution of methyl 2-(iodomethyl)but-3-enoate in CD_3OD after exposure to ^{60}Co γ-rays at 77 K and annealing to *ca.* 130 K, showing features assigned to the radical $CH_2 = CHCH(CO_2Et)^{\cdot}CH_2$. The intense central absorption is due to the solvent. 1 G = 10^{-4} T.

et al. 1985). These observations can be explained by postulating light-induced homolysis of the Co–C bond in each cobaloxime, to give a radical (**4*a*** from **2*a***, **4*b*** from **2*b***, etc.), which equilibrates with its isomers (**4*a*** $\rightleftharpoons$ **4*c*** or **4*d*** $\rightleftharpoons$ **4*b***). At equilibrium, the predominant radical is **4*a***. This reacts with cobaloxime(II) to give hydrido(pyridine)cobaloxime and 1-ethoxycarbonylbuta-1,3-diene. Recombination of these affords allylcobaloxime **2*f***, which by oxygen insertion into its Co–C bond gives the peroxyalkylcobaloxime **2*e***.

Three independent pieces of evidence have shown the ability of radicals **4*a*–4*d*** to be interconverted readily. On the contrary, the corresponding alkylcobaloximes **2*d*–3*d*** are not interconverted easily. It is inferred that α-methyleneglutarate mutase operates via the protein-bound radicals **4*e*–4*g***, the butenyl species being formed by reaction between the adenosyl radical and a substrate molecule. The deoxyadenosine formed in this process prevents the combination of an organic radical with cob(II)alamin. This is undesirable because the resulting alkylcobalamins are presumed *not* to be readily interconverted. Dowd *et al.* (1984) have reported the conversion (in darkness, 230 h, pH 8.4, 25°) of the alkylcobalamin **1*j*** into α-methyleneglutaric acid (*ca.* 30 %), amongst other products. This result was interpreted in support of a mechanism for α-methyleneglutarate mutase via organocorrinoid intermediates. The possibility of rearrangement via radicals, either *before* (by the electron-transfer pathway in the formation of alkylcobalt compounds) or *after* formation of alkylcobalamin **1*j*** (by homolysis of the Co–C σ-bond) was not excluded.

4. Cleavage of the Co–C bond of AdoCbl

The Co–C_α–C_β bond angle was measured as 125° ± 3 in the crystal structure analysis of AdoCbl (Lenhert 1968; Glusker 1982). This has been taken to imply strain in the molecule, making the Co–C bond intrinsically weak and ready to be cleaved on demand by an enzyme (Glusker 1982; Pratt 1982). To assess whether or not 125° is an especial angle, relevant to the mode of action of AdoCbl, it is necessary to have data for other alkylcobalamins and model compounds. We have therefore determined the crystal structures of the diastereoisomeric 2,3-dihydroxypropylcobalamins (**1*c*** and **1*d***). The *S* isomer **1*d*** has a Co–C_α–C_β angle of 113.6° ± 2.1, whereas that for the *R* isomer **1*c*** is 119.6° ± 1.7. Cobalamins **1*c*** and **1*d*** have Co–C bond lengths that do not differ significantly from Co–C for AdoCbl [2.03 (6) Å]. A very interesting feature of the structures of **1*c***, **1*d*** and AdoCbl is that in each case the group σ-bonded to Co lies in a channel between rings C and D defined by the sentinel groups at C-12 (β-methyl) and C-17 (C-54 methyl). For **1*d*** this is a 'comfortable' arrangement that benefits from an intramolecular hydrogen bond (C_β–OH...O=C (of *c*-acetamido group)). AdoCbl is isostructural with **1*d*** (same chirality at C_β), but adopts a different rotamer about the C_α–C_β bond (N.B. it cannot form an analogous H-bond to **1*d***). This enables the ribose of adenosyl to slot perpendicularly into the C–D channel, while the adenine is roughly parallel to the corrin and partly over ring C. The β-face of ring C is probably the sterically least demanding region of the corrin and the arrangement described may profit from hydrophobic bonding, providing the adenine does not come too close to the β-methyl group at C-12 (N-9 ... C-46 = 3.71 Å). Minimization of steric interactions may necessitate an increase in the Co–C_α–C_β angle. This may weaken the Co–C σ-bond and make AdoCbl intrinsically strained. Nevertheless, although the bond dissociation energy of AdoCbl is only 132 kJ mol^{-1}, it has been estimated that diol dehydratase must accelerate the rate of cleavage of the Co–C bond by at least 10^{10}-fold (Finke

& Hay 1984). The crystal structure of cobalamin **1c** indicates that too close contact between C_β-OH and H-19 is avoided by increasing the angle Co–C_α–C_β. Cobalamin **1c** cannot adopt the same conformation as ***1d***, with a hydrogen bond from C_β-OH to the *c*-acetamido because this would press the CH_2OH group of the side chain into ring C. For further discussion of these results see Alcock *et al.* (1985).

Conclusions

The claim (Golding & Radom 1976) that a mechanistic scheme in the reactions catalysed by diol dehydratase, ethanolamine ammonia lyase, the other aminomutases and methylmalonyl CoA mutase 'is consistent with all experimental data to date' still holds true today. Our solution to the 'B_{12} mystery' is based on what we regard as the best assessment of the available evidence from experiments with the enzymes, theoretical calculations and model studies. We have been especially influenced by theory and models and trust that we do not suffer from what Henry James called the 'perversity (of) an innate preference for the represented subject over the real one'.

We thank Dr N. W. Alcock (University of Warwick) for collaboration on the crystal structure determinations and Professor M. C. R. Symons (University of Leicester) for assistance with e.s.r. measurements. Financial support from SERC is gratefully acknowledged.

References

Abeles, R. H. & Dolphin, D. 1976 The vitamin B_{12} coenzyme. *Acct. chem. Res.* **9**, 114–120.

Alcock, N. W., Dixon, R. M. & Golding, B. T. 1985 The crystal structures of (*R*)- and (*S*)-2,3-dihydroxypropylcobalamin; comparison with the structure of adenosylcobalamin. *J. chem. Soc. chem. Commun.*, 603–605.

Alcock, N. W., Golding, B. T. & Mwesigye-Kibende, S. 1985 Routes of formation and crystal structure of an alkylperoxycobalamin: ethyl (*E*)-4-[(pyridine)cobaloximeperoxy]pent-2-enoate. *J. chem. Soc. Dalton Trans.* (In the press.)

Alworth, W. L. 1972 Stereochemistry and its application in biochemistry. New York: Wiley Interscience.

Atkins, M. P., Golding, B. T., Bury, A., Johnson, M. D. & Sellars, P. J. 1980 $^3\eta$-Homoallylcobalt complexes in the intramolecular rearrangements of but-3-enylcobaloximes. *J. Am. chem. Soc.* **102**, 3630–3632.

Babior, B. M. 1970 Mechanism of action of ethanolamine ammonia-lyase, B_{12}-dependent enzyme. *J. biol. Chem.* **245**, 6125–6133.

Bansal, K. M., Gratzel, M., Henglein, A. & Janata, E. 1973 Polarographic and optical absorption studies of radicals produced in the pulse radiolysis of aqueous solutions of ethylene glycol. *J. phys. Chem.* **77**, 16–19.

Bouma, W. J., Nobes, R. H. & Radom, L. 1983 Unusual low-energy isomers of the ethanol and dimethyl ether radical cations. *J. Am. chem. Soc.* **105**, 1743–1746.

Chemaly, S. & Pratt, J. M. 1976 Carbon-skeleton rearrangement of the alkyl ligand in a vitamin B_{12} derivative. *J. chem. Soc. chem. Commun.*, pp. 988–989.

Cockle, S. A., Hill, H. A. O., Williams, R. J. P., Davies, S. P. & Foster, M. A. 1972 The detection of intermediates during the conversion of propane-1,2-diol to propionaldehyde by glycerol dehydrase, a coenzyme B_{12}-dependent reaction. *J. Am. chem. Soc.* **94**, 275–277.

Dowd, P., Shapiro, M. & Kang, K. 1975 The mechanism of action of vitamin B_{12}. *J. Am. chem. Soc.* **97**, 4754–4757.

Dowd, P., Shapiro, M. & Kang, J. 1984 The mechanism of action of vitamin B_{12}. *Tetrahedron* **40**, 3069–3086.

Effio, A., Griller, D., Ingold, K. U., Beckwith, A. L. J. & Serelis, A. K. 1980 Allylcarbinyl-cyclopropylcarbinyl rearrangement. *J. Am. chem. Soc.* **102**, 1734–1736.

Eggerer, H., Overath, P., Lynen, F. & Stadtmann, E. R. 1960 Mechanism of the cobamide coenzyme-dependent isomerisation of methylmalonyl CoA to succinyl CoA. *J. Am. chem. Soc.* **82**, 2643–2644.

Elliott, C. M., Hershenhart, E., Finke, R. G. & Smith, B. L. 1981 Coenzyme B_{12} model studies: an electrochemical comparison of cobaloxime and $Co[C_2(DO)(DOH)_{pn}]$ complexes to coenzyme B_{12}. *J. Am. chem. Soc.* **103**, 5558–5566.

Finke, R. G. & Hay, B. P. 1984 Thermolysis of adenosylcobalamin: a product, kinetic, and Co-C5′ bond dissociation energy study. *Inorg. Chem.* **23**, 3041–3043.

Finke, R. G., Schiraldi, D. A. & Mayer, B. J. 1984 Towards the unification of coenzyme B_{12}-dependent diol dehydratase stereochemical and model studies: the bound radical mechanism. *Coord. Chem. Rev.* **54**, 1–22.

Finlay, T. H., Valinsky, J., Sato, K. & Abeles, R. H. 1972 Mechanism of action of coenzyme B_{12}. Formation of 5′-deoxyadenosine and $B_{12(r)}$ in the reaction of dioldehydrase with chloroacetaldehyde. *J. biol Chem.* **247**, 4197–4207.

Gilbert. B. C., Larkin, J. P. & Norman, R. O. C. 1972 Evidence for heterolytic and homolytic transformations of radicals from 1,2-diols and related compounds. *J. chem. Soc. Perkin Trans.* II, pp. 794–802.

Glusker, J. P. 1982 X-ray crystallography of B_{12} and cobaloximes. In B_{12}, vol. 1 (ed. D. Dolphin), pp. 23–106. New York: Wiley-Interscience.

Golding, B. T., Holland, H. L., Horn, U. & Sakrikar, S. 1970 Solvolysis of σ-2-acetoxyalkylbis(biacetyldioximato)-pyridinecobalt: evidence for a novel intermediate olefinic complex. *Angew. Chem. int. Edn Engl.* **9**, 959–960.

Golding, B. T. & Radom, L. 1973 Facilitation of intramolecular 1,2-shifts in radicals by protonation, and the mechanism of reactions catalysed by 5′-deoxyadenosylcobalamin. *J. chem. Soc. chem. Commun.*, pp. 939–941.

Golding, B. T., Kemp, T. J., Nocchi, E. & Watson, W. P. 1975 Conversion of 1,2-diols into aldehydes induced by photolysis of alkylcobaloximes: a model system for dioldehydrase. *Angew. Chem. int. Edn Engl.* **14**, 813–814.

Golding, B. T. & Radom, L. 1976 On the mechanism of action of adenosylcobalamin. *J. Am. chem. Soc.* **98**, 6331–6338.

Golding, B. T. 1979 Vitamin B_{12}. In *Comprehensive organic chemistry*, vol. 5 (ed. D. Barton & W. D. Ollis), pp. 549–584. Oxford: Pergamon.

Golding, B. T., Sell, C. S. & Sellars, P. J. 1980 Model experiments relevant to the mechanism of adenosylcobalamin-dependent diol dehydrase: further investigations of isomerisations of dihydroxyalkyl radicals. *J. chem. Soc. Perkin Trans.* II, pp. 961–970.

Golding, B. T. 1982 Mechanisms of action of the B_{12} coenzyme: theory and models. In B_{12}, vol 1 (ed. D. Dolphin), pp. 543–582. New York: Wiley-Interscience.

Golding, B. T. & Mwesigye-Kibende, S. 1983 Synthesis and rearrangements of ethoxycarbonyl-substituted but-3-enyl and cyclopropylmethyl(pyridine)-cobaloximes. *J. chem. Soc. chem. Commun.*, pp. 1103–1105.

Halpern, J. 1982 Determination and significance of transition-metal-alkyl bond dissociation energies. *Acct. chem. Res.* **15**, 238–244.

Lenhert, P. G. 1968 The X-ray analysis of the vitamin B_{12} coenzyme. *Proc. R. Soc. Lond.* A **303**, 45–84.

Okabe, M. & Tada, M. 1982 The reaction of bis(dimethylglyoximato)-(pyridine)cobalt(I), cobaloxime(I), with 2-(allyloxy)ethyl halides and photolysis of the resulting organocobaloximes. *Bull. chem. Soc. Japan* **55**, 1498–1503.

Pratt, J. M. 1982 Coordination chemistry of the B_{12} dependent isomerisation reactions. In B_{12}, vol. 1 (ed. D. Dolphin), pp. 325–392, New York: Wiley Interscience.

Retey, J. 1982 Vitamin B_{12}: stereochemical aspects of its biological functions and of its biosynthesis. In *Stereochemistry* (ed. Ch. Tamm), p. 249–282. Amsterdam: Elsevier Biomedical Press.

Rooney, J. J. 1984 The mechanism of the vicinal interchange reactions mediated by coenzyme B_{12}. *J. chem. Res.* S, pp. 48–49.

Schrauzer, G. N. 1968 Organocobalt chemistry of vitamin B_{12} model compounds (cobaloximes). *Acct. chem. Res.* **1**, 97–103.

Silverman, R. B., Dolphin, D. & Babior, B. M. 1972 A model for the mechanism of action of coenzyme B_{12} dependent enzymes. Evidence for $\sigma \rightleftharpoons \pi$ rearrangements in cobaloximes. *J. Am. chem. Soc.* **94**, 4028–4030.

Summers, M. F., Marzilli, L. G., Bresciani-Pahor, N. & Randaccio, L. 1984 Unusual bond lengths, conformations, and ligand exchange rates in B_{12} models with the bis(salicylidene)-*o*-phenylenediamine equatorial ligand. *J. Am. chem. Soc.* **106**, 4478–4485.

Weisle, T., Halim, H. & Schwarz, H. 1985 *Chem. Ber.* (In the press.)

Whitlock, H. W. 1963 Concerning the mechanism of a reaction catalysed by coenzyme B_{12}. *J. Am. chem. Soc.* **85**, 2343–2344.

Zagalak, B. 1982 Vitamin B_{12} as a biologically active model compound. *Naturwissenschaften* **69**, 63–74.

Appendix

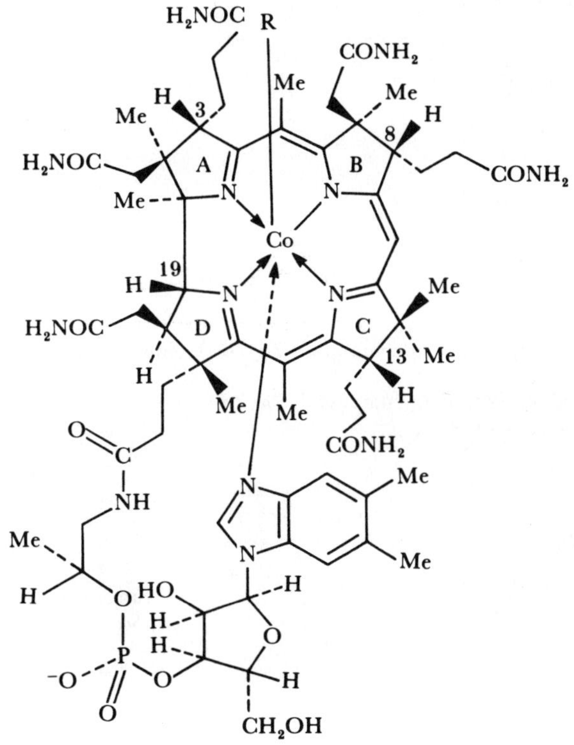

RCbl

1*a* R =

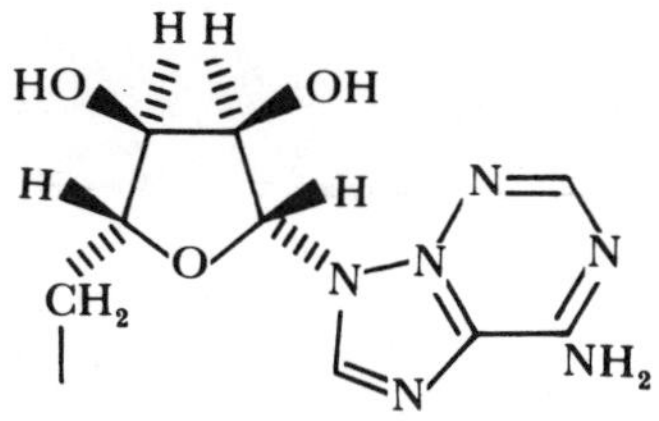

1*b* R = single electron
1*c* R = *R*-2,3-dihydroxypropyl
1*d* R = *S*-2,3-dihydroxypropyl
1*e* R = H
1*f* R = OH
1*g* R = $HOCH_2CHOH(CH_2)_3$
1*h* R = $HOCH_2CHOH(CH_2)_4$
1*i* R = $HOCH_2CHOHCH_2CHMe$
1*j* R = $CH_2CH(CO_2R^1)C(=CH_2)CO_2R^1$ (R^1 = tetrahydropyranyl)

2	$RCo(dmgH)_2py$ (dmgH = monoanion of dimethylglyoxime; py = pyridine)
2a	$R = CH(CO_2Et)CH_2CH{=}CH_2$
2b	$R = CH_2CH(CO_2Et)CH{=}CH_2$
2c	R = [cyclopropane bearing CH_2 and CO_2Et]
2d	R = [cyclopropane bearing CH_2 and CO_2Et]
2e	*t*-$O_2CHMeCH{=}CHCO_2Et$
2f	*t*-$CHMeCH{=}CHCO_2Et$
3a–3d	alkyl halides corresponding to R in compounds ***2a–2d*** (halide = Br in ***3a***, I in ***3b–3d***)
4a–4d	alkyl radical corresponding to R in compounds ***2a–2d***
4e	$HO_2C\dot{C}HCH_2C({=}CH_2)CO_2H$
4f	[cyclopropane bearing $\dot{C}H_2$, CO_2^- and ^-O_2C] (*cis* or *trans*)
4g	$HO_2CCH(\dot{C}H_2)C({=}CH_2)CO_2H$

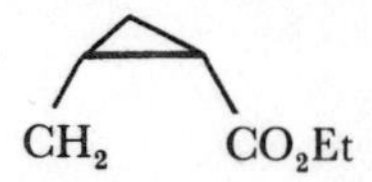

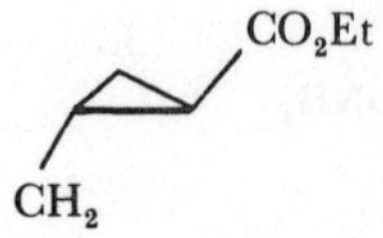

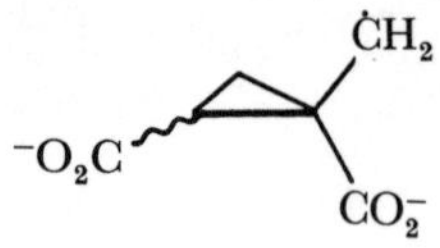

Phil. Trans. R. Soc. Lond. B **311**, 545–563 (1985)
Printed in Great Britain

Organic free radicals and proteins in biochemical injury: electron- or hydrogen-transfer reactions?

By R. L. Willson, Christina A. Dunster, L. G. Forni, Carolyn A. Gee and Katherine J. Kittridge

Biochemistry Department, Brunel University of West London, Uxbridge, Middlesex UB8 3PH, U.K.

The reactions of organic free radicals, acting as either reductants or oxidants, have been studied by pulse radiolysis in neutral aqueous solution at room temperature. Many hydroxyl-substituted aliphatic carbon-centred radicals and one-electron adducts have been shown to be good one-electron reductants, while several oxygen-, sulphur- and nitrogen- (but not carbon-) centred free radicals have been shown to be good one-electron oxidants. Several carbon-centred radicals can be reduced rapidly by hydrogen transfer, from undissociated thiol compounds which can thus act as catalysts facilitating the overall reduction of a carbon-centred radical by an electron-donating molecule.

Kinetic considerations influenced by the one-electron redox potentials of the radical–molecule couples involved, determine whether a particular reaction predominates.

In this paper examples of such reactions, involving a water-soluble derivative of vitamin E (Trolox C) and the coenzyme NADH, are described, together with studies showing (*a*) that even in complex multi-solute systems some organic peroxy radicals can inactiviate alcohol dehydrogenase under conditions where the superoxide radical does not, and (*b*) the superoxide radical can be damaging if urate is also present, and this damage can be reduced by the presence of superoxide dismutase.

Introduction

Although free radicals have long been accepted as important intermediates in a large number of chemical processes of industrial importance, there is still a considerable reluctance to believe that they have any role in biology, let alone medicine. Radiation chemistry, likewise, is still often regarded as a somewhat esoteric subject, relevant only to areas such as nuclear safety, radiation sterilization and radiation therapy. Indeed, it is still widely believed that when a system is exposed to radiation a host of more or less random reactions takes place and that the intermediates involved, even if they are free radicals, are somewhat different from those that might be formed by conventional chemistry. This is just not so. It cannot be overstressed that, in most instances (indeed in all the experiments of the type referred to here), radiation-induced free radicals are of thermal energy and follow the same laws of chemistry as other free radicals. We hope this presentation will demonstrate that radiation effects, free radicals and contemporary biology, even medicine, can provide an exciting multidisciplinary area of study for physical organic chemists, biochemists, biologists and clinicians alike.

Clearly, when a biological system is exposed to free radicals a host of molecules will be affected. Some of these molecules may be important, for example DNA, or perhaps a key protein whose normal rate of synthesis is comparatively slow. The repercussions may then be serious. Other molecules effected may be less important. For example, destruction of a small amount of glucose is unlikely to effect a biological system *per se*, although there does remain the possibility that products derived from such innocuous molecules may themselves be toxic. We must not forget that in organic chemistry, when one free radical reacts with another molecule, another free radical is generally formed and the reactions of this radical must be taken into consideration (see, for example, Adams *et al.* 1970, 1972; Willson 1976, 1977*a*, *b*, 1978, 1979, 1982, 1983).

Let us consider, for example, the many questions that can arise when we consider how free radicals react with a protein. For example, if we irradiate an enzyme in solution what happens?

(*a*) Is its activity lost?

(*b*) If so, by how much?

(*c*) Which free radicals are involved?

(*d*) What chemical changes occur?

(*e*) Can only one type of radical initiate a certain type of damage or can other radicals initiate the same or different damage?

(*f*) Are certain amino-acid residues attacked preferentially or is the free-radical attack simply random?

(*g*) Can some residues be destroyed without loss of activity?

(*h*) Does the damage remain on one particular residue or can it be transferred to another?

(*i*) If so, how fast are these processes?

(*j*) Do they occur intramolecularly or intermolecularly?

(*k*) What type of compounds can prevent these reactions?

(*l*) If such reactions can be prevented, do the compounds act by loosely binding to the enzyme, so modifying the active site and making it less susceptible to damage?

(*m*) Or, do they block the active site, shielding it from free-radical attack?

(*n*) Do other protective compounds act indirectly by scavenging free radicals generally?

(*o*) Or, do they act by repairing damage that has already been caused?

(*p*) If so, what reaction mechanisms are involved: do the processes occur, for example, by electron- or hydrogen-transfer reactions?

Clearly the number of possible reactions that can take place in such complex systems is enormous. Furthermore, any other inorganic or organic compounds present, even in relatively small concentrations, may have a significant effect on the overall processes that take place. To minimize such complications, radiation chemists traditionally have worked in systems which are as clean as possible. Materials have often been highly purified. Water, not just distilled, but often triply distilled, has been used for making up solutions and in many cases oxygen has been rigorously excluded. The fact that the biological relevance of information obtained in such sterile *in vitro* systems has sometimes been viewed with scepticism is perhaps, therefore, not surprising.

It is in this light that we have been undertaking experiments not only in pure chemical systems but also in 'dirty' but defined model biological systems containing an assortment of foreign compounds ('Analar dirt') and natural substances such as α-tocopherol (vitamin E), a water-soluble derivative of vitamin E (Trolox C), ascorbate (vitCH), and NADH.

When a neutral aqueous solution is irradiated at room temperature, whether it be pulsed radiation from an electron linear accelerator or continuous radiation from a ^{60}Co source, most chemical changes that take place can be attributed to initial reactions of the solvated electron and hydroxyl radicals formed after the ionization of water. A small amount of hydrogen atoms (*ca.* 10%) is also formed, but this can often, but not always, be neglected

$$H_2O \rightarrow e^-_{aq} + OH^{\cdot} + H_3O^+. \quad (1)$$

The solvated electron, a strongly reducing free radical, and the hydroxyl radical, a strongly oxidizing species, will react rapidly either with themselves or with other substances present, the extent depending on the relative concentrations and the respective rate constants (k) involved:

$$e^-_{aq} + A \xrightarrow{k_1} A^{-\cdot}; \quad (2)$$

$$e^-_{aq} + B \xrightarrow{k_2} B^{-\cdot}; \quad (3)$$

$$e^-_{aq} + C \xrightarrow{k_3} C^{-\cdot}; \quad (4)$$

$$[A^{-\cdot}]:[B^{-\cdot}]:[C^{-\cdot}] = k_1[A]:k_2[B]:k_3[C].$$

The product radicals from these reactions may, however, enter into further reactions with the other solutes present, the relative probabilities again depending on the respective concentrations and rate constants. Thus, although damage may occur initially with a compound present in high concentration, it may nevertheless be transferred subsequently to a compound present only in minor amounts. Such electron-transfer (or 'hole'-transfer) processes, taking place either intermolecularly or intramolecularly, can be readily envisaged when one considers the following cycling processes which have been characterized by pulse radiolysis (Forni & Willson 1984; Willson, 1970, 1971; Asmus *et al.* 1979; Mahood *et al.* 1980):

(i) radical as a reducing agent:

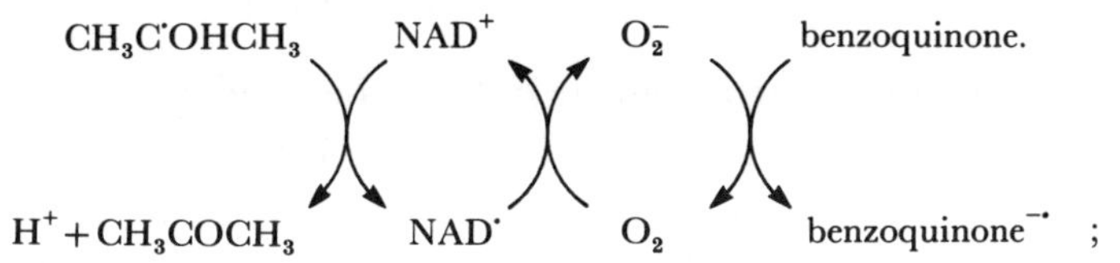

$CH_3C^{\cdot}OHCH_3$ / $H^+ + CH_3COCH_3$ ⇄ NAD^+ / $NAD^{\cdot}$ ⇄ O_2^- / O_2 ⇄ benzoquinone. / benzoquinone$^{-\cdot}$;

(ii) radical as an oxidizing agent:

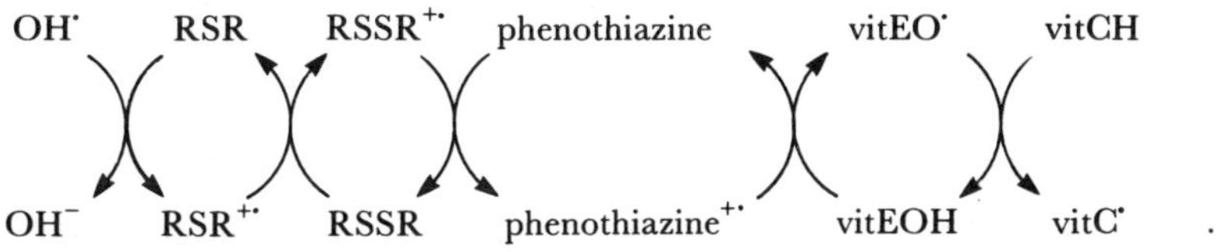

$OH^{\cdot}$ / OH^- ⇄ RSR / $RSR^{+\cdot}$ ⇄ $RSSR^{+\cdot}$ / RSSR ⇄ phenothiazine / phenothiazine$^{+\cdot}$ ⇄ $vitEO^{\cdot}$ / vitEOH ⇄ vitCH / $vitC^{\cdot}$.

Free-radical inactivation of lysozyme

Let us consider the irradiation of the enzyme lysozyme, in doubly distilled water containing a little phosphate buffer to adjust the solution to neutral pH. If we measure the extent of inactivation after a certain dose of radiation, we find that it is reduced by approximately half (table 1). These effects can be attributed to the reaction of the water free radicals $OH^{\cdot}$ and

Table 1. Percentage remaining activity of a lysozyme solution[a] after irradiation in solution[b] in the presence of an increasing assortment of other solutes

	additional solutes	percentage remaining activity
	air	40
plus	thymine (100 μM) leucine (100 μM) acetate (100 μM) glucose (100 μM) lactate (100 μM) citrate (100 μM) isocitrate (100 μM) NAD^+ (100 μM)	60
plus	Br^- (5 mM)	45
plus	C_2H_5OH (0.5 M)	90
plus	$CHCl_3$ (5 mM)	25
plus	acetone–isopropanol (1 M)	90
plus	CCl_4 (10 mM)	40
plus	NADH (1 mM)	90
	or Trolox C (1 mM)	85
	or tryptophan (1 mM)	85
	or glutathione (1 mM)	55

[a] Concentration 50 μg ml^{-1}.
[b] With 150 Gy (*ca.* 100 μM free radicals).

e^-_{aq} with the enzyme and oxygen respectively, but with the superoxide radical formed having no effect on the enzyme activity (Adams *et al.* 1970).

$$OH^{\cdot} + \text{lysozyme} \rightarrow \text{inactivation}, \tag{5}$$

$$e^-_{aq} + O_2 \rightarrow O_2^{\cdot}, \tag{6}$$

$$O_2^{\cdot} + \text{lysozyme} \rightarrow \text{no inactivation}. \tag{7}$$

We may then repeat the experiment, but include in the irradiated solution an assortment of biological molecules. For example, if in addition to the DNA base thymine, we include an amino acid such as leucine, a fatty acid such as acetate, a monosaccharide such as glucose, some intermediates of the citric acid cycle such as citrate and isocitrate, as well as lactate and the coenzyme NAD^+, all at 100 μM concentration, the remaining activity is now increased (see figure 1). This can be attributed to the scavenging of the hydroxyl free radicals by the additional substances that have been included in the irradiated solution:

$$OH^{\cdot} + S \rightarrow \text{'products'}, \tag{8}$$

$$\text{'products'} + \text{lysozyme} \rightarrow \text{no inactivation}. \tag{9}$$

If bromide (5 mM) is added before irradiation, the extent of damage is increased. The extent is decreased again, however, if ethanol (100 mM) in high concentration is also included. The effects can be attributed to the action of bromide ion scavenging the hydroxyl radical, and the resulting bromine radical-anion reacting with the enzyme and inactivating it (Adams *et al.* 1969, 1972; Posner *et al.* 1976):

$$OH^{\cdot} + Br^{\cdot} \rightarrow Br^{\cdot}, \tag{10}$$

$$Br^{\cdot} + Br^- \rightarrow Br_2^{-\cdot}, \tag{11}$$

$$Br_2^{-\cdot} + \text{lysozyme} \rightarrow \text{inactivation}. \tag{12}$$

In the further presence of ethanol, the hydroxyl radicals are now selectively scavenged by the alcohol. Because the ethanol radical or radicals derived from it, such as the peroxy radical or superoxide radical, do not inactivate lysozyme, considerable protection is observed.

$$OH^{\cdot} + CH_3CH_2OH \rightarrow CH_3C^{\cdot}HOH + H_2O, \tag{13}$$

$$CH_3C^{\cdot}HOH + O_2 \rightarrow CH_3CH(O_2^{\cdot})OH, \tag{14}$$

$$CH_3CH(O_2^{\cdot})OH + OH^- \rightarrow CH_3CH(O_2^{\cdot})O^- + H_2O, \tag{15}$$

$$CH_3CH(O_2^{\cdot})O^- \rightarrow CH_3CHO + O_2^- + H_2O. \tag{16}$$

If chloroform (5 mM) is now included before irradiation, the solvated electrons are preferentially scavenged by the halocarbon to form the chloroform peroxy radical, $CHCl_2O_2^{\cdot}$. This peroxy radical can react with the enzyme and inactivate it (Willson 1982).

$$e_{aq}^- + CHCl_3 \rightarrow CHCl_2 + Cl^-, \tag{17}$$

$$CHCl_2 + O_2 \rightarrow CHCl_2O_2^{\cdot}, \tag{18}$$

$$CHCl_2O_2^{\cdot} + \text{lysozyme} \rightarrow \text{inactivation}. \tag{19}$$

However, if isopropanol and acetone in high concentration (1 M) are also included, the water free radicals are now selectively scavenged by these compounds. The resulting isopropanol radical, or the corresponding peroxy radical or superoxide radical derived from it, do not react with the enzyme and therefore protection is again observed.

$$e_{aq}^- + CH_3COCH_3 \rightarrow CH_3COCH_3^{-\cdot}, \tag{20}$$

$$CH_3COCH_3^{-\cdot} + H_2O \rightarrow CH_3C^{\cdot}OHCH_3 + OH^-, \tag{21}$$

$$OH^{\cdot} + (CH_3)_2CHOH \rightarrow (CH_3)_2C^{\cdot}OH + H_2O, \tag{22}$$

$$(CH_3)_2C^{\cdot}OH + O_2 \rightarrow (CH_3)_2C(O_2^{\cdot})OH, \tag{23}$$

$$(CH_3)_2C(O_2^{\cdot})OH + OH^- \rightarrow (CH_3)_2C(O_2^{\cdot})O^- + H_2O, \tag{24}$$

$$(CH_3)_2C(O_2^{\cdot})O^- \rightarrow (CH_3)_2CO + O_2^-. \tag{25}$$

Finally, if carbon tetrachloride (10 mM) is added before irradiation, considerable inactivation again occurs. However, the extent is again reduced if any of the electron donors NADH, Trolox C (TxOH, a water soluble derivative of vitamin E), glutathione (GSH), tryptophan (trp), or the sulphur-containing amino acid, methionine, are also included. In the presence of carbon tetrachloride, the isopropanol radical is oxidized to acetone and the radical $CCl_3O_2^{\cdot}$ is formed (Koster & Asmus 1971; Willson & Slater 1975). This peroxy radical then reacts with lysozyme and inactivates it (Broadhurst *et al.* 1982; Willson 1983; Hiller *et al.* 1983):

$$CH_3C^{\cdot}OHCH_3 + CCl_4 \rightarrow CH_3COCH_3 + CCl_3^{\cdot} + Cl^- + H^+, \tag{26}$$

$$CCl_3^{\cdot} + O_2 \rightarrow CCl_3O_2^{\cdot}, \tag{27}$$

$$CCl_3O_2^{\cdot} + \text{lysozyme} \rightarrow \text{inactivation}. \tag{28}$$

When NADH or any of the other electron-donating substances described are present at the time of radiation, $CCl_3O_2^{\cdot}$ is scavenged by these compounds to form a species which does not inactivate the enzyme appreciably, if at all.

$$CCl_3O_2^{\cdot} + NADH \rightarrow CCl_3O_2H + NAD^{\cdot}; \quad (29)$$

$$CCl_3O_2^{\cdot} + TxOH \rightarrow CCl_3O_2H + TxO^{\cdot}. \quad (30)$$

It is important to state that in none of these systems, nor in the systems containing alcohol dehydrogenase which will be described later, did the presence of the high concentrations of organic compounds used affect the activity of the enzymes significantly over the time course of the experiment, *ca.* 30 min.

Clearly, a host of chemical reactions are going on in these systems. Unfortunately, we cannot go into them in detail here. Let us concentrate, therefore, on the free radicals derived from isopropanol and carbon tetrachloride, and look at their reactions with NADH, Trolox C, GSH, tryptophan and tyrosine.

Reactions of organic peroxy radicals with Trolox C

Previous studies have shown that when aerated solutions containing vitamin E (α-tocopherol, vitEOH), isopropanol, acetone and carbon tetrachloride are pulse-irradiated, a transient absorbing species, ($\lambda_{max} = 435$ nm) can be observed (figure 1). The absorption

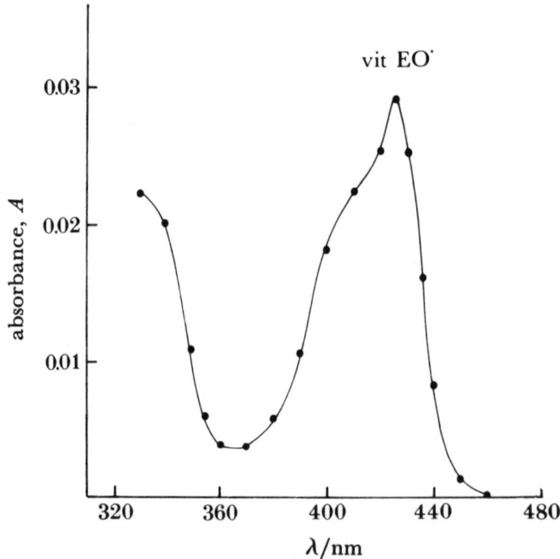

FIGURE 1. Absorption spectrum of the phenoxyl free radical from α-tocopherol observed by following the reaction of $CCl_3O_2^{\cdot}$; 25 ml solution contains 12 ml i-PrOH, 2.5 ml acetone, 0.04 M CCl_4, 8.8×10^{-4} M α-tocopherol. Structure of α-tocopherol:

HO, CH_3, CH_3, CH_3 (ring substituents); O; CH_3; $-CH_2-(CH_2-CH_2-\underset{|}{\overset{CH_3}{C}}H-CH_2)_2-CH_2-CH_2-\overset{CH_3}{C}H-CH_3$.

is similar to the absorption recorded on flash photolysis of various phenols and has been attributed to the corresponding phenoxy free radical (Packer *et al.* 1979). No such absorptions are observed in the absence of CCl_4 or O_2 in agreement with the lack of any reaction of $CCl_3^{\cdot}$ or $CH_3C^{\cdot}OHCH_3$ with the vitamin (figure 2).

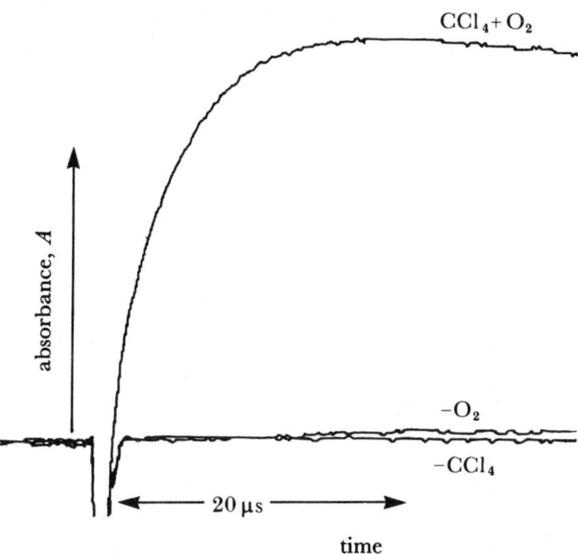

FIGURE. 2. Changes in absorption at 435 nm observed on pulse radiolysis of solutions containing isopropanol and acetone (1 M) and α-tocopherol (100 μm), in the absence and presence of air or CCl_4 (10^{-2} M).

$$CH_3C^{\cdot}OHCH_3 + vitEOH \rightarrow CH_3CHOHCH_3 + vitEO^{\cdot} \text{ (relatively slow if at all);} \quad (31)$$

$$CCl_3^{\cdot} + vitEOH \rightarrow CCl_3H + vitEO^{\cdot} \text{ (relatively slow if at all);} \quad (32)$$

$$CCl_3O_2^{\cdot} + vitEOH \rightarrow CCl_3O_2H + vitEO^{\cdot} \text{ (fast).} \quad (33)$$

Similar results have recently been obtained with Trolox C in agreement with:

$$CCl_3O_2^{\cdot} + TxOH \rightarrow CCl_3O_2H + TxO^{\cdot}. \quad (34)$$

Although (34) is written above as a hydrogen-transfer reaction it was of interest to know whether the reaction did occur in this manner, or was in fact an electron-transfer reaction, i.e.:

$$CCl_3O_2^{\cdot} + TxOH \rightarrow CCl_3O_2^- + TxOH^{+\cdot} \xrightarrow{H_2O} TxO^{\cdot} + H_3O^+. \quad (35)$$

The peroxy radical $CCl_3O_2^{\cdot}$ and related peroxy radicals can also be formed on pulse radiolysis of aerated aqueous solutions of the halocarbon and excess t-butanol, according to:

$$OH^{\cdot} + (CH_3)_3COH \rightarrow {}^{\cdot}CH_2(CH_3)_2COH + H_2O, \quad (36)$$

$${}^{\cdot}CH_2(CH_3)_2COH + O_2 \rightarrow {}^{\cdot}O_2CH_2(CH_3)_2COH, \quad (37)$$

$$e_{aq}^- + CCl_4 \rightarrow CCl_3^{\cdot} + Cl^-. \quad (38)$$

On pulse radiolysis of solutions containing t-butanol and Trolox C no formation of the $TxO^{\cdot}$ could be observed. In the presence of CCl_4, however, an exponential formation of the absorbtion was again apparent, in agreement with reaction (34) and reaction (39) being relatively slow.

$$^{\bullet}O_2CH_2(CH_3)_2COH + TxOH \rightarrow HO_2CH_2(CH_3)_2COH + TxO^{\bullet}. \tag{39}$$

On performing experiments with solutions of deuterated Trolox C, (TxOD), prepared from an alkaline solution of Trolox C ($pK_a = 11.9$) in D_2O, a similar absolute rate constant was obtained, in agreement with the occurrence of an electron-transfer (35) rather than a hydrogen-transfer reaction (34). Further support of an electron-transfer reaction comes from other studies, which have shown that $CCl_3O_2^{\bullet}$ reacts rapidly with a variety of one-electron donors (table 2) and that Trolox C reacts rapidly with a variety of other electrophilic radicals (Willson & Slater 1975; Packer *et al.* 1978, 1979, 1980, 1981; Bahnemann *et al.* 1981; Forni *et al.* 1983*b*).

TABLE 2. ABSOLUTE RATE CONSTANTS[a] OF FREE-RADICAL REACTIONS[b]

radical	TxOH	NADH	vitC$^-$	UO$^-$	PZ	cytII *c*	GSH	tyr	trp	MV^{2+}	O_2
$CH_3C^{\bullet}OHCH_3$	< 0.01	< 0.01	0.01	—	< 0.01	< 0.01	1.8	—	—	*ca.* 10	4.1
$OH^{\bullet}$	68	80	64	71	97	> 100	> 100	105	80		
$N_3^{\bullet}$	5	28	—	49	46	13	—	1	41		
$Br_2^{-\bullet}$	3.8	9.0	8.7	9	—	9.7	—	0.2	7.7		
$GS^{\bullet}$	—	2.5	6.0	0.14	0.33	2.5	0.14	—	—	—	6
$CCl_3O_2^{\bullet}$	3.7	5.6	2	7.0	4.5						
$CHCl_2O_2^{\bullet}$	1.1	3.3	2.6	—	0.67						
$NO_2^{\bullet}$	< 0.01[c]	—	0.3	—	0.3						
$TxO^{\bullet}$	—	—	0.4								
$tyrO^{\bullet}$	3.2	0.61									
$C_6H_5O^{\bullet}$	4.1										
m-$CH_3C_6H_4O^{\bullet}$	2.8										
p-$CH_3C_6H_4O^{\bullet}$	0.95										
o-$CH_3C_6H_4O^{\bullet}$	< 0.01										

[a] Units: $10^8\ M^{-1}\ s^{-1}$. [b] At *ca.* pH 7. [c] Takes value 5.0 at pH 11.

Reactions of peroxy radicals with NADH

The strong absorption of NADH at 340 nm provides a convenient handle for determining the rate of reaction of free radicals with the coenzyme (Swallow 1953, 1955; Land & Swallow 1971; Bielski & Chan 1973, 1974). On pulse radiolysis of aqueous solutions containing t-butanol or acetone, and isopropanol, carbon tetrachloride and NADH, an exponential bleaching at 340 nm was observed (figure 3), in agreement with the reactions (26) and (27) or (36), (37) and (38) followed by

$$CCl_3O_2^{\bullet} + NADH \rightarrow CCl_3O_2H + NAD^{\bullet}. \tag{40}$$

Again, although formerly written above as a hydrogen-transfer reaction, it was of interest to know whether the reaction occurs in this manner or was in fact an electron-transfer reaction:

$$CCl_3O_2^{\bullet} + NADH \rightarrow CCl_3O_2^{-} + NADH^{+\bullet} \xrightarrow{H_2O} NAD^{\bullet} + H_3O^{+}. \tag{41}$$

With solutions containing t-butanol and monodeuterated NADH, i.e. NADD (prepared by incubating C_2D_5OD and NAD^+ with YADH), no difference in absolute rate constants was obtained (figure 4). This is again in agreement with an electron-transfer reaction. Further

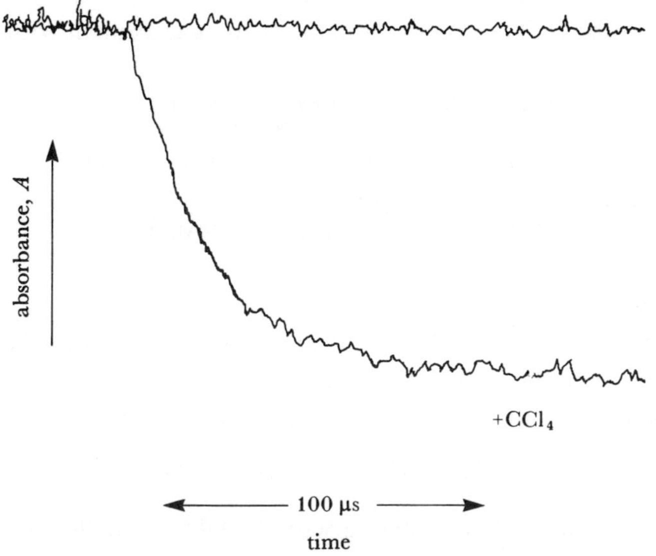

FIGURE. 3. Decrease in absorption at 340 nm observed on pulse radiolysis of an air-saturated solution containing t-butanol and NADH (100 µm), in the absence and presence of CCl_4 (10^{-2} M).

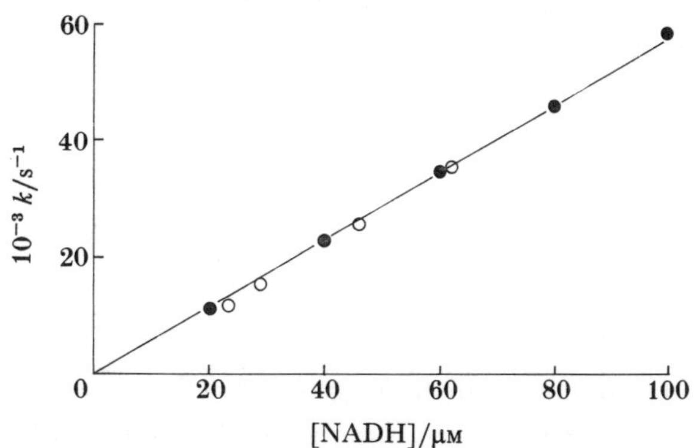

FIGURE 4. Plot of first-order rate constant against NADH (●) or NADD (○) concentration for the loss of absorption at 340 nm owing to reaction of CCl_3O_2 with the coenzyme (see text); $k = 5.6 \times 10^8\ \text{M}^{-1}\ \text{s}^{-1}$.

support for such a mechanism also comes from other studies, showing that NADH reacts rapidly with a variety of electrophilic radicals (table 2). Furthermore, as with Trolox C, no reaction was apparent in the absence of O_2 or CCl_4, again indicating that the reaction of $CCl_3^{\cdot}$ or the carbon-centred t-butanol or isopropanol radicals with NADH, occur relatively slowly if at all.

$$CH_3C^{\cdot}OHCH_3 + NADH \rightarrow CH_3CHOHCH_3 + NAD^{\cdot} \text{ (relatively slow if at all);} \quad (42)$$

$$CCl_3^{\cdot} + NADH \rightarrow CCl_3H + NAD^{\cdot} \text{ (relatively slow if at all).} \quad (43)$$

Reactions of thiyl radicals with NADH

The fact that thiyl radicals such as $GS^{\cdot}$ can also react rapidly with NADH again comes from pulse radiolysis studies of solutions containing isopropanol, acetone and NADH (Forni & Willson 1983). It was found that the exponential loss of the NADH absorption could also occur if GSH rather than CCl_4 was present in the solution. Thiols have been shown to repair organic radicals by a comparatively rapid hydrogen-transfer reaction (Adams *et al.* 1967, 1968, 1969; Wolfenden & Willson 1983). The observations are then in agreement with the cycling sequence

$CH_3C^{\bullet}OHCH_3$ GSH $NAD^{\bullet}$

hydrogen transfer electron transfer

$CH_3CHOHCH_3$ $GS^{\bullet}$ NADH

If methyl viologen (paraquat), MV^{2+}, is also present in this system, a slow appearance of the characteristic absorption of the $MV^{+\bullet}$ radical cation is apparent. If NADH is not included in the irradiated solution the absorption is absent in agreement with the above reactions followed by the reaction (Anderson 1980; Farrington *et al.* 1980)

$$NAD^{\bullet} + MV^{2+} \rightarrow NAD^{+} + MV^{+\bullet}. \tag{44}$$

Again, further support that $GS^{\bullet}$ was reacting with NADH by an electron-transfer mechanism comes from other pulse-radiolysis studies, showing that thiyl radicals can react rapidly with a variety of electron donors (table 2) (Forni *et al.* 1983*a*; Forni & Willson 1983; Willson 1983). Furthermore, on pulse radiolysis of GSSG in the presence of t-butanol, bleaching of the NADH absorption was again observed and the same absolute rate constant was obtained whether NADH or NADD was used.

$$e^{-}_{aq} + RSSR \rightarrow RSSR^{-\bullet} \rightarrow RS^{\bullet} + RS^{-}, \tag{45}$$

$$RS^{\bullet} + NADH \rightarrow RS^{-} + NAD^{+\bullet} \xrightarrow{H_2O} NAD^{\bullet} + H_3O^{+}. \tag{46}$$

Reactions of tyrosine radicals with NADH *and intramolecular hydrogen transfer*

In the context of protein damage our finding that NADH can also be readily oxidized by phenoxyl free radicals, $\phi O^{\bullet}$, is particularly interesting. Previous studies on pulse radiolysis of solutions containing azide ion and tyrosine (Prutz 1979) have shown that the tyrosine phenoxyl radical can be formed rapidly:

$$OH^{\bullet} + N_3^{-} \rightarrow N_3^{\bullet} + OH^{-}, \tag{47}$$

$$N_3^{\bullet} + tyrOH \rightarrow N_3^{-} + tyrOH^{+\bullet} \xrightarrow{H_2O\,(N_3^{-})} tyrO^{\bullet} + H_3O^{+}\ (N_3H). \tag{48}$$

In the additional presence of NADH the phenoxyl radical absorption decayed rapidly, with a parallel decrease in absorption at 340 nm. This is in agreement with

$$tyrO^{\bullet} + NADH \rightarrow tyrO^{-} + NADH^{+\bullet} \tag{49}$$

$$tyrO^{-} \xrightarrow{H_2O} OH^{-} + tyrOH \qquad NADH^{+\bullet} \xrightarrow{H_2O} NAD^{\bullet} + H_3O^{+}.$$

Furthermore, previous studies have shown that when $N_3^{\bullet}$ or $CCl_3O_2^{\bullet}$ reacts with the peptide tryptophanyl-tyrosine, the tryptophanyl radical ($trp^{\bullet}$) is soon formed, but a rapid intramolecular transfer process subsequently takes place in which the free-radical centre becomes located on the tyrosine (Prutz *et al.* 1980, 1981; Packer *et al.* 1981):

$$N_3^{\cdot} + \text{trpH-tyrOH} \rightarrow \text{trpH}^{+\cdot}\text{-tyrOH}, \qquad (50)$$

$$\text{tryH}^{+\cdot}\text{-tyrOH} + N_3^- \rightarrow \text{trp}^{\cdot}\text{tyrOH} + N_3H, \qquad (51)$$

$$\text{trp}^{\cdot}\text{-tyrOH} \rightarrow \text{trpH-tyrO}^{\cdot}. \qquad (52)$$

When similar experiments were undertaken with NADH also present, the tryptophanyl radical, trp˙, was soon observed, but this subsequently decayed with the formation of the tyrosine phenoxyl absorption at 410 nm. This in turn decayed and after 3 ms only a bleaching of the NADH absorption at 340 nm was apparent (figure 5). This can be represented:

$$\text{trpH-tyrO}^{\cdot} + \text{NADH} \rightarrow \text{trpH-tyrO}^- + \text{NADH}^{+\cdot} \xrightarrow{H_2O} \text{NAD}^{\cdot} + H_3O^+. \qquad (53)$$

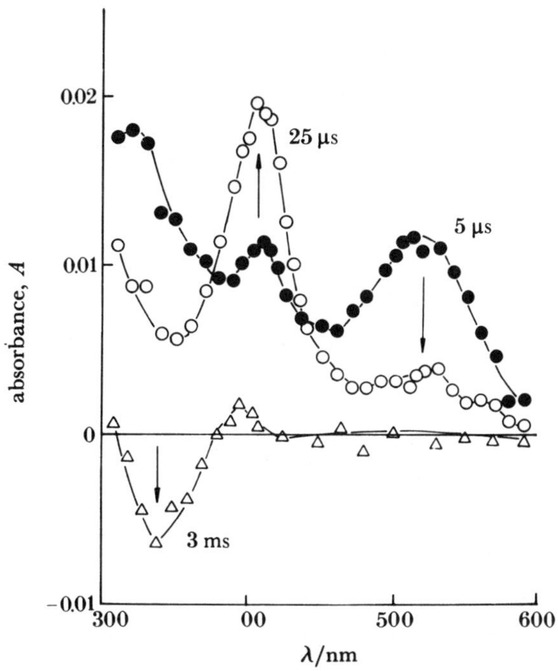

FIGURE 5. Absorption spectra observed on pulse radiolysis of N_2O-saturated aqueous solutions containing azide (50 mM), tryptophanyl-tyrosine (3 mM), and NADH (25 μM); ●, after 5 μs; ○, after 25 μs; △, after 3 ms.

Further support for an electron-transfer mechanism comes from other pulse-radiolysis studies showing that phenoxy radicals can react rapidly with a variety of electron donors, including ascorbate and various ionized phenols (Schuler 1977; Packer *et al.* 1979; Steenken & Neta 1979, 1982). Some rate constants are shown (table 2).

Reactions of peroxy radicals $RO_2^{\cdot}$ *with* NADH

Although reactions of $CCl_3O_2^{\cdot}$ and related halocarbon peroxy radicals with NADH have been characterized by pulse radiolysis, no reactions of the thymine hydroxyl radical-adduct peroxy radicals nor of aliphatic peroxy radicals ($RO_2^{\cdot}$) have been observed directly. However, stationary-state studies in which solutions of NADH saturated with a nitrous oxide–air mixture have been irradiated, have shown that such reactions do occur, albeit relatively slowly (table 3). In studies in which formate or ethanol were irradiated with the coenzyme, little loss of NADH absorption at 340 nm was apparent at pH = 8. With thymine at pH = 5–6 or at pH = 8 or

Table 3. Destruction of NADH by organic peroxy free radicals but not superoxide[a]

scavenger	pH 5–6	*ca.* pH 8
formate	90	100
ethanol	55	95
isopropanol	79	88
thymine (1 mM)	26	45

[a] Values given are percentage of coenzyme remaining after irradiation in $N_2O:O_2$ (4:1) saturated solutions in the presence of hydroxyl radical scavenger. Initial NADH concentration is 10^{-4} M; radiation dose is 100 Gy (*ca.* 60 μM free radicals).

with ethanol at pH = 5.9, however, some destruction was obtained, which is in agreement with O_2^- reacting only slowly, if at all, with the coenzyme and the thymine and ethanol peroxy radical reacting to a significant extent.

$$e_{aq}^- + N_2O \rightarrow N_2 + OH^{\cdot} + OH^-, \tag{54}$$

$$OH^{\cdot} + T \rightarrow T(OH)^{\cdot}, \tag{55}$$

$$T(OH^{\cdot}) + O_2 \rightarrow T(OH)O_2^{\cdot}, \tag{56}$$

$$T(OH)O_2^{\cdot} + NADH \rightarrow T(OH)O_2H + NAD^{\cdot}, \tag{57}$$

$$OH^{\cdot} + HCO_2^- \rightarrow CO_2^{-\cdot} + H_2O, \tag{58}$$

$$CO_2^{-\cdot} + O_2 \rightarrow CO_2 + O_2^-, \tag{59}$$

$$O_2^- + NADH \rightarrow HO_2^- + NAD^{\cdot} \text{ (relatively slow if at all).} \tag{60}$$

The slow reaction of O_2^- with NADH is in agreement with previous studies where the rate constant $k_{60} = 10^4\ M^{-1}\ s^{-1}$ was determined when the enzyme was irradiated in the presence of lactate dehydrogenase, but was only $k_{60} = 28\ M^{-1}\ s^{-1}$ in the absence of the enzyme (Bielski & Chan 1973, 1974). It was suggested that this was a result of the enzyme affecting the conformation of NADH, making it more susceptible to oxidation. However, one wonders whether this large difference in rate constant may be partly because of the effect of the protein environment on the equilibrium concentration of the acid form of superoxide, $HO_2^{\cdot}$, in the vicinity of the coenzyme, with $k_{61} \gg k_{60}$:

$$HO_2^{\cdot} + NADH \rightarrow H_2O_2 + NAD^{\cdot}. \tag{61}$$

Such an effect of protein may have been observed with urate (see later).

Reaction of peroxy radicals with alcohol dehydrogenase and protection by SOD

Recent studies have shown that, in addition to $CCl_3O_2^{\cdot}$ and related halocarbon peroxy radicals, peroxy radicals from a variety of aromatic compounds can inactivate yeast alcohol dehydrogenase (YADH) under conditions where O_2^- is non-damaging (Kittridge & Willson 1984; Gee *et al.* 1985). Indeed, the extent of damage is often greater than that observed with a similar concentration of hydroxyl radicals, $OH^{\cdot}$. This is in agreement with a lower oxidizing ability conferring a greater degree of selectivity (Adams *et al.* 1969). When the enzyme was irradiated in the presence of both formate and thymine, the extent of inactivation observed depended on the relative formate and thymine concentrations, in agreement with the reactions (55) and (58) competing for hydroxyl radicals. Thus, in the presence of excess formate, little inactivation of the enzyme occurred.

Damage could also be prevented by the presence at the time of irradiation of a variety of electron donors (antioxidants) in relatively low concentration (100 μM). These included Trolox C, NADH, GSH and cysteamine, as well as cysteine, tryptophan and methionine. Other common amino acids did not protect the enzyme, perhaps indicating that the above residues were likely to be the vulnerable sites in YADH (Gee *et al.* 1985).

A further series of enzyme experiments in which YADH was irradiated in solutions of increasing complexity, hopefully serves to illustrate how, if a detailed biochemical description of a system is available, we may be able to predict the biological repercussions more accurately (table 4). Many of the reactions involved are similar to those in the lysozyme system. When chloroform is added, the peroxy radical $CHClO_2^{\cdot}$ is formed according to (17) and (18). However, unlike reaction with carbon tetrachloride, the reaction of $CHCl_3$ with the isopropanol radical is relatively slow and it cannot compete effectively with oxygen for this radical: hence the observed sensitization on the addition of chloroform followed by protection on the addition of acetone.

Table 4. Percentage remaining activity of a yeast alcohol dehydrogenase solution[a] after irradiation[b] in the presence of an increasing assortment of other molecules known to scavenge free radicals

additional solutes	concentration	radical	10^{-9} rate constant/ $M^{-1}\,s^{-1}$	percentage remaining activity
oxygen (air)	250 μM	e_{aq}^-	12	41
plus formate	100 μM	$OH^{\cdot}$	4	72
plus thymine	500 μM	$OH^{\cdot}$	5	3
plus ethanol	5 mM	$OH^{\cdot}$	2	43
plus bromide	50 mM	$OH^{\cdot}$	1	0
plus isopropanol	1 M	$OH^{\cdot}$	2	55
plus chloroform	5 mM	e_{aq}^-	30	1
plus acetone	1 M	e_{aq}^-	6	53
plus CCl_4	10 mM	$CH_3C^{\cdot}OHCH_3$	0.2	0
plus methyl viologen	10 mM	$CH_3C^{\cdot}OHCH_3$	1	43
plus urate	1 mM	O_2^-/HO_2?	—	17
plus SOD	10 μg ml^{-1}	O_2^-	—	71

[a] Concentration 20 μg ml^{-1}.
[b] Dose 15 Gy (*ca.* 10 μM free radicals).

$$CH_3C^{\cdot}OHCH_3 + CHCl_3 \rightarrow CHCl_2O_2^{\cdot} + CH^{\cdot}COCH_3 + Cl^- + H^+ \text{ (relatively slow if at all).} \quad (62)$$

The appropriate absolute rate constants are shown (table 4), together with the extent of enzyme inactivation observed.

Uric acid, iron and the reactions of H_2O_2 *and* O_2^- *with enzymes*

In the irradiated system (table 4), it will have been noticed that the presence of urate (UO^-) increases the extent of inactivation, contrary to the suggestion that uric acid (UOH) is one of man's important antioxidants (Ames *et al.* 1981).

Even in simpler systems, the purine, like the pyrimidine thymine, does not protect alcohol dehydrogenase against inactivation. Indeed the extent of inactivation is again considerably increased (Kittridge & Willson 1984). What is particularly surprising is that extensive inactivation still occurs even when excess formate is present. This is in spite of the fact that when urate is irradiated in the presence of formate alone, little destruction of the purine occurs as measured spectrophotometrically at 292 nm. In the presence of excess formate, O_2^- is likely to be the only free radical able to react with urate or the enzyme and similar effects are also observed when O_2^- is generated in systems containing isopropanol, acetone (1 M) and methyl viologen (paraquat, 10 mM):

$$CH_3C^{\cdot}OHCH_3 + MV^{2+} \rightarrow CH_3COCH_3 + MV^{+\cdot} + H^+, \quad (63)$$

$$MV^{+\cdot} + O_2 \rightarrow MV^{2+} + O_2^-. \quad (64)$$

In both superoxide-generating systems the effect of urate is strongly pH-dependent, decreasing as the pH was increased from 6 to 8. No inactivation of YADH is apparent when the enzyme is added to a previously irradiated solution containing urate and excess formate.

Damage to the enzyme is therefore attributed to the radiation-induced formation of a short-lived toxin when O_2^- is generated in the presence of the protein and urate. The acid dissociation constant of urate is $pK_a = 5.5$ and that of the hydroperoxy radical, $HO_2^{\cdot}$, is $pK_a = 4.9$. Thus the damage may be due to $HO_2^{\cdot}$, a urate radical ($UO^{\cdot}$) or subsequent permanent product formed from urate in the presence of the enzyme:

$$UOH \rightarrow UO^- + H^+, \quad (65)$$

$$O_2^- + UOH \rightarrow UO^- + HO_2^{\cdot}, \quad (66)$$

$$HO_2^{\cdot} + UO^- \rightarrow HO_2^- + UO^{\cdot}, \quad (67)$$

$$UO^{\cdot} + YADH \rightarrow \text{inactivation}, \quad (68)$$

$$UO^{\cdot} \rightarrow \text{`toxin'}, \quad (69)$$

$$\text{`toxin'} + YADH \rightarrow \text{inactivation}, \quad (70)$$

$$\text{or} \quad HO_2^{\cdot} + YADH \rightarrow \text{inactivation}. \quad (71)$$

Interestingly, the above damaging action is again prevented by including Trolox C, ascorbate or NADH (100 μM) in the irradiated system. Superoxide dismutase also protects to a considerable extent. Whether these compounds are reacting with O_2^- in the presence of the enzyme or whether they are reacting directly with the urate toxin or with the damaged enzyme, remains to be elucidated.

At the present time, the possibility that a Fenton reaction involving hydrogen peroxide and trace iron contaminants also play some role, cannot be ruled out completely:

$$Fe^{II} + H_2O_2 \rightarrow OH^{\cdot} + OH^- + Fe^{III}. \quad (72)$$

Hydrogen peroxide does inactivate such enzyme preparations in the absence of radiation. However, the concentrations of hydrogen peroxide and ferrous ion required for an appreciable effect between the end of the radiation and time of activity measurement (less than 30 min) are unlikely to be sufficiently high, unless a long chain reaction involving urate or the enzyme, or both, takes place. Furthermore, in the irradiated systems containing the enzyme and thymine, or the enzyme, formate and urate, the inclusion of ferrous ions (20 μM), like SOD ($10\,\mu g\ ml^{-1}$), has a protective rather than a sensitizing effect.

Significance of in vitro studies and free-radical-induced protein damage

The vast wealth of free-radical reaction rate constants now available, and the steady accumulation of oxidation–reduction potentials of many free-radical–molecule couples are providing a very useful framework for the design of *in vitro* model experiments (Patel & Willson 1973; Wardman & Clark 1976; Steenken & Neta 1979, 1982; Butler *et al.* 1982; Anderson 1980; Farrington *et al.* 1980). Some examples of redox potentials are given (table 5). By

TABLE 5. REDOX POTENTIAL OF SOME FREE-RADICAL COUPLES[a]

couple	E/V
$OH^{\cdot}/OH^-$	2.00 (1.77)
$N_3^{\cdot}/N_3^-$	1.9
$Br_2^{-\cdot}/2Br^-$	1.7 (1.45)
$I^{\cdot}/I^-$	1.31
$I_2^{-\cdot}/2I^-$	1.0
$RS^{\cdot}$, H^+/RSH	*ca.* 1.0
trp$^{\cdot}$, H^+/trpH	0.98
promethazine$^{+\cdot}$/promethazine	0.896 (0.865)
$C_6H_5O^{\cdot}$, H^+/C_6H_5OH	> 0.8
tyrO$^{-\cdot}$/H^+, tyrOH	*ca.* 0.65
p-$CH_3OC_6H_5O^{\cdot}$, H^+/*p*-$CH_3OC_6H_5OH$	0.60
catechol $O^{\cdot}$, H^+/catechol OH	0.53
Trolox $O^{\cdot}$, H^+/Trolox OH	0.48
ascorbate$^{\cdot}$/ascorbate$^-$	0.30
$NAD^{\cdot}$, $H^+/NADH$	0.30
cyt(III) *c*/cyt(II) *c*	0.255
benzoquinone/benzoquinone$^{-\cdot}$	0.10
O_2/O_2^-	−0.16
methyl viologen^{2+}/methyl viologen$^{+\cdot}$	−0.46
$NAD^+/NAD^{\cdot}$	−0.93
CH_3CHO, $H^+/CH_3C^{\cdot}HOH$	−1.1
CH_3COCH_3, $H^+/CH_3C^{\cdot}OHCH_3$	−1.8
$CO_2/CO_2^{-\cdot}$	−2.0

[a] At pH 7.

ranking the couples in this manner it can be readily seen whether particular reactions are thermodynamically favourable: species on the left-hand side of a couple may oxidize those on the right-hand side of couples lower in the series, although kinetic factors must also be taken into consideration. Whether such studies will be relevant or have application *in vivo* remains to be seen. In the context of cell damage, it is clear that lysozyme or YADH are hardly likely to be key proteins. Nonetheless, the dehydrogenase enzyme, in particular, may serve as a model demonstrating what might occur when other more critical proteins are subject to free-radical attack. For example, some polymerase enzymes are known to be sensitive to oxidation and any serious depletion of these when they are required to repair damage to nucleic acid may be seriously detrimental. Various chemotactic factors and protease inhibitors are also known to be sensitive to oxidation, and again any damage to these at critical times within the cell cycle may have important repercussions. Clearly when proteins are exposed to free radicals, whether they are formed chemically, biochemically, or by exposure to radiation, the nature and extent of any damage will depend strongly on the local environment of the site of radical generation. Factors such as radical reactant concentrations, the polarity of the local medium and the presence of basic or acidic residues, not to mention simple steric constraints and the facility for intramolecular electron or hydrogen transfer, will all have a bearing on the biological outcome.

We hope that further stationary- and pulse-radiolysis studies will continue to provide information, enabling us to describe more fully the various criteria involved.

None of the above experiments described would have been undertaken at Brunel without the enthusiastic collaboration of Professor T. F. Slater and many others, in particular, Professor Dieter Asmus, Dr D. Bahneman and Dr J. Monig from the Hahn Meitner Institute of Berlin and of Professor J. Packer from the Department of Chemistry, University of Auckland.

We are also much indebted to the Cancer Research Campaign for the upgrading and running costs of the linear accelerator, to the Medical Research Council for the provision of computer equipment and the Cobalt Gamma source and to the National Foundation for Cancer Research for studentships for LGF and KJK.

Finally, we are grateful to Professor R. O. C. Norman, F.R.S., for his part in helping us to acquire a linear accelerator, and to the late Professor W. A. Waters, F.R.S., whose papers on hydroxyl radicals and on the oxidation potentials of free radicals inspired much of this work.

References

Adams, G. E., Aldrich, J. E., Bisby, R. H., Cundall, R. B., Redpath, J. L. & Willson, R. L. 1972 Selective free radical reactions with proteins and enzymes: reactions of inorganic radical anion with amino acids. *Radiat. Res.* **49**, 278–289.

Adams, G. E., Armstrong, R. C., Charlesby, A., Michael, D. B. & Willson, R. L. 1969 Pulse radiolysis of sulphur compounds. *Trans. Faraday Soc.* **65**, 732.

Adams, G. E., Cundall, R. B. & Willson, R. L. 1970 Pulse radiolysis: an oxidative reductive probe in general biochemical and enzyme studies. In *Chemical reactivity of biological role of functional groups in enzymes* (ed. R. M. S. Smellie), pp. 171–182. London and New York: Academic Press.

Adams, G. E., McNaughton, G. S. & Michael, B. D. 1967 The pulse radiolysis of sulphur compounds, part 1 Cysteamine and Cystamine. In *Excitation and ionisation* (ed. G. Scholes & G. R. A. Johnson), pp. 281–293. London: Taylor & Francis.

Adams, G. E., McNaughton, G. S. & Michael, B. D. 1968 Pulse radiolysis of sulphur compounds. *Trans. Faraday Soc.* **64**, 902.

Adams, G. E. & Willson, R. L. 1969 Pulse radiolysis studies on the oxidation of organic radicals in aqueous solution. *Trans. Faraday Soc.* **65**, 2981–2987.

Adams, G. E., Willson, R. L., Aldrich, J. E. & Cundall, R. B. 1969 On the mechanism of the radiation-induced inactivation of lysozyme in dilute aqueous solution. *Int. J. radiat. Biol.* **16**, 333–342.

Aldrich, J. E., Cundall, R. B., Adams, G. E. & Willson, R. L. 1969 Identification of essential residues in lysozyme. A pulse radiolysis method. *Nature, Lond.* **221**, 1049–1050.

Ames, B. N., Cathcart, R., Schwiers, E. & Hochstein, P. 1981 Uric acid provides an antioxidant defence in humans against oxidant- and radical-caused aging and cancer: a hypothesis. *Proc. natn. Acad. Sci. U.S.A.* **78**, 6858.

Anderson, R. F. 1980 Energetics of the one-electron steps in the $NAD^{\cdot}$/NADH redox couple. *Biochim. biophys. Acta* **590**, 277–281.

Asmus, K.-D., Bahnemann, D., Monig, J., Searle, A. & Willson, R. L. 1979 Electrophilic free radicals and phenothiazines: a pulse radiolysis study. In *Structures in physical and theoretical chemistry*, vol. 6 (ed. H. E. Edwards, S. Navaratnam, B. J. Parson & G. O. Phillips), pp. 39–47. Amsterdam: Elsevier Scientific.

Bahnemann, D., Asmus, K.-D. & Willson, R. L. 1981 Free radical reactions of the phenothiazine, metiazinic acid. *J. Chem. Soc. Perkins Trans.* II, pp. 890–895.

Bahnemann, D., Asmus, K.-D. & Willson, R. L. 1983 Phenothiazine radical-cations: electron transfer equilibria with iodide ion and the determination of one-electron redox potential by pulse radiolysis. *J. chem. Soc. Perkin Trans.* III, pp. 1669–1673.

Bielski, H. J. B. & Chan, P. C. 1975 Enzyme-catalyzed free radical reactions with nicotinamide-adenine nucleotides. 1. Lactate dehydrogenase-catalyzed chain oxidation of bound NADH by superoxide radicals. *Arch. Biochem. Biophys.* **159**, 873–879.

Bielski, H. J. B. & Chan, P. C. 1974 Kinetic study by pulse radiolysis of the lactate dehydrogenase-catalyzed chain oxidation of nicotinamide adenine dinucleotide by $HO_2^{\cdot}$ and $O_2^{-\cdot}$ radicals. *J. biol. Chem.* **250**, 318.

Bisby, R. H., Ahmed, S. & Cundall, R. B. 1984 Repair of amino acid radicals by a vitamin E analogue. *Biochem. biophys. Res. Commun.* **119**, 245–251.

Breitenkamp, M., Henglein, A. & Lilie, J. 1976 Mechanism of the reduction of lead ions in aqueous solution (a pulse radiolysis study). *Ber. Bunsen Ges. phys. Chem.*, pp. 973–979.

Broadhurst, A., Packer, J. E., Taylor, J. J., Schaefer, M., Searle, A. J. F. & Willson, R. L. 1981 Reaction of radiation-induced peroxy radicals with amino acids and enzymes (abstract). *Radiat. Res.*, **87**, 507.

Butler, J., Land, E. J., Prutz, W. A. & Swallow, A. J. 1982 Charge transfer between tryptophan and tyrosine in proteins. *Biochem. biophys. Acta* **705**, 150–162.

Farrington, J. A., Land, E. J. & Swallow, A. J. 1980 The one-electron reduction potentials of NAD. *Biochim. Biophys. Acta* **590**, 273–276.

Forni, L. G., Monig, J., Mora-Arellano, V. O., Willson, R. L. 1983*a* Thiyl free radicals: direct observation of electron transfer reactions with phenothiazines and ascorbate. *J. chem. Soc. Perkin Trans.* II, pp. 961–965.

Forni, L. G., Packer, J. E., Slater, T. F. & Willson, R. L. 1983*b* Reactions of the trichloromethyl and halothane peroxy radicals with unsaturated fatty acids: a pulse radiolysis study. *Chem. biol. Interact.* **45**, 171–177.

Forni, L. G. & Willson, R. L. 1983 Vitamin C and consecutive hydrogen atom and electron transfer reactions in free radical protection: a novel catalytic role for glutathione. In *Protective mechanisms in cancer* (ed. D. C. H. McBrien & T. F. Slater), pp. 159–173. London & New York: Academic Press.

Forni, L. G. & Willson, R. L. 1984 Electron and hydrogen atom transfer reactions: determination of free radical redox potentials by pulse radiolysis. In *Methods in enzymology*, vol. 105 (ed. L. Packer), pp. 170–188. Orlando: Academic Press.

Gee, C. A., Kittridge, K. S. & Willson, R. L. 1985 Peroxy free radicals, enzymes and radiation damage: sensitisation by oxygen and protection by superoxide dismutase and antioxidants. *Br. J. Radiol.* **58**, 251–256.

Henglein, A. 1980 Energetics of reactions of O^- transfer reactions between radicals. *Radiat. Phys. Chem.* **15**, 151–158.

Hiller, K.-O., Hodd, P. L. & Willson, R. L. 1983 Anti-inflammatory drugs: protection of a bacterial virus as an *in vitro* biological measure of free radical activity. *Chem. Biol. Interact.* **47**, 293–305.

Kittridge, K. & Willson, R. L. 1984 Uric acid substantially enhances the free radical induced inactivation of alcohol dehydrogenase. *FEBS Lett.* **170**, 162–164.

Koppenol, W. H. & Liebman, J. F. 1984 The oxidizing nature of the hydroxyl radical. A comparison with the ferryl ion (FeO^{2+}) *J. phys. Chem.* **88**, 99–101.

Koster, R. & Asmus, K.-D. 1971 Die reduction von tetrachlorkohlenstoff durch hydratisete elektronen, H-atome und reduzierende radikale. *Z. Naturforsch* **266**, 1104–1108.

Land, E. J. & Swallow, A. J. 1971 One-electron reactions in biochemical systems as studied by pulse radiolysis. IV. Oxidation of hydronicotinamide-adenine dinucleotide. *Biochim. biophys. Acta* **234**, 34–42.

Mahood, J. S., Packer, J. E., Searle, A. J. F., Willson, R. L. & Wolfenden, B. S. 1980 Chlorpromazine, vitamin C, vitamin E and catecholamines: direct observations of free radical interactions. In *Proc. 4th Int. Symp. on Phenothiazines and Related Drugs.* (Ed. E. Usdin, H. Eckert & S. Forrest), pp. 103–106. Amsterdam: Elsevier.

Merz, J. H. & Waters, W. A. 1949*a* The oxidation of aromatic compounds by means of the free hydroxyl radical. *J. chem. Soc.*, pp. 2427–2433.

Merz, J. H. & Waters, W. A. 1949*b* Some oxidations involving the free hydroxyl radical. *J. chem. Soc.*, pp. 515–524.

Packer, J. E., Mahood, J. S., Willson, R. L. & Wolfenden, B. 1981 Reactions of the trichloromethylperoxy free radical (Cl_3COO) with tryptophan, tryptophanyl-tyrosine and lysozyme. *Int. J. radiat. Biol.* **39**, 135–141.

Packer, J. E., Mahood, J. S., Mora-Arellano, V. O., Slater, T. F., Willson, R. L. & Wolfenden, B. S. 1981 Free radicals and singlet oxygen scavengers: reaction of a peroxy radical with -carotene, diphenylfuran and 1,4-diazobicyclo-(2,2,2)-octane. *Biochem. biophys. Res. Commun.* **98**, 901–906.

Packer, J. E., Slater, T. F. & Willson, R. L. 1978 Reactions of the carbon tetrachloride-related peroxy free radical ($CCl_3O_2^{\cdot}$) with amino acids: pulse radiolysis evidence. *Life Sci.* **23**, 2617–2620.

Packer, J. E., Slater, T. F. & Willson, R. L. 1979 Direct observation of a free radical interaction between vitamin E and vitamin C. *Nature, Lond.* **278**, 737–738.

Packer, J. E., Willson, R. L., Bahnemann, D. & Asmus, K.-D. 1980 Electron transfer reactions of halogenated aliphatic peroxy radicals: measurements of absolute rate constants by pulse radiolysis. *J. chem. Soc. Perkin Trans.* II, pp. 296–299.

Patel, K. B. & Willson, R. L. 1973 Semiquinone free radicals and oxygen: a pulse radiolysis study of one electron transfer equilibria. *J. chem. Soc. Faraday Trans.* I **69**, 814–825.

Posener, M. L., Adams, G. E., Wardman, P. & Cundall, R. B. 1976 Mechanism of tryptophan oxidation by some inorganic radical-ions: a pulse radiolysis study. *J. chem. Soc. Faraday Trans.* **72**, 2231–2239.

Prutz, W. A., Butler, J., Land, E. J. & Swallow, A. J. 1980 Direct demonstration of electron transfer between tryptophan and tyrosine in proteins. *Biochem. biophys. Res. Commun.* **96**, 408–414.

Prutz, W. A. & Land, E. J. 1979 Charge transfer in peptides. Pulse radiolysis investigation of one-electron reactions in dipeptides of tryptophan and tyrosine. *Int. J. radiat. Biol.* **36**, 513–520.

Prutz, W. A., Land, E. J. & Sloper, R. W. 1981 Charge transfer in peptides. *J. chem. Soc. Faraday Trans.* **77**, 281–292.

Schuler, R. H. 1977 Oxidation of ascorbate anion by electron transfer to phenoxyl radicals. *Radiat. Res.* 417–433.

Steenken, S. & Neta, P. 1979 Electron transfer rates at equilibria between substituted phenoxide ions and phenoxyl radicals. *J. phys. Chem.* **83**, 1134–1137.

Steenken, S. & Neta, P. 1982 One-electron redox potentials of phenols: hydroxyl- and amino phenols and related compounds of biological interest. *J. phys. Chem.* **86**, 3661.

Swallow, A. J. 1953 The radiation chemistry of ethanol and disphosphopyridine nucleotide and its bearing on dehydrogenase action. *Biochem. J.* **54**, 253–257.

Swallow, A. J. 1955 The reduction of diphosphopyridine nucleotide by ethanol under the influence of -irradiation. *Biochem. J.* **61**, 197–203.

Wardman, P. & Clarke, E. D. 1976 One-electron reduction potential of substituted nitroimidazoles measured by pulse radiolysis. *J. chem. Soc.* **72**, 1377–1390.

Waters, W. A. 1971 In *Proceedings of 23rd International Congress of Pure and Applied Chemistry*, vol. 4, pp. 307–324. London: Butterworth.

Willson, R. L. 1971 Pulse radiolysis of electron transfer in aqueous quinone solutions. *Trans. Faraday Soc.* **67**, 3020–3029.

Willson, R. L. 1976 Electrophilic free radicals and nucleic acid damage: pulse radiolysis studies. *Panminerva Med.* **18**, 391–402.

Willson, R. L. 1977*a* 'Free?' radicals and electron transfer in biology and medicine. *Chemy Ind., Lond.*, 183–193.

Willson, R. L. 1977*b* Iron, zinc, free radicals and oxygen in tissue disorders and cancer control. In *Iron metabolism* (*Ciba Foundation symp. 51*) (ed. R. Porter), pp. 333–354. Amsterdam: Elsevier (Excerpta Medica).

Willson, R. L. 1978 Free radicals and tissue damage: mechanistic evidence from radiation studies. In *Biochemical mechanisms of liver injury* (ed. T. F. Slater), pp. 123–224.

Willson, R. L. 1979 Hydroxyl radicals and biological damage *in vitro*: what relevance *in vivo*? In *Oxygen free radicals and tissue damage.* (*Ciba Foundation Symp. 65*), pp. 19–42. Amsterdam: Elsevier (Excertpa Medica).

Willson, R. L. 1982 Iron and hydroxyl free radicals in enzyme inactivation and cancer. In *Free radicals, lipid peroxidation and cancer* (ed. D. C. H. McBrien & T. F. Slater), pp. 275–303. London: Academic Press.

Willson, R. L. 1983 Free radical protection: why vitamin E, not vitamin C, β-carotene or glutathione? In *Biology of vitamin E* (*Ciba Foundation Symp. 101*) (ed. R. Porter & J. Whelan), pp. 19–44. London: Pitman.

Willson, R. L. & Slater, T. F. 1975 Carbon tetrachloride and biological damage: pulse radiolysis studies of associated free radical reactions. In *Fast processes in radiation chemistry and biology* (ed. G. E. Adams *et al.*), pp. 147–161. Chichester: The Institute of Physics/Wiley.

Wolfenden, B. S. & Willson, R. L. 1982 Radical-cations as reference chromogens in kinetic studies of one electron transfer reactions: pulse radiolysis studies of 2,2-azinobis-(3-ethylbenzthiazoline-6-sulphonate). *J. chem. Soc. Perkin Trans.* II, pp. 805–812.

Discussion

H. Sies (*Institut für Physiologische Chemie I, Universität Dusseldorf, F.R.G.*). The reaction of $GS^{\cdot}$ with NADH to form $NAD^{\cdot}$ is quite interesting *in vitro*. I wonder whether this reaction may occur to a significant extent in cells? The NADH concentration is buffered to 10^{-6} M by specialized binding proteins. On the other hand, there is a considerable capacity to reduce GSSG by GSSG reductase. Given that there are several reaction pathways to yield GSSG from $GS^{\cdot}$, I suppose that the latter pathway is more likely to occur in the cell than the reaction with NADH?

R. L. Willson. Surely we just don't know. The experiments I described were indeed undertaken *in vitro*, where we do know other reactions of $GS^{\cdot}$, for example with O_2 or GS^-, could occur if the reaction concentrations were favourable. But I am not at all sure whether we really know what the concentrations of the various reactants are at any given time in the cell, either overall or, more importantly, in particular intracellular micro-environments. All we can say at present is that such reactions can occur *in vitro* according to homogeneous kinetics. What occurs *in vivo* seems very much an open question.

G. Scott (*Department of Molecular Sciences, Aston University*). In several of Professor Willson's reaction schemes he had an alkyl radical reacting with a hydrogen donor. For a thiol I can accept this, but the reaction is known to be very slow for phenolic antioxidants. Would he indicate what was the partial oxygen pressure in the system because this determines the relative importance of hydrogen abstraction by alkylperoxyl as opposed to alkyl.

R. L. Willson. The reactions I have described were generally undertaken under deoxygenated conditions. As Professor Scott correctly infers, the reaction of carbon-centred radicals with hydrogen donors will occur in competition with the reaction of the radicals with oxygen. Such reactions will therefore only predominate at relatively high hydrogen-donor to oxygen concentrations. Nevertheless such reactions may occur within the cell: indeed this has long been accepted as one of the mechanisms by which oxygen sensitizes cells to radiation damage and by which thiol compounds often afford protection. With respect to phenolic antioxidants, we have not yet observed any reaction with carbon-centred radicals by pulse radiolysis. This is in agreement with Professor Scott's statement that such reactions are known to be very slow.

Phil. Trans. R. Soc. Lond. B **311**, 565–578 (1985)
Printed in Great Britain

Biological antioxidants

By G. W. Burton,[1] D. O. Foster[2], B. Perly[3], T. F. Slater[4], I. C. P. Smith[2] and K. U. Ingold[1]

[1] *Division of Chemistry, National Research Council of Canada, Ottawa, Canada K1A 0R6*
[2] *Division of Biosciences, National Research Council of Canada, Ottawa, Canada K1A 0R6*
[3] *Departement de Physico-chimie, Centre d'Etudes Nucleaires de Saclay, Gif-sur-Yvette, France*
[4] *Department of Biochemistry, Brunel University, Uxbridge, Middlesex UB8 3PH*

The mechanism of lipid peroxidation and the ways in which the rate of this reaction can be reduced by small quantities of certain specific chemicals, called antioxidants, are described. The types and roles of the different antioxidants found in living systems are considered. Vitamin E (α-tocopherol) has long been recognized as an important lipid-soluble, chain-breaking antioxidant. It has an unexpectedly high reactivity towards peroxyl radicals, which can be understood only after detailed consideration of its structure. It is the major antioxidant of its class in human blood and its effectiveness in plasma is greatly improved by a synergistic interaction with water-soluble reducing agents such as ascorbic acid. Experiments designed to locate vitamin E within phospholipid bilayers and to discover the origin of the different biopotencies of stereoisomers of α-tocopherol are also described.

Introduction

Many foods, even when stored in a refrigerator, will eventually become rancid, which is a sign that the lipid material (i.e. the fats) in the food have undergone a chemical reaction with atmospheric oxygen. Such more or less spontaneous oxidation processes, which occur under mild conditions at ambient and sub-ambient temperatures are called autoxidations or, more commonly in biological circles, lipid peroxidations, which may be non-enzymic or enzyme-catalysed reactions. The process is a free-radical chain reaction which can be represented by four elementary reactions:

initiation: production of $R^{\cdot}$ (rate $= R_i$); (1)

propagation: $$R^{\cdot} + O_2 \rightarrow ROO^{\cdot}; \quad (2)$$

$$ROO^{\cdot} + RH \rightarrow ROOH + R^{\cdot}; \quad (3)$$

termination $$ROO^{\cdot} + ROO^{\cdot} \rightarrow \text{molecular products}. \quad (4)$$

In this scheme, RH represents the lipidic material and $R^{\cdot}$ the carbon-centred radical derived from it by removal of a hydrogen atom. The radical $R^{\cdot}$ reacts very rapidly with oxygen to form the peroxyl radical $ROO^{\cdot}$ which, in a subsequent much slower step, attacks the lipid to form a molecule of hydroperoxide, ROOH, and a new $R^{\cdot}$ radical. The propagation sequence of reactions 2 and 3 is eventually broken when two of the chain-carrying $ROO^{\cdot}$ radicals react together to give molecular products. For every radical that initiates a new chain there are therefore many molecules of RH that are oxidized to ROOH. The overall rate of autoxidation can be represented by

$$-\mathrm{d}[O_2]/\mathrm{d}t = k_3[[RH]R_i^{\frac{1}{2}}/(2k_4)^{\frac{1}{2}}, \tag{5}$$

where R_i is the rate of chain initiation and the *k*s are the rate constants for the indicated reactions.

Living organisms are exposed to much more severe oxidative stress than is food in a refrigerator. Nevertheless, they do not become rancid until they, in their turn, become food. What this means is that living organisms have some mechanism or mechanisms by which they protect themselves against autoxidation. The materials that are most readily autoxidized and hence are in most need of protection are the polyunsaturated fatty acids. Like other fatty acids, these form a part of various lipid materials within the organism including, in particular, biomembranes. Autoxidation of a biological membrane will breach its integrity and this can have disastrous consequences for the organism, because one of the principal functions of a membrane is to act as a dividing wall that compartmentalizes biochemical processes into specific cells and into specific regions within an individual cell.

Antioxidants

How does Nature protect cell membranes against peroxidation?

To answer this question it is helpful to consider the ways in which man protects certain products of his technology such as lubricating oils, rubber, and plastics, against oxidative degradation. The protection is provided by the addition of fairly small quantities of certain specific compounds called antioxidants. Such compounds are divided into two broad classes, referred to as preventive antioxidants and chain-breaking antioxidants.

Preventive antioxidants reduce the rate of chain initiation. The molecular precursor for the initiation process is generally the hydroperoxidic product, ROOH, of the oxidation. Most commercial preventive antioxidants function by converting hydroperoxides to molecular products that are not potential sources of free radicals. Similarly, most biological preventive antioxidants are also peroxide decomposers. The most important are certain enzymes that are capable of reducing hydroperoxides: for example, catalase reduces H_2O_2 to H_2O and glutathione peroxidase can reduce H_2O_2 to H_2O and also lipid hydroperoxides to the corresponding alcohol, i.e.

$$\mathrm{ROOH} \xrightarrow[\text{glutathione peroxidase}]{[2H]} \mathrm{ROH} + H_2O. \tag{6}$$

A second class of preventive antioxidants are those which inhibit photo-induced autoxidations. One such process involves the dye-sensitized formation of singlet oxygen by visible light and its 'ene' reaction with unsaturated fatty acids to form hydroperoxide.

$$\text{dye} \xrightarrow[\text{light}]{\text{visible}} \text{dye}^* \xrightarrow{^3O_2} \text{dye} + {}^1O_2, \tag{7}$$

$${}^1O_2 + \mathrm{RCH{=}CHCH_2R'} \rightarrow \mathrm{RCH(OOH)CH{=}CHR'}. \tag{8}$$

The compound β-carotene is an extremely efficient quencher of singlet oxygen. It can therefore also be classified as a preventive antioxidant.

$${}^1O_2 + \beta\text{-carotene} \rightarrow {}^3O_2 + \beta\text{-carotene} + \text{heat}. \tag{9}$$

Commercial chain-breaking antioxidants are generally phenols or aromatic amines. They owe their antioxidant activity to their ability to trap peroxyl radicals. For a phenol, such as one of the four tocopherols (α-(see figure 1), β-, γ- and δ-) that together constitute vitamin E, the initial step involves a very rapid transfer of the phenolic hydrogen atom (Burton & Ingold 1981).

$$ROO^{\cdot} + ArOH \rightarrow ROOH + ArO^{\cdot}. \tag{10}$$

The phenoxyl radical is resonance-stabilized and is relatively unreactive toward RH and O_2, and therefore it does not continue the chain. It is eventually either destroyed by reaction with a second peroxyl radical,

$$ROO^{\cdot} + ArO^{\cdot} \rightarrow \text{molecular products} \tag{11}$$

or, in certain systems, it may be 'repaired', that is, reduced to the starting phenol by reaction with a water-soluble reducing agent such as the ascorbate anion, AH^-, vitamin C (Packer *et al.* 1979).

$$ArO^{\cdot} + AH^- \rightarrow ArOH + A^{-\cdot}. \tag{12}$$

Such water-soluble reducing agents may also trap any chain-carrying peroxyl radicals that are present in the aqueous phase (Doba *et al.* 1984), whereas vitamin E, being lipid-soluble, traps $ROO^{\cdot}$ radicals in the lipid phase.

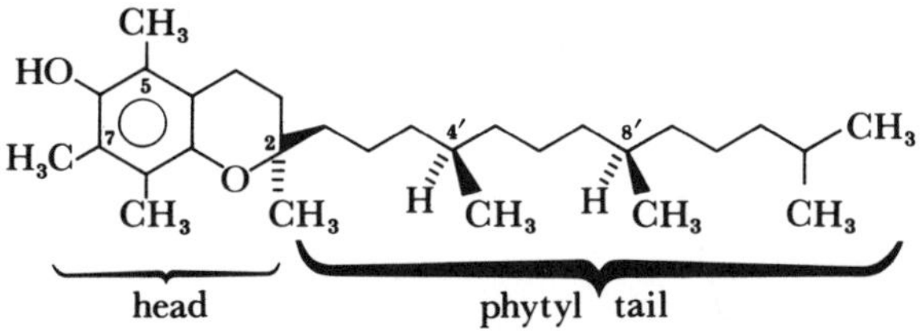

FIGURE 1. Natural (2*R*, 4′*R*, 8′*R*) α-tocopherol. Only those positions that are specifically referred to in the text have been numbered.

The main radical species that is likely to be present in the aqueous phase of aerobically respiring cells is the superoxide anion. Though not itself particularly reactive, O_2^- is in equilibrium with $HOO^{\cdot}$, its conjugate acid.

$$O_2^- + H^+ \rightleftharpoons HOO^{\cdot}. \tag{13}$$

The $HOO^{\cdot}$ radical is generally more reactive than O_2^-. However, at physiological pH this equilibrium favours O_2^- by a factor of about 100. The superoxide dismutase enzymes are the biological chain-breaking antioxidants that control the O_2^- threat to the organism.

Finally, let us reconsider the roles of β-carotene. Epidemiological studies have sparked a growing interest in this compound because they have indicated a lower incidence of certain types of cancer among individuals with an above-average intake of β-carotene and other carotenoids (Peto *et al.* 1980; Shekelle *et al.* 1981). These studies led us to investigate the chain-breaking antioxidant activity of β-carotene. We found that in addition to its efficient quenching of singlet oxygen, $^1O_2$, it belongs to a previously unknown class of biological antioxidants that exhibit good radical-trapping behaviour only at partial pressures of oxygen significantly less than the 150 Torr† of normal air (Burton & Ingold 1984). Such low O_2

† 1 Torr = 101325/760 Pa.

pressures are found in most tissues under physiological conditions (e.g. 15 Torr in the capillaries of actively working muscles). The antioxidant activity of β-carotene arises because this compound is highly reactive towards ROO˙ radicals yielding a resonance-stabilized carbon-centred radical. The reaction of this carbon-centred radical with oxygen is reversible at ambient temperatures.

$$\text{ROO}^\bullet + \beta\text{-carotene} \rightarrow \beta\text{-carotenyl radical,} \quad (14)$$

$$\beta\text{-carotenyl}^\bullet + \text{O}_2 \rightleftharpoons \beta\text{-caroteneperoxyl radical.} \quad (15)$$

At low O_2 partial pressures the β-carotenyl radical acts as an efficient trap for a second peroxyl radical.

$$\text{ROO}^\bullet + \beta\text{-carotenyl}^\bullet \rightarrow \text{molecular products.} \quad (16)$$

Absolute reactivities of lipid-soluble phenolic antioxidants *in vitro*

In model systems a phenolic antioxidant traps two peroxyl radicals, thereby breaking two oxidation chains and so reducing the chain length and the rate of oxidation. Naturally, the phenol is being continuously consumed and when it has all been used up the rate will return to the value it would have had if no antioxidant had been added (see figure 2). The time during which the rate of oxidation is suppressed is known as the induction period and its duration, τ, is very simply related to the concentration of phenol and rate of chain initiation by the equation

$$\tau = 2[\text{ArOH}]/R_\text{i}. \quad (17)$$

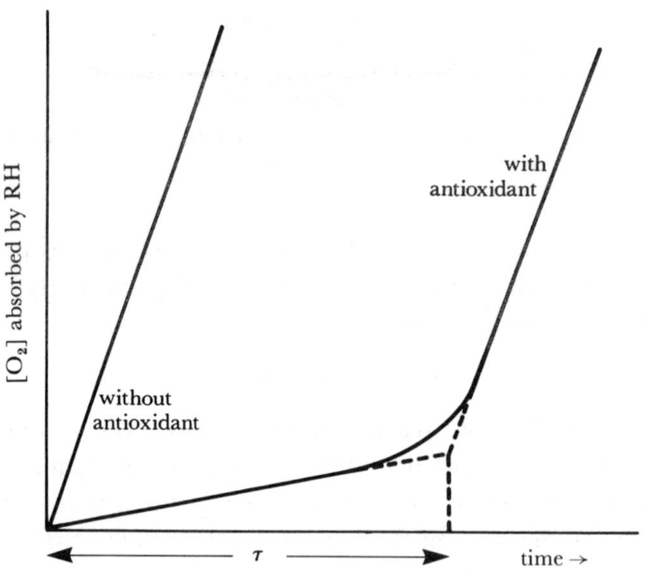

Figure 2. Effect of a chain-breaking antioxidant.

The initial rate of a phenol-inhibited oxidation in a model system can be represented by

$$-\text{d}[\text{O}_2]/\text{d}t = k_3[\text{RH}]\, R_\text{i}/2k_{10}[\text{ArOH}]. \quad (18)$$

The activity of the phenolic antioxidant is directly related to the magnitude of k_{10}. This quantity is determined by measuring the inhibited rate of autoxidation of some substrate RH, for which

k_3 has been measured, at a constant and known rate of chain initiation. We have used this technique to measure k_{10} values for a variety of phenols at 30 °C (Burton & Ingold 1981; Burton *et al.* 1983, 1984). Our results, some of which are summarized in table 1, show that the order of antioxidant activity of the tocopherols (α- > β- ⩾ γ- > δ-) is the same as that of their biological activities (Century & Horwitt 1965). Furthermore, α-tocopherol (α-T) and the structurally related model compound, pentamethylhydroxychroman (**3**), are considerably better traps for peroxyl radicals than the major phenolic antioxidants used in commerce, 2,6-di-t-butyl-4-methyl (**5**) and 4-methoxyphenol (**6**).

TABLE 1. VALUES OF k_{10} AT 30 °C AND SOME θ VALUES FOR CERTAIN PHENOLIC ANTIOXIDANTS

phenol	$10^{-4}\,k_{10}/(\mathrm{M^{-1}\,s^{-1}})$	θ
δ-T[a]	65	—
γ-T[a]	130	—
α-T[a]	320	—
(**1**)[b]	39	90°
(**2**)[a]	250	38°
(**3**)[a]	380	17°
(**4**)[a]	570	⩽ 5°
(**5**)[c]	1.4	—
(**6**)[c]	11	—

	R_2	$R_{3,4}$	R_5	R_7
δ-T	$C_{16}H_{33}$	CH_2CH_2	H	H
γ-T	$C_{16}H_{33}$	CH_2CH_2	H	CH_3
α-T	$C_{16}H_{33}$	CH_2CH_2	CH_3	CH_3
(**2**)	CH_3	$CH{=}CH$	CH_3	CH_3
(**3**)	CH_3	CH_2CH_2	CH_3	CH_3
(**4**)	CH_3	CH_2	CH_3	CH_3

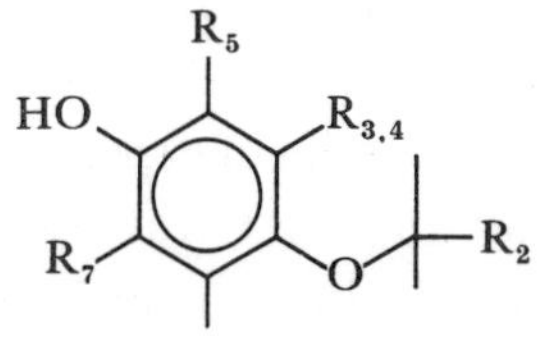

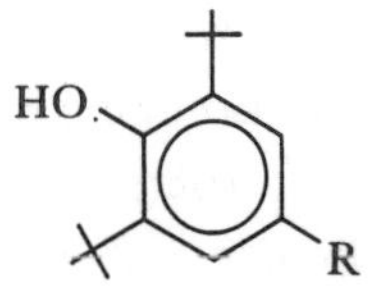

(**5**) R = CH_3
(**6**) R = CH_3O

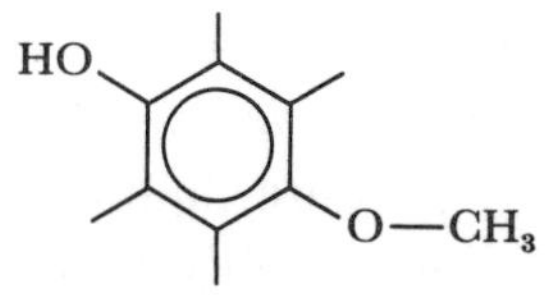

We have invoked stereoelectronic factors to explain the fact that 4-methoxytetramethylphenol (**1**) has only about 10% of the reactivity of α-T or **3** (Burton & Ingold 1981). In (**1**) the methoxy group is perpendicular to the plane of the aromatic ring (i.e. $\theta = 90°$; see figure 3). In this position, the p-type lone pair on the ethereal oxygen atom lies in the plane of the aromatic ring and so cannot stabilize the corresponding phenoxyl radical. However, in α-T and in (**3**) the second ring adopts a half-chair conformation and holds the ethereal oxygen in such a position that $\theta \approx 17°$. The lone pair orbital can therefore overlap with the orbital containing the unpaired electron and stabilize the phenoxyl radical. The more stable the phenoxyl radical, the weaker will be the O—H bond in the parent phenol, and the weaker this bond, the more readily it will be cleaved by an attacking ROO$^\cdot$, i.e. the more effective it will be as an antioxidant.

This stereoelectronic description of antioxidant activity leads to the prediction that for structurally related 4-alkoxyphenols, k_{10} will increase as θ decreases from 90° (**1**) towards 0°.

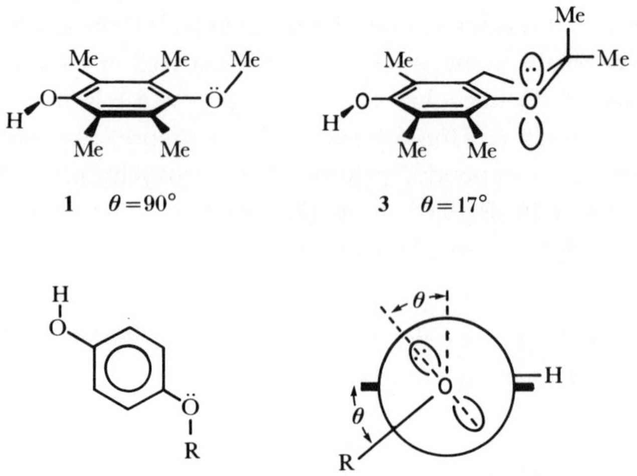

FIGURE 3. Structures of **1** and **3** and their θ values. The definition of θ is shown at the lower right for the 4-alkoxyphenol shown at the lower left.

This prediction was confirmed by showing that the pentamethylhydroxychromenol (**2**), with $\theta = 38°$, is less reactive than **3**, while the pentamethylhydroxydihydrobenzofuran (**4**), with $\theta \leqslant 5°$, is more reactive, see table 1 (Burton *et al.* 1983, 1984).

PROOF THAT VITAMIN E IS THE MAJOR LIPID-SOLUBLE, CHAIN-BREAKING ANTIOXIDANT IN HUMAN BLOOD

Vitamin E functions as a protective agent against lipid peroxidation *in vivo* (Machlin 1980; Tappel 1980). As we have demonstrated, it is also an extremely potent (though not the most potent) chain-breaking phenolic antioxidant *in vitro*. We have performed a series of experiments (Burton *et al.* 1983) in which high pressure liquid chromatography (h.p.l.c.) was used to measure the concentrations of the four tocopherols (mainly α-T) in lipid extracts of human blood plasma and human erythrocyte ghost membranes. We also added a small portion of each of these lipid extracts to styrene that was undergoing autoxidation at 30 °C with a constant and known rate of chain initiation. The lipid extracts produced induction periods and from the length of these induction periods we could calculate, by using (17), the total concentration of all chain-breaking antioxidants present in the lipid extract. Comparison with the total concentration of tocopherols, measured by h.p.l.c. (that equals, of course, the total concentration of vitamin E), showed that most, if not all, of the chain-breaking antioxidant in the plasma and ghost membrane extracts was vitamin E. We have therefore concluded that vitamin E is the major lipid-soluble, chain-breaking antioxidant in human blood (Burton *et al.* 1983).

A tetramethylhydroxydihydrobenzofuran having a $C_{16}H_{33}$ phytyl 'tail' attached at position 2 (i.e. **4** but with $R_2 = C_{16}H_{33}$) has recently been synthesized in our laboratory (L. Hughes 1984, unpublished results). We hope that the phytyl tail will give this compound appropriate lipophilicity and other desirable characteristics for its incorporation into biomembranes *in vivo*. We have recently started to measure the vitamin E activity of this compound on vitamin-E-depleted rats (C. D. O. Foster and coworkers 1984, unpublished results) by using one of the standard bioassay procedures (Machlin *et al.* 1982).

Kinetic studies on the autoxidation of phospholipid bilayers

The kinetic rate law for uninhibited, liquid-phase, autoxidation (5) has frequently been shown to apply to homogeneous organic systems. However, it is not self-evident that it should also apply to the heterogeneous types of systems that would be involved in lipid peroxidation *in vivo*, nor even for the autoxidation of phospholipid bilayers dispersed in water. The first quantitative kinetic studies on the latter type of system used egg lecithin phosphatidylcholine bilayers dispersed as multilamellar vesicles (Barclay & Ingold 1981). Oxidation was initiated by the thermal decomposition of a lipid-soluble azo compound, and the rate of chain initiation that it produced was determined (by means of (17)) by using α-T as the lipid-soluble, chain-breaking antioxidant. Both the azo initiator and the α-T had to be added to a methylene chloride solution of the egg lecithin. The CH_2Cl_2 was then removed by evaporation and this was followed by the dispersion of the residue in water. Despite the somewhat inelegant experimental approach, (5) was shown to describe the oxidation kinetics of this system.

It was clear that this experimental approach would have to be modified if we were to study the autoxidation kinetics of biologically interesting systems. The breakthrough came when we discovered that not only could the oxidation of phospholipid bilayers be initiated with a lipid-soluble initiator, but also with a water-soluble initiator (Barclay *et al.* 1984). Furthermore, whichever initiator was used, the reaction could be inhibited by both α-T and a water-soluble analogue of α-T (Trolox(−), which has a modified structure **3** with R_2 as CO_2^-). The rate constant ratio, $k_3/(2k_4)^{\frac{1}{2}}$ of (5) (which is generally referred to as the oxidizability of RH) had, as it should, essentially the same value irrespective of the phase in which the initiator and inhibitor were present. Clearly peroxyl radicals generated in the aqueous phase can diffuse into the bilayer to start oxidation chains, while Trolox(−) can get sufficiently close to the peroxyl radicals that carry the chain within the bilayer to intercept them (via reaction 10).

Vitamin E–vitamin C interaction

It has been known for many years that vitamin C is, by itself, an ineffective antioxidant but that in many systems it will greatly extend the induction period produced by vitamin E (Golumbic & Mattill 1941). That is to say, vitamin E–vitamin C mixtures are synergistic because their joint effect is greater than the sum of their individual effects. As Tappel (1968) originally suggested, vitamin C may, in living organisms, 'repair' (i.e. regenerate) vitamin E by reducing the tocopheroxyl radicals formed in reaction (10). Pulse radiolytic experiments have subsequently shown that in homogeneous solution the ascorbate anion can indeed reduce the α-tocopheroxyl radical, reaction (12) (Packer *et al.* 1979).

Although there have been several studies of vitamin E–vitamin C synergistic antioxidant effects in, for example, micellar oxidations (Barclay *et al.* 1983; Yamamoto *et al.* 1984) only some of our own recent measurements appear to be relevant to the probable situation in living organisms (Doba *et al.* 1985). By using multilamellar vesicles of dilinoleoylphosphatidylcholine (DLPC) as the oxidizable bilayer we have found that for oxidations initiated by the thermal decomposition of a lipid-soluble azo compound, vitamin C produces no induction period, though it will extend the induction period due to α-T. (Based on the fact that α-T traps two peroxyl radicals per molecule α-T, the increase in τ produced by the addition of ascorbate indicates an effective trapping of *ca.* 0.4 peroxyl radicals per molecule ascorbate in the vitamin E–vitamin C system). However, in reactions initiated by thermal decomposition of a water-soluble azo-compound, vitamin C produces an excellent induction period, the length of which implies that each molecule of ascorbate can trap 0.6 peroxyl radicals. Therefore, although the ascorbate anion (unlike Trolox(−)) cannot penetrate sufficiently deeply into the phospholipid bilayer (see also Schreier-Mucillo *et al.* 1976) to intercept the chain-carrying ROO˙ radicals, it can and does trap peroxyl radicals that are formed in the aqueous phase before they can diffuse into the bilayer. It is therefore highly probable that in living organisms vitamin C functions both as a direct chain-breaking antioxidant for water-soluble peroxyl radicals (e.g. HOO˙) and as an indirect chain-breaking antioxidant by regenerating vitamin E.

Kinetic studies on the autoxidation of blood plasma

The discovery that water-soluble azo compounds can be used in heterogeneous systems to initiate lipid peroxidation has allowed us to commence investigation of the autoxidation of human blood plasma. We have already shown (vide supra) that vitamin E is essentially the only lipid-soluble antioxidant in plasma. However, preliminary comparison of vitamin E concentration measured by h.p.l.c. with the length of the induction period indicate that the 'effective' antioxidant level in plasma is approximately 15–20 times greater than can be accounted for by the vitamin E alone (Wayner *et al.* 1986). The relative contributions to this extended induction period of ascorbate, urate (see Ames *et al.* 1981) and superoxide dismutase are currently being explored (D. Wayner 1984, unpublished results).

Comparison of natural *RRR*-α-T with unnatural *SRR*-α-T

Tocopherols have three chiral carbon atoms and the natural materials have the 2*R*, 4′*R*, 8′*R*-configuration (see figure 1). Natural *RRR*-α-T is the most bioactive form of vitamin E having, for example, about three times the bioactivity of the synthetic isomer with inverted stereochemistry at position 2, i.e. *SRR*-α-T (Weiser & Vecchi 1982). The magnitude of this difference in bioactivity between stereoisomers would appear to be far too small to be a result of some enzyme-mediated chemical or physical (such as transport) process. There appears to have been no real attempt to explain and few attempts to explore the origin of the difference in bioactivity of these two stereoisomers (Weber *et al.* 1964*a*, *b*). We therefore undertook the following experiment.

Male rats were raised from weaning on a standard vitamin-E-free diet to which we had added 36 mg of natural *RRR*-α-T acetate per kilogram of chow. After four weeks the diet was changed to one based on the same chow but in which the natural α-T acetate was replaced by an equimolar mixture of *RRR*-α-T 5,7-$(CD_3)_2$ acetate and *SRR*-α-T 5-CD_3 acetate, in a total quantity identical to that of the *RRR*-α-T acetate present in the initial diet. Blood samples were taken at various times by heart puncture under anaesthetic and tissue samples after sacrifice. Lipids were extracted and the α-tocopherol-containing fraction was separated by h.p.l.c. Subsequentl analysis by g.c.–m.s. allowed the direct measurement of the *RRR*:*SRR* ratio of the two deuterated isomers without interference by the natural (i.e. unlabelled) *RRR*-α-T. Figure 4 shows this ratio as a function of time for blood plasma and red blood cell membranes. Measurements of this ratio at times less than one day are clearly desirable (see Weber *et al.* 1964*b*) and will soon be undertaken. However, in the meantime we tentatively suggest that the initial (time$\rightarrow$0) ratio of 1.35 in plasma is a result of chiral discrimination (favouring the natural stereochemistry) on the transport from gut to plasma. The initial ratio of 1.8 in the red blood cell membranes suggests a second chiral discrimination favouring the absorption into this membrane of the natural stereoisomer, again by a factor of *ca.* 1.35 (i.e. $(1.35)^2 \approx 1.8$). If we assume that α-T in the lipoproteins of the plasma and in the red blood cell membranes are in equilibrium on the timescale of our experiment, as seems highly probable (Silber *et al.* 1969; Poukka & Bieri 1970; Bjornson *et al.* 1975), then the rate of transfer of *RRR*-α-T from plasma to membrane is larger than that for *SRR*-α-T, or the transfer rate of *RRR*-α-T in the reverse direction is smaller than for *SRR*-α-T, or both. We believe that this rather small chiral discrimination by the red blood cell membrane simply reflects the fact that membranes are themselves composed of chiral molecules that 'recognize' (i.e. absorb) one stereoisomer of vitamin E preferentially. Passage through a series of membranes could lead to a fairly high discrimination in favour of the *RRR*-isomer in certain tissues, which could explain the origin of the observed differences in bioactivity. Experiments are currently underway to measure the magnitude of this chiral discrimination, by using *in vitro* (rather than *in vivo*) model systems and both natural and synthetic membranes.

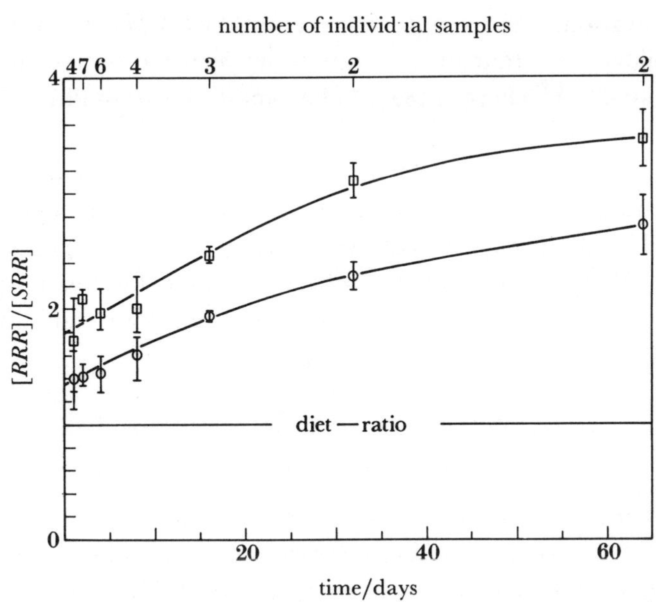

FIGURE 4. Ratio of [*RRR*]/[*SRR*]α-tocoperhol of rats receiving a 1:1 ratio of these stereoisomers in their diet; ○, plasma; □, red blood cell membranes. Number of samples refers to the number of separate rats. Error bars represent the total spread in the ratio.

Location of vitamin E in phospholipid bilayers

Knowledge of the preferred location of vitamin E within a biomembrane would aid in our understanding of its interaction with the membrane derived peroxyl radicals, of the synergistic interaction with vitamin C and urate, and even of the origin of the chiral discrimination described above. A relatively simple ^{13}C-n.m.r. technique that can yield information on the position of α-T in a phospholipid bilayer has been developed by using *RRR*-α-T [5-^{13}C]H_3

(B. Perly 1984, unpublished results). Unilamellar egg lecithin phosphatidylcholine vesicles containing relatively high concentrations of this labelled compound show a single peak from the ^{13}C label. Addition of the shift reagent, Pr^{3+}, to the liposome suspension causes this signal to split into two distinct peaks. The position of one peak remains unchanged, while the other is shifted downfield by an amount dependent on the Pr^{3+} concentration. We interpret this result as indicating that the α-T cannot be randomly distributed throughout the bilayer but is, instead, located in two definite and distinct regions in the liposome bilayer with slow exchange (on the n.m.r. timescale) between the regions. The unshifted signal must be due to α-T in the inner monolayer of the bilayer, no Pr^{3+} penetrating into the aqueous centre of the liposome. The other signal must be due to α-T in the outer monolayer. Because it is fairly sharp and is also well shifted, the hydroxyl group of α-T must be fairly close to the surface of the bilayer.

In a second experiment, varying concentrations of Gd^{3+} were added to the *RRR*-α-T [5-^{13}C]H_3-containing liposomes dispersed with sufficient Pr^{3+} to split the ^{13}C line. Values for T_1 of the two ^{13}C peaks were measured at varying concentrations of Gd^{3+}. (For a given Gd^{3+} concentration, T_1 will vary as the sixth power of the distance between the Gd and ^{13}C nuclei.) For the 'inner' ^{13}C signal, T_1 is unchanged up to the maximum Gd^{3+} concentration that could be used (10^{-2} M). For the outer signal, however, T_1 had decreased by more than 90% at the maximum Gd^{3+} concentration. These results confirm that the 5-CH_3 group of *RRR*-α-T is close to the surface of the bilayer and that the separation between the 'inner' and 'outer' 5-CH_3 groups is greater than 40–50 Å† (Perly 1984). (The overall thickness of the bilayer should be *ca.* 60 Å.)

Deuterium n.m.r. experiments aimed at determining both the precise shape of the heterocyclic ring of α-T and the orientation of its phytyl tail when the α-T is present in a phospholipid bilayer (I. C. P. Smith & I. H. Ekiel 1984, unpublished results) are being performed. Numerous specifically deuterated *RRR*-α-tocopherols have been prepared for this purpose (L. Hughes 1984, unpublished results). These experiments will also yield information about the mobility of various portions of the α-T molecule within the bilayer.

Conclusion

Our overall defences against free-radical damage have been represented in figure 5. It can be seen that α-tocopherol plays a vital role in these defences. Application of the techniques and, more particularly, the powerful methodology of physical organic chemistry have revealed many facts about α-tocopherol. The most important are (i) its high antioxidant activity and the stereoelectronic explanation for this activity; (ii) its presence as the major (or only) lipid-soluble, chain-breaking antioxidant in human blood and the 'extension' of its antioxidant activity by other water-soluble compounds that are present in the plasma (e.g. ascorbate) and (iii) possible thermodynamic reasons (absorption by chiral membranes) for the varying bioactivities of its different stereoisomers. Similarly, the techniques of biophysics are now providing preliminary information about the location, conformation, and mobility of α-T in phospholipid bilayers and eventually biomembranes. Because radical damage to living organisms has been implicated in many pathological processes (such as heart disease, cancer and ageing), a better understanding of the role of vitamin E should have both scientific and practical benefits. As an example, our

† 1 Å = 10^{-10} m = 10^{-1} nm.

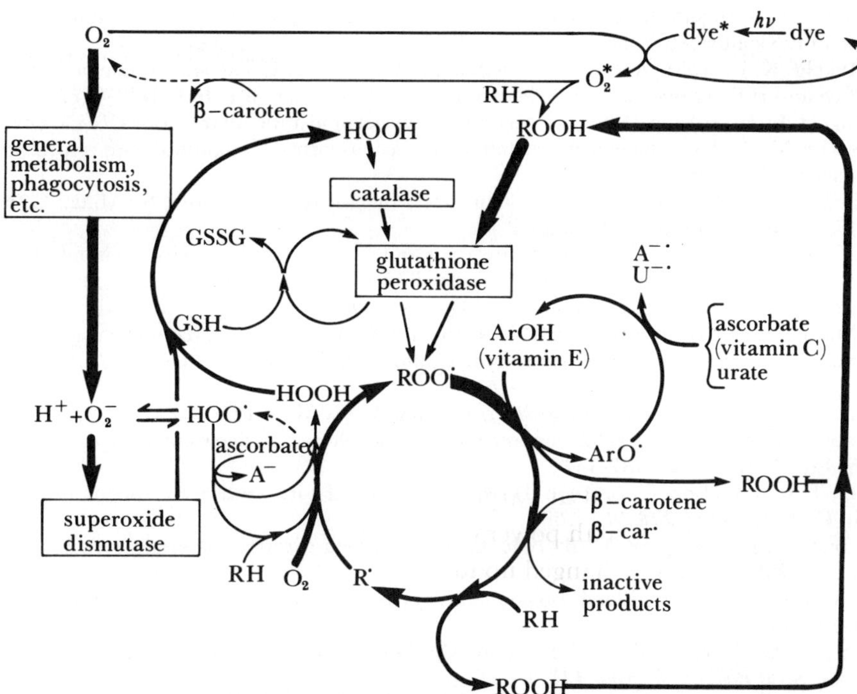

FIGURE 5. Biological autoxidation and antioxidants: the global picture (1984). For simplicity, thin arrows leading from an enzyme are meant to represent that portion of the substrate that is not destroyed by the enzyme.

discovery that the lipids present in various cancerous tissues (Slater *et al.* 1984) and in other rapidly multiplying cells (Cheeseman *et al.* 1985) contain significantly enhanced levels of vitamin E can be cited.

Thanks are due to all those colleagues and collaborators whose names are in the references. Without their unstinting efforts and imagination such a complete investigation of vitamin E would never have been possible. We also thank the National Foundation for Cancer Research and the Association for International Cancer Research for their generous financial support for much of this work.

References

Ames, B. N., Cathcart, R., Schwiers, E. & Hochstein, P. 1981 Uric acid provides an antioxidant defence in humans against oxidant- and radical-caused aging and cancer: a hypothesis. *Proc. natn. Acad. Sci. U.S.A.* **78**, 6858–6862.

Barclay, L. R. C. & Ingold, K. U. 1981 Autoxidation of biological molecules. 2. The autoxidation of a model membrane. A comparison of the autoxidation of egg lecithin phosphatidylcholine in water and in chlorobenzene. *J. Am. chem. Soc.* **103**, 6478–6485.

Barclay, L. R. C., Locke, S. J. & MacNeil, J. M. 1983 The autoxidation of unsaturated lipids in micelles. Synergism of inhibitors vitamin C and E. *Can. J. Chem.* **61**, 1288–1290.

Barclay, L. R. C., Locke, S. J., MacNeil, J. M., VanKessel, J., Burton, G. W. & Ingold, K. U. 1984 Autoxidation of micelles and model membranes. Quantitative kinetic measurements can be made by using either water-soluble or lipid soluble initiators with water-soluble or lipid-soluble chain-breaking antioxidants. *J. Am. chem. Soc.* **106**, 2479–2481.

Bjornson, L. K., Gniewkowski, C. & Kayden, H. J. 1975 Comparison of exchange of α-tocopherol and free cholesterol between rat plasma lipoproteins and erythrocytes. *J. Lipid Res.* **16**, 39–53.

Burton, G. W., Doba, T., Gabe, E., Hughes, L., Lee, F. L., Prasad, L. & Ingold, K. U. 1984 Autoxidation of biological molecules. 4. Maximizing the antioxidant activity of phenols. (In preparation.)

Burton, G. W., Hughes, L. & Ingold, K. U. 1983 Antioxidant activity of phenols related to vitamin E. Are there chain-breaking antioxidants better than α-tocopherol? *J. Am. chem. Soc.* **105**, 5950–5951.

Burton, G. W. & Ingold, K. U. 1981 Autoxidation of biological molecules. 1. The antioxidant activity of vitamin E and related chain-breaking phenolic antioxidants in vitro. *J. Am. chem. Soc.* **103**, 6472–6477.

Burton, G. W. & Ingold, K. U. 1984 β-Carotene: An unusual type of lipid antioxidant. *Science, Wash.* **224**, 569–573.

Century, B. & Horwitt, M. K. 1965 Biological availability of various forms of vitamin E with respect to different indices of deficiency. *Fed. Proc., Pt, 1*, **24**, 906–911.

Cheeseman, K. H., Benedetto, C., Burton, G. W., Collins, M., Ingold, K. U., Maddix, S., Milia, A., Proudfoot, K. & Slater, T. F. 1985 *Biochem. J.* (Submitted.)

Doba, T., Burton, G. W. & Ingold, K. U. 1985 Antioxidant and co-antioxidant activity of vitamin C. The effect of vitamin C, either alone or in the presence of vitamin E or a water-soluble vitamin E analogue, upon the peroxidation of aqueous multilamellar phospholipid liposomes. *Biochim. biophys. Acta* (In the press.)

Golumbic, C. & Mattill, H. A. 1941 Antioxidants and the autoxidation of fats. XIII. The antioxigenic action of ascorbic acid in association with tocopherols, hydroquinones and related compounds. *J. Am. chem. Soc.* **63**, 1279–1280.

Machlin, L. J. (ed.) 1980 *Vitamin E: a comprehensive treatise*. New York: Marcel Dekker.

Machlin, L. J., Gabriel, E. & Brin, M. 1982 Biopotency of α-tocopherols as determined by curative myopathy bioassay in the rat. *J. Nutr.* **112**, 1437–1440.

Packer, J. E., Slater, T. F. & Willson, R. L. 1979 Direct observation of a free radical interaction between vitamin E and vitamin C. *Nature, Lond.* **278**, 737–738.

Peto, R., Doll, R., Buckley, J. D. & Sporn, M. B. 1980 Can dietary beta-carotene materially reduce human cancer rates? *Nature, Lond.* **290**, 201–208.

Poukka, R. K. H. & Bieri, J. G. 1970 Blood α-tocopherol: erythrocyte and plasma relationships in vitro and in vivo. *Lipids* **5**, 757–761.

Schreier-Mucillo, S., Marsh, D. & Smith, I. C. P. 1976 Monitoring the permeability profile of lipid membranes with spin probes. *Archs Biochem. Biophys.* **172**, 1–11.

Shekelle, R. B., Liu, S., Raynor, J., Jr, Lepper, M., Maliza, C., Rosoff, A. H., Paul, O., Shryock, A. M. & Stamler, J. 1981 Dietary vitamin A and risk of cancer in the Western Electric study. *Lancet* (ii), pp. 1185–1190.

Silber, R., Winter, R. & Kaydon, H. J. 1969 Tocopherol transport in the rat erythrocyte. *J. clin. Invest.* **48**, 2089–2095.

Slater, T. F., Benedetto, C., Burton, G. W., Cheeseman, K. H., Ingold, K. U. & Nodes, J. T. 1984 Lipid peroxidation in animal tumours: a disturbance in the control of cell division? In *Icosanoids and cancer* (ed. H. Thaler-Dao), pp. 21–29. New York: Raven Press.

Tappel, A. L. 1968 Will antioxidant nutrients slow ageing process? *Geriatrics* **23**, 97–105.

Tappel, A. L. 1980 Measurement of and protection from *in vivo* lipid peroxidation. In *Free radicals in biology*, vol. 4 (ed. W. A. Pryor), pp. 1–47. New York: Academic Press.

Wayner, D., Burton, G. W., Ingold, K. U. & Locke, S. J. 1986 Quantitative measurement of total peroxyl radical-trapping antioxidant capability of human blood plasma by controlled peroxidation. The important contribution made by plasma proteins. (In preparation.)

Weber, F., Gloor, U., Wùrsch, J. & Wiss, O. 1964*a* Synergism of *d*- and *l*-α-tocopherol during absorption. *Biochem. biophys. Res. Commun.* **14**, 186–188.

Weber, F., Gloor, U., Wùrsch, J. & Wiss, O. 1964*b* Studies on the absorption, distribution and metabolism of *l*-α-tocopherol in the rat. *Biochem. biophys. Res. Commun.* **14**, 189–192.

Weiser, H. & Vecchi, M. 1982 Stereoisomers of α-tocopherol acetate. II. Biopotencies of all eight stereoisomers, individually or in mixtures, as determined by rat resorption–gestation tests. *Int. J. Vit. Nutr. Res.* **52**, 351–370.

Yamamoto, Y., Haga, S., Niki, E. & Kamiya, Y. 1984 Oxidation of lipids. V. Oxidation of methyl linoleate in aqueous dispersion. *Bull. chem. Soc. Japan* **57**, 1260–1264.

Discussion

R. L. Willson (*Department of Biochemistry, Brunel University, Uxbridge, Middlesex*). We have actually found a compound that has very much the structure Professor Ingold is suggesting; 6-hydroxy-1,4-dimethylcarbazole (HDC).

HO, N H, CH_3, CH_3

HDC

In our *in vitro* microsomal lipid peroxidation system this is a much better antioxidant than vitamin E; indeed, it is the most potent antioxidant we have yet found. Would not its structure be in agreement with Professor Ingold's electron orbital requirements? (See Willson 1983.)

Reference

Willson, R. L. 1983 In *Ciba Symposium: Biology of vitamin E*, pp. 19–44. London: Pitman.

K. U. Ingold. Yes.

G. Scott (*Department of Molecular Sciences, Aston University, Gosta Green, Birmingham B4 7ET*). I have a comment and a question. The comment is that BHT is by no means the most active of the synthetic antioxidants, and I wonder whether Professor Ingold has done full justice to the products that man has already produced. Chain-breaking antioxidants are known that are very much more effective than BHT because they give rise to much more stable aryloxyls on reaction with alkylperoxyl. Perhaps the most studied is the reduced form of the stable radical galvinoxyl,

hydrogalvinoxyl $\xrightarrow{ROO^{\cdot}}$ galvinoxyl

This system is particularly potent because galvinoxyl is itself an effective antioxidant (chain-breaking electron acceptor) under conditions of limited oxygen availability. This is because it can oxidize an alkyl radical by hydrogen abstraction from the α-position with the regeneration of hydrogalvinoxyl. This amounts to a catalytic antioxidant process in which one molecule of antioxidant can knock out many chain-propagating radicals (Scott 1984).

My question is whether Professor Ingold has any explanation for the fact reported by several workers that, *in vivo*, the concentration of α-tocopherol does not reduce to zero as required by the mechanism he has outlined under conditions of oxidative stress? Indeed there is some evidence that in iron overload, the α-tocopherol concentration is actually higher than it is normally (Green *et al.* 1967). We wonder whether the α-tocopheroloxyl radical may not be reduced by the substrate radicals as in the case of galvinoxyl.

References

Green, J., *et al.* 1967 *Br. J. Nutr.* **21**, 725.
Scott, G. (ed.) 1984 *Developments in polymer stabilisation*, vol. 7, p. 65. Amsterdam: Elsevier.

K. U. Ingold. In biological systems there may be some 'regeneration' of the α-tocopheroxyl radicals by the mechanism Professor Scott suggests under certain conditions. However, Nature appears to have arranged things so that the α-tocopheroxyl formed in the biomembranes is generally reduced back to α-tocopherol by reducing agents that are present in the surrounding aqueous medium (urate, ascorbate, etc.).

There is no requirement in our mechanism that α-tocopherol levels should fall to zero under conditions of *in vivo* 'oxidative stress'. One would probably have to reduce the concentration of all the aqueous reducing agents (some of which were referred to above) to zero before the α-tocopherol level would reach zero. I am confident that the *in vivo* experimental object would be dead long before this limit was reached. The possibility that *in vivo* 'oxidative stress' may induce certain biological antioxidants must not be ignored.

Phil. Trans. R. Soc. Lond. B **311**, 579–591 (1985)
Printed in Great Britain

Cytochrome P_{450}: substrate and prosthetic-group free radicals generated during the enzymatic cycle

By D. Dolphin

Department of Chemistry, The University of British Columbia, Vancouver, British Columbia V6T 1Y6, Canada

During the enzymatic cycle of the cytochromes P_{450}, dioxygen binds to the ferrous haemprotein when the resting ferric haemprotein has undergone a one-electron oxidation after substrate binding. A further one-electron reduction generates an intermediate that is isoelectronic with a peroxide dianion coordinated to a ferric iron. Heterolytic cleavage of the O—O bond generates water and a species which is formally an oxene (oxygen atom) coordinated by iron(III). However, on the basis of model reactions and by analogy to the catalases and peroxidases, this active oxidizing intermediate is formulated as an oxo–Fe^{IV} porphyrin π-cation radical. The radical is stabilized by delocalization on the porphyrin macrocycle and the high oxidation state is achieved by oxidizing both the metal and the porphyrin ring of the haemprotein.

Hydrogen atom abstraction from a saturated hydrocarbon substrate generates a substrate free radical, constrained by the protein binding site, and the equivalent of a hydroxyl radical bound to iron(III). Coupling of the 'hydroxy' and substrate radicals generates hydroxylated product and resting protein.

For olefins an initial electron transfer to oxidized haemprotein gives a substrate cation radical. Further reaction of this radical can give the epoxide, the principal product; an aldehyde or ketone by rearrangement; or an alkylated haemprotein resulting in suicide inhibition.

Two classes of radicals occur in Nature: those which are destructive and unwelcome in living cells and for which complex and elegant protective mechanisms exist and those which are generated and controlled enzymatically (Boschke 1983). Many examples of both types occur in the reductive chemistry of the naturally occurring stable free radical, triplet dioxygen. Scheme 1 outlines some of the radical species generated from the reduction and subsequent reactions of dioxygen.

Much of the chemistry of dioxygen and its reduction products is controlled, naturally, by haemproteins. The classic example is the catalase-catalysed destruction of hydrogen peroxide (Hewson & Hager 1979). As we shall see, this decomposition of hydrogen peroxide occurs without formation of hydroxyl radicals; however, stable radicals are generated on the haem moiety (Dolphin & Felton 1974). A closely related family of enzymes, the peroxidases, also use hydrogen peroxide as an oxidant but then generate substrate (aromatic amines, phenols) free radicals which may have biological significance (D. H. R. Barton, this symposium).

We plan to show in this paper that the cytochromes P_{450}, currently the most widely studied of all enzymes, function in a manner analogous to the catalases and peroxidases and generate both haem and substrate radicals under carefully controlled conditions.

Interest in these haemproteins arose from our studies on the redox properties of metallo-porphyrins containing a redox-inert metal (Fajer *et al.* 1970). Thus both zinc and magnesium

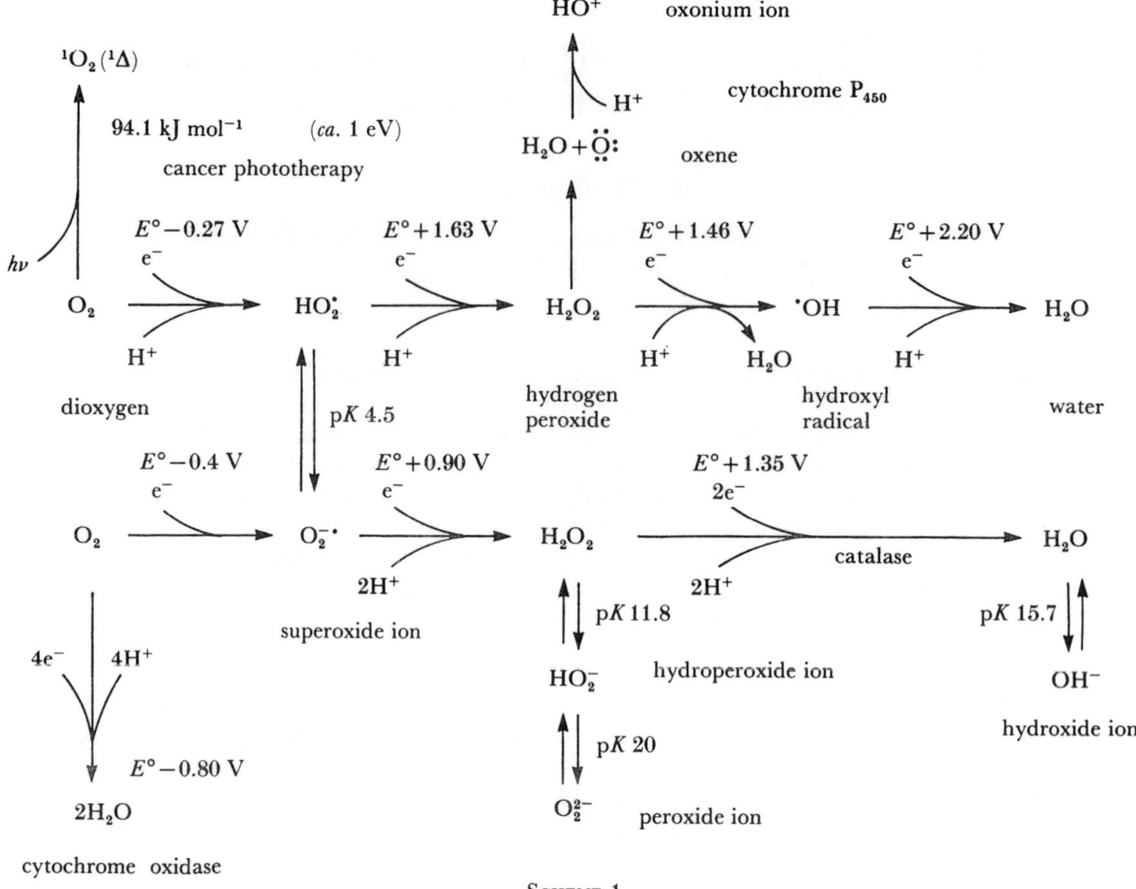

Scheme 1

porphyrins exhibit two reversible one-electron oxidations. The first oxidation generates a cation radical and e.p.r. showed that the unpaired electron was delocalized within the aromatic π-system of the porphyrin macrocycle. However, two classes of these π-cation radicals were observed. The highest filled π-orbitals of porphyrins and metalloporphyrins are degenerate with a_{1u} and a_{2u} symmetries (Hanson *et al.* 1976). Removal of a single electron from one or other of these orbitals, determined by the porphyrin peripheral substituents and the metal and its axial coordination, generates either a $^2A_{1u}$ or $^2A_{2u}$ ground state for these π-cation radicals. The $^2A_{1u}$ states are characterized by low unpaired spin density on the *meso*-carbon and nitrogen atoms, while in the $^2A_{2u}$ states, spin density appears on these atoms (Fajer *et al.* 1973).

The two ground states also exhibit quite different, but characteristic optical spectra. Figure 1 shows the optical spectra of the π-cation radicals of the magnesium complexes of octaethylporphyrin (OEP) ($^2A_{1u}$) and *meso*-tetraphenylporphyrin (TPP) ($^2A_{2u}$) ground states. Figure 2 shows the π-cation radicals derived from the one-electron oxidation of Co^{III}OEP. In this case, the species $Co^{III}(OEP)^{+\cdot}$ occupies a $^2A_{1u}$ state when coordinated by bromide but a $^2A_{1u}$ state when the bromide is replaced by perchlorate. Of special note is part *b* of figure 2; this shows the optical spectra of catalase and peroxidase after the resting ferric haemproteins have been oxidized, with hydrogen peroxide, by two electrons. Clearly these high oxidation states, known as compounds I, are porphyrin π-cation radicals (Dolphin *et al.* 1971). Mössbauer spectroscopy (Moss *et al.* 1969) had already suggested an oxidation of the iron,

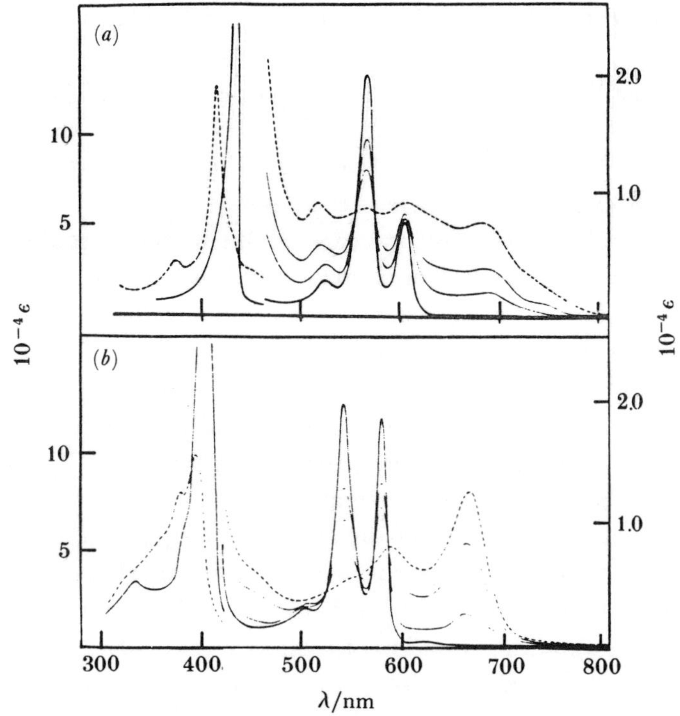

FIGURE 1. Optical changes during the electrochemical oxidation of magnesium porphyrins. (a) $Mg^{II}OEP$ (solid line); $Mg^{II}(OEP)^{+\cdot}$ (dotted line). (b) $Mg^{II}TPP$ (solid line); $Mg^{II}(TPP)^{+\cdot}$ (dotted line).

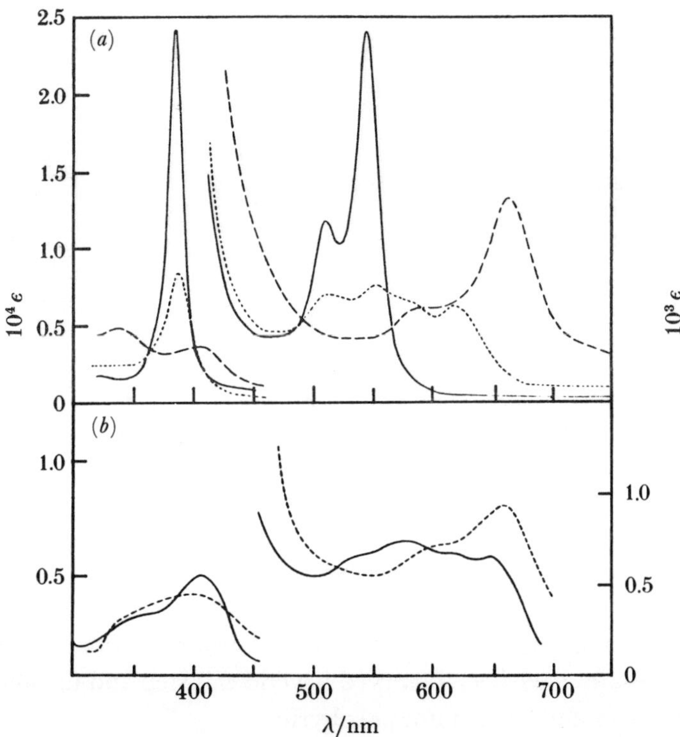

FIGURE 2. (a) $Co^{III}(OEP)Br$ (solid line); $Co^{III}(OEP)^{+\cdot}(Br)_2$ (broken line); $Co^{III}(OEP)^{+\cdot}(ClO_4)_2$ (dotted line). (b) Catalase compound I (solid line); horseradish peroxidase compound I (broken line).

and this coupled with the porphyrin oxidation and the stoichiometry allows the reaction to be represented as in (1).

$$Fe^{III}porphyrin \xrightarrow{H_2O_2} \underset{\mathbf{1}}{O{=}Fe^{IV}(porphyrin)^{+\cdot}} + H_2O. \qquad (1)$$

The coordination of the oxygen to the iron was established (Hager *et al.* 1972) and elegantly confirmed by ENDOR experiments (Roberts *et al.* 1981).

While the simple porphyrin π-cation radicals exhibit textbook-quality e.p.r. spectra, the same is not true for the compounds I of catalase and horseradish peroxidase. Indeed, only recently has the e.p.r. signal of the Fe^{IV} porphyrin π-cation radical been observed, a difficult task considering it spans thousands of gauss (Schultz *et al.* 1979).

Now that the oxo-Fe^{IV} π-cation radical is well established (excepting for some discussion as to the specific doublet ground state of some haemproteins (Rutter *et al.* 1983) for catalase and the peroxidases we would like to suggest that the cytochromes P_{450} use the 'same' high oxidation state intermediate during their enzymatic cycle.

Scheme 2 shows the enzymatic cycle for the cytochrome P_{450}. Binding of CO to the ferrous intermediate gives the characteristic absorption at *ca.* 450 nm (Hanson *et al.* 1976) for which this family of enzymes are named (Omura & Sato 1962). This atypical haem spectrum was shown to result from the axial coordination of a thiolate (cysteinyl) anion to the ferrous iron (Collman & Sorrell 1975; Chang & Dolphin 1975). The thiolate remains coordinated when the natural ligand, dioxygen, binds (Dolphin *et al.* 1980).

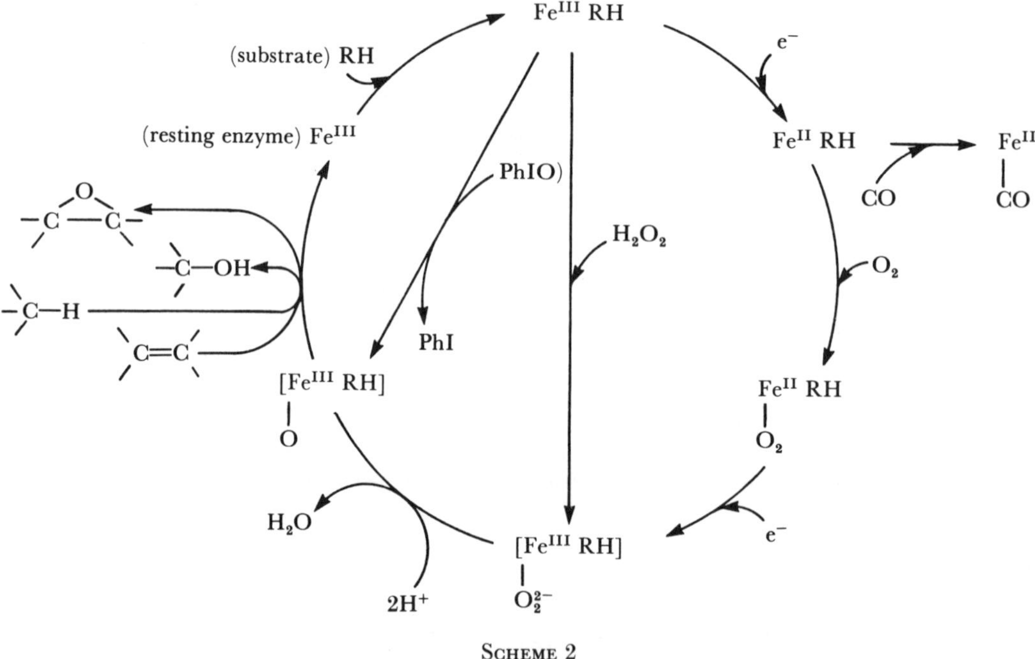

SCHEME 2

The rate-limiting enzymic step is the second one-electron reduction (Coon & White 1980). No enzymatic intermediates have been observed beyond this step and further discussion relates to studies with model iron and ruthenium porphyrins.

The resting ferric enzyme may be shunted by using hydrogen peroxides or peracids, suggesting that the final one-electron reduction product may be a ferric peroxide complex. A one-electron electrochemical reduction of oxygenated ferrous OEP (Welborn *et al.* 1981) gives the η^2-ferric peroxide (McCandlish *et al.* 1980) (**2**, scheme 3) suggesting that the same type of intermediate may occur naturally. Heterolytic cleavage of the O—O bond in **2** (and its enzymic equivalent) would, via protonation or peracid formation (Sligar *et al.* 1980) generate a ferric oxene complex, **3** (scheme 4).

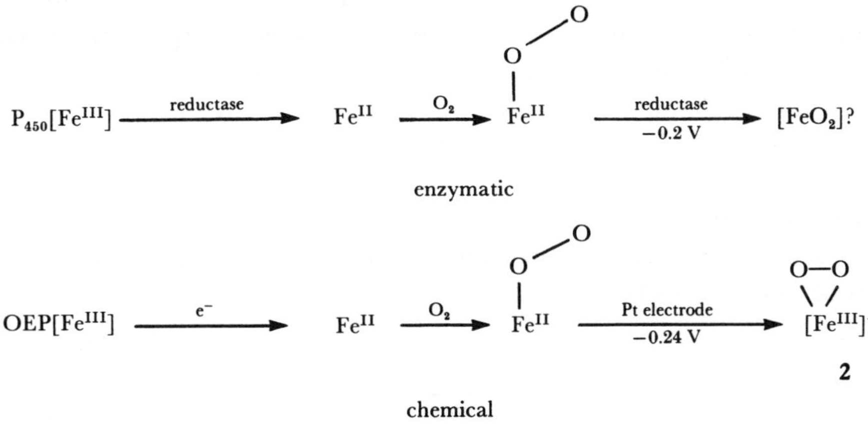

$$P_{450}[Fe^{III}] \xrightarrow{\text{reductase}} Fe^{II} \xrightarrow{O_2} Fe^{II}\text{–O–O} \xrightarrow[-0.2\ V]{\text{reductase}} [FeO_2]?$$

enzymatic

$$OEP[Fe^{III}] \xrightarrow{e^-} Fe^{II} \xrightarrow{O_2} Fe^{II}\text{–O–O} \xrightarrow[-0.24\ V]{\text{Pt electrode}} [Fe^{III}(O\text{—}O)]^- \ \mathbf{2}$$

chemical

SCHEME 3

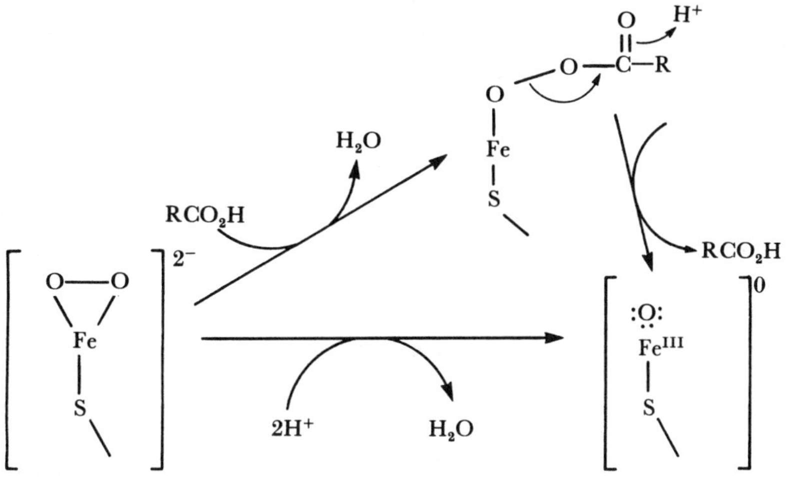

SCHEME 4

If an oxene were to react like a carbene (Kirmse 1964) or nitrene (Lwowski 1970), then direct insertion into a C—H bond or concerted addition across a double bond might be anticipated. Such chemistry, proceeding as it would without formation of intermediates, might seem appropriate for the known P_{450} chemistry. However, the oxene formulation accounts for none of the observed P_{450} chemistry.

Numerous metal-catalysed oxidations using hydrogen peroxide or dioxygen and a reductant are known. Groves and his colleagues (Groves & McClusky 1976) have shown that Fenton-like reactions of ferrous perchlorate and hydrogen peroxide hydroxylated cyclohexanol (Groves & Van der Puy 1976) possess both regio- and stereoselectively, while at the same time exhibiting a pronounced K_H/K_D isotope effect of 3.2. This reaction was described as occurring via an initial hydrogen-atom abstraction by the oxidizing species to generate the organic substrate free radical. Recombination of this radical with a hydroxyl radical, stabilized by coordination to iron, generated the hydroxylated organic product. This stepwise process involving a substrate radical was described as an 'oxygen rebound' mechanism (Groves *et al.* 1978). Deuterium isotope effects and the preference for tertiary hydroxylations (Frommer *et al.* 1970) suggest that direct insertion and carbanionic intermediates were not involved. Similarly, the absence of rearranged products during hydroxylation discounts the intermediacy of any free carbonium ions. Moreover, the generation of charged carbon species in an active site designed to accommodate hydrophobic hosts could be energetically unfavourable.

The elegant studies on the P_{450}-mediated hydroxylation of tetradeuteronorbornane (Groves *et al.* 1978), where retention of deuterium at a hydroxylated carbon (scheme 5) as well as formation of both endo- and exo-alcohols, clearly demonstrate the occurrence of an intermediate in the reaction, and all of the evidence (including an intramolecular K_H/K_D of 11) point to the radical nature of this intermediate.

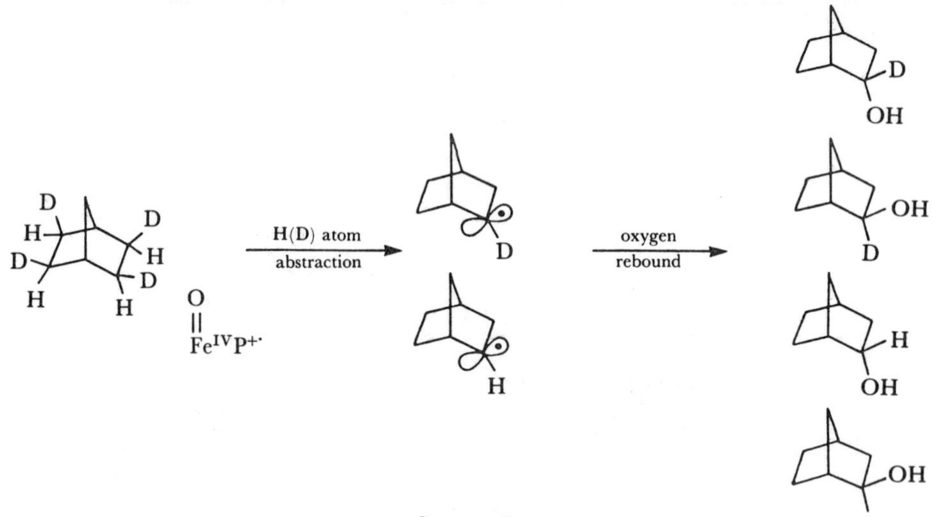

H(D) atom abstraction

oxygen rebound

$O=Fe^{IV}P^{+\cdot}$

Scheme 5

How then can an oxene, generated by the routes described above, account for the observed enzymic chemistry? There are several other resonance structures, perhaps even discrete electronic configurations, that can be written for **3** (scheme 6). One of these, **4**, in which the oxygen contains an octet of electrons, is the $O=Fe^{IV}$ porphyrin π-cation radical, which we

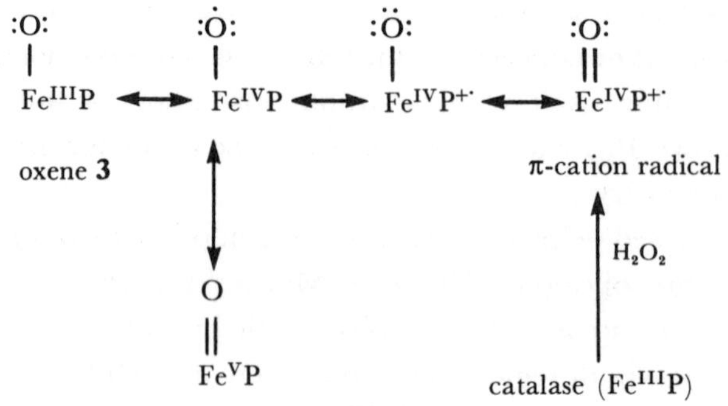

$:O:-Fe^{III}P \longleftrightarrow \dot{O}-Fe^{IV}P \longleftrightarrow \ddot{O}-Fe^{IV}P^{+\cdot} \longleftrightarrow :O:=Fe^{IV}P^{+\cdot}$

oxene **3**

π-cation radical

$O=Fe^{V}P$

H_2O_2

catalase ($Fe^{III}P$)

Scheme 6

have shown is the 'active intermediate' in catalase and peroxidases (Dolphin *et al.* 1971). Groves and his colleagues have found that model iron porphyrins and the oxygen atom transfer reagent iodosylbenzene will hydroxylate aliphatic hydrocarbons in reactions that closely parallel the enzymic ones, including an isotope effect of *ca.* 13 for cyclohexane hydroxylation (Groves & Nemo 1983). At −80 °C, using the sterically 'protected' *meso*-tetramesityl haemin (TMP, **5**) an intermediate has been observed and characterized as the oxoferryl Fe^{IV} porphyrin π-cation radical (Groves *et al.* 1981).

We have taken a parallel approach to the characterization of such high-oxidation-state intermediates by using ruthenium instead of iron porphyrins. The inorganic chemistry of ruthenium is dominated by much slower rates of ligand exchange compared with iron (Basolo & Pearson 1958), and this experience is observed with ruthenium porphyrins (Dolphin *et al.* 1983). While slow ligand exchange may make the preparation of ruthenium complexes more difficult, it should also stabilize products once they are formed. This is so; scheme 7 outlines the chemistry we have observed with RuOEP (Dolphin *et al.* 1983). The most significant observation, within the current context, is the preparation and characterization of **6** (scheme 7), i.e. $O{=}Ru^{IV}(OEP)^{+\cdot}Br^{-}$, which is 'stable' at room temperature and which hydroxylates

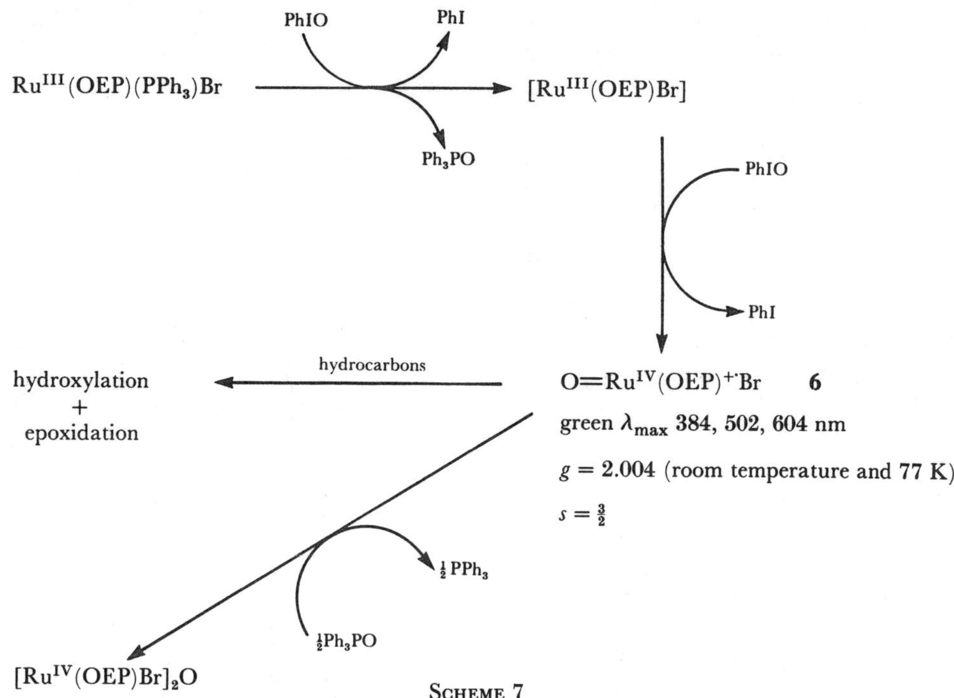

SCHEME 7

hydrocarbons and epoxidises alkenes in manners analogous to the corresponding iron and manganese systems (table 1). As expected, the catalytic activity of ruthenium porphyrins is lower than that of the corresponding iron complexes. Nevertheless Ru(TMP) does exhibit yields and catalytic turnover numbers comparable with the other metalloporphyrins (T. Leung, B. R. James & D. Dolphin 1985, unpublished results). We find, however, that the benzylic methyl groups of TMP are very reactive (M. Camenzind, D. Dolphin & B. R. James 1985, unpublished results). This accounts for the loss of oxidizing power of the iodosylbenzene and destruction of the catalyst. Replacement of the methyl groups of **5** by chlorines to give **7**

TABLE 1. HYDROXYLATION OF CYCLOHEXANE AND EPOXIDATION OF CYCLOHEXENE BY USING METALLOPORPHYRINS AND IODOSOBENZENE; YIELDS ARE BASED ON CONSUMED OXIDANT

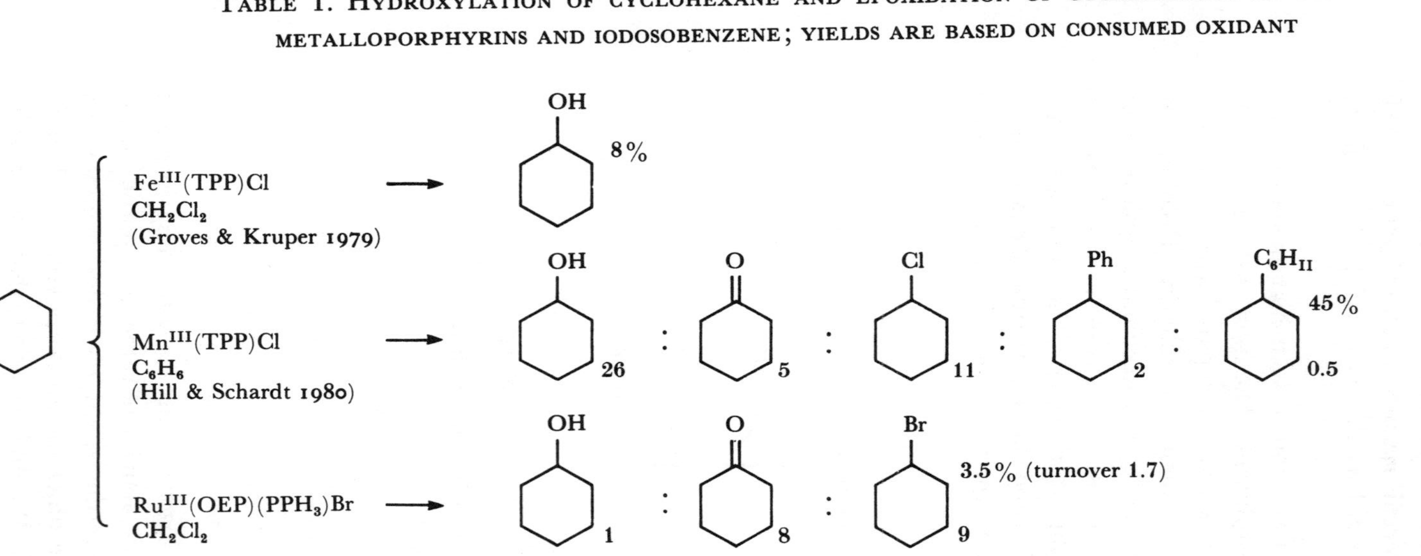

for $Ru^{III}(TMP)(PPh_3)Br$ the turnover was 100 with a 50% yield

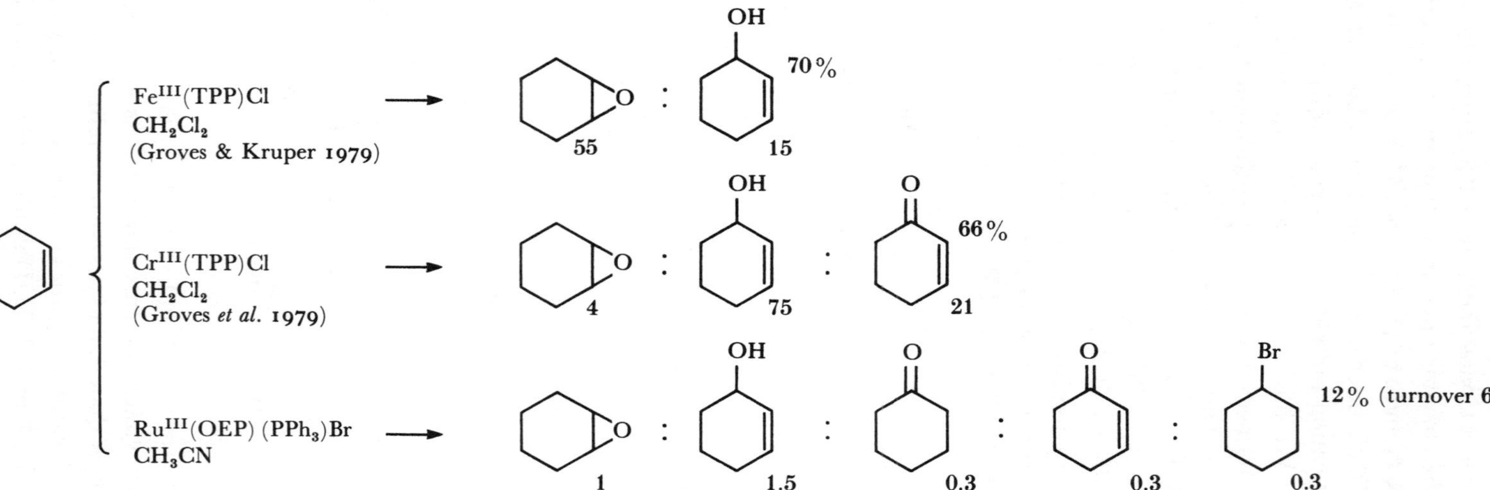

for $Ru^{III}(TMP)(PPh_3)Br$ the turnover was 120 with a 78% yield (scheme 10) dramatically enhances the stability of the catalyst and turnover numbers of more than 10000 have been observed (Traylor *et al.* 1984).

Scheme 8 shows how the oxoferryl π-cation radical formulation and oxygen rebound mechanism can be incorporated into a P_{450} mechanism. Clearly such a mechanism requires that electrons can readily be transferred from porphyrin and metal to ligand. Our studies on

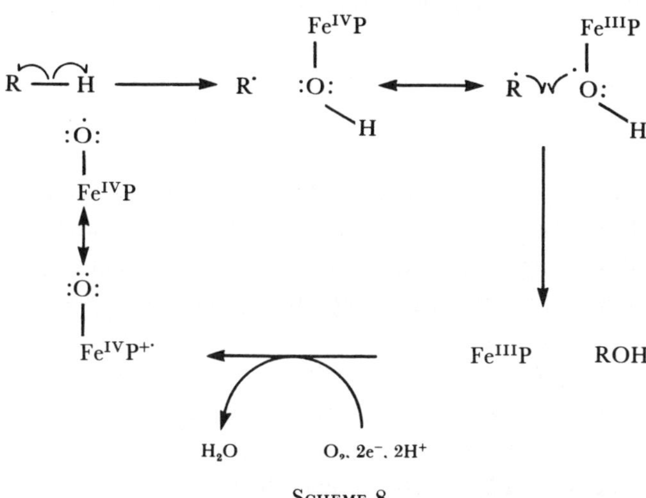

Scheme 8

ruthenium porphyrins show that intramolecular transfer between porphyrin and metal is in fact a facile process which can be controlled by axial ligation (Barley *et al.* 1984). Two examples demonstrate the point. Thus oxidation of a ruthenium(II) porphyrin carbonyl complex **8** generates the corresponding $Ru^{II}CO$ porphyrin π-cation radical **9**. Removal of the CO is then accompanied (Barley *et al.* 1984) by an intramolecular electron rearrangement with the electron moving to the periphery to give the Ru^{III} porphyrin (**10**, scheme 9). Conversely the bis-triphenylarsine complex **11** is stable as the oxidized Ru^{III} porphyrin. Replacement of one of the arsine ligands by CO is accompanied by a reversible migration of an electron from the porphyrin to the metal to give the corresponding porphyrin π-cation radical **12**.

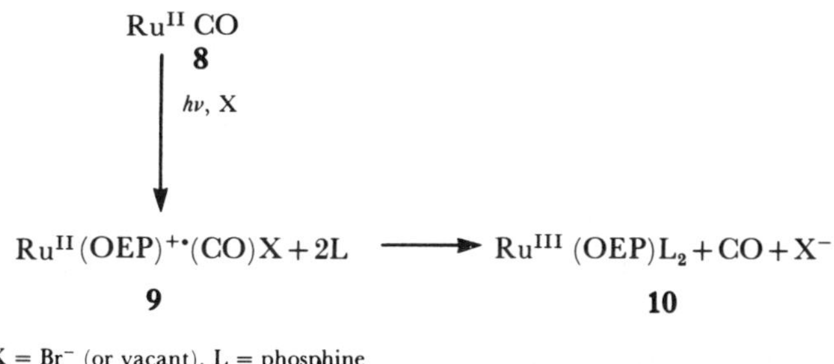

Scheme 9

In addition to hydroxylations, cytochromes P_{450} also effect epoxidations (Groves 1980). Although the epoxide often reflects the stereochemistry of the alkene from which it was derived, additional metabolic products require the intervention of an intermediate. While the principle product from *trans*-1-phenyl-1-butene was the *trans* epoxide, both 1-phenyl-1-(and -2-) butanones were produced (Liebler & Guengerich 1983). These ketones were not further metabolites of the epoxide. Moreover, it has been reported (J. T. Groves, 1985, personal communication), for the specific case of propylene, that deuterium can be regio- and stereospecifically incorporated into the recovered alkene when the enzymic reaction is run in D_2O. Furthermore, deuterium can be lost from deuterated propylene when the reaction is performed in water. Groves has suggested that the reaction proceeds via a metallocycle which generates a metallocarbene upon deprotonation. Be that as it may, the experiment clearly demonstrates the production of an intermediate in these reactions. Of even greater mechanistic and biochemical significance are the observations that both terminal alkenes and alkynes may cause suicide inhibition of cytochromes P_{450} by *N*-alkylation of the prosthetic haem group. The *N*-alkyl group contains the oxygen atom that normally generates the epoxide, but *N*-alkylation does not occur via initial epoxide formation (Ortiz de Montellano *et al.* 1982). We have recently shown that the P_{450} model system using **7** and iodosopentafluorobenzene gives rise to some inactivation and *N*-alkylation (Mashiko *et al.* 1985). Thus 4,4-dimethyl-1-pentene gives principally the corresponding epoxide; however, the 21-(4,4-dimethyl-2-oxidopentyl)-10,15,20-tetrakis (2,6-dichlorophenyl) haemin (**13**) is also slowly formed, from which the corresponding *N*-4,4-dimethyl-2-hydroxypentylporphyrin (**14**, scheme 10) can be isolated in greater than

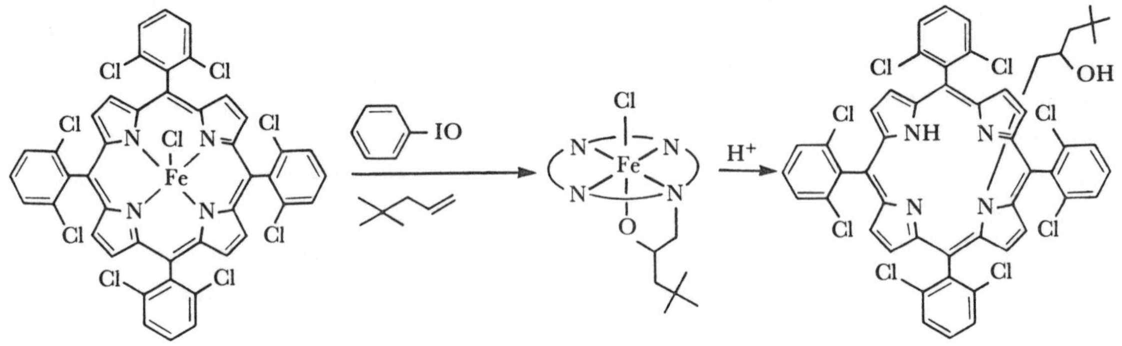

SCHEME 10

60% yield from the original reaction mixture. This observation establishes even more firmly the use of haemin–iodosobenzene systems as effective models for cytochrome P_{450}, while at the same time confirming the suicide inhibition as a basic reaction of these systems which is not substantially influenced by the apoprotein. Though the exact nature of the intermediates involved in olefin (and alkyne) metabolism are not firmly established (Guengerich & Macdonald 1984), an initial electron transfer to generate a cation radical of the substrate seems the most plausible pathway. Such a radical intermediate can account for all of the observed enzymic reactions (as outlined in scheme 11).

We suggest, in conclusion, that those haemproteins oxidized by hydrogen peroxide including the catalases and peroxidases, and those which generate 'peroxide' by a two-electron reduction of dioxygen (cytochromes P_{450} and the cytochrome oxidases (Blair *et al.* 1983)) all function in a common manner. The key reaction that Nature has learned to achieve and control is the

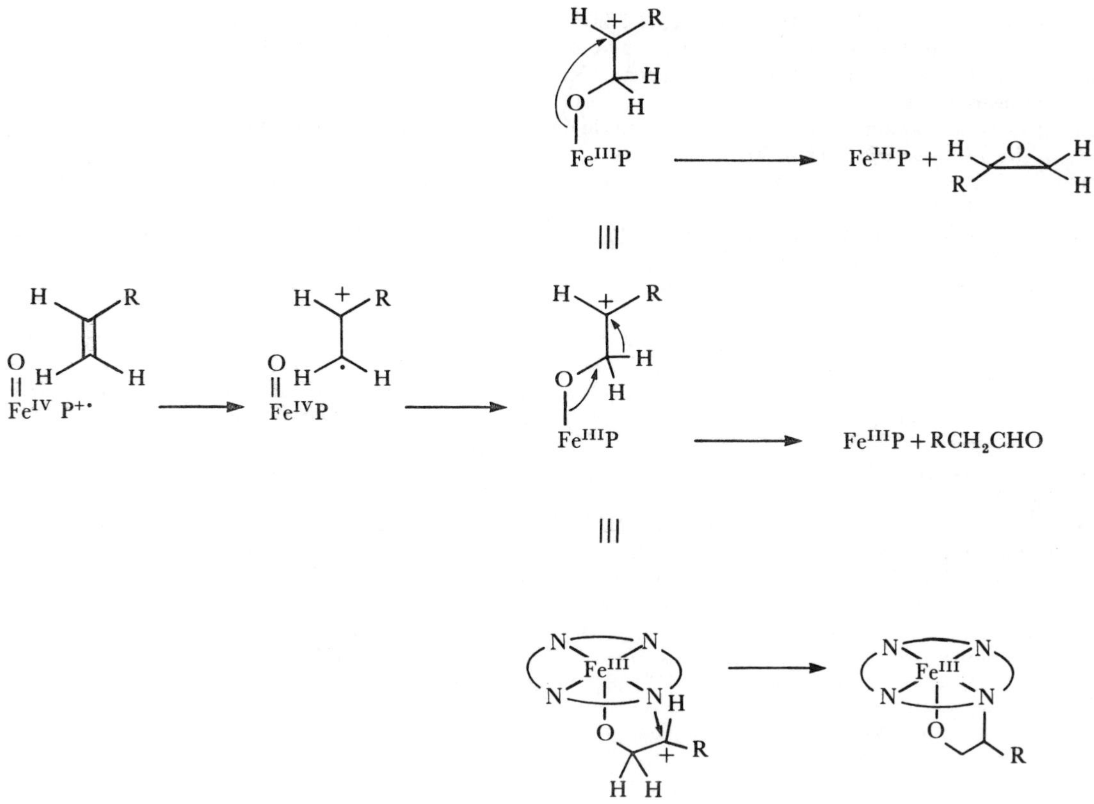

SCHEME 11

heterolytic cleavage of the O—O bond of the coordinated peroxide to give water and a two-electron oxidation product of the resting enzymes, namely the oxo-iron(IV) porphyrin π-cation radical, **1**. The specific chemistry attainable by these high oxidation states is then controlled by the steric and electronic (including the sixth axial ligand) constraints imposed by the protein. For cytochrome P_{450} (cam) the preliminary X-ray crystallographic results (Poulos 1984) suggest how the subsequent generation of substrate free radicals can be achieved and controlled. It is equally clear that any one specific class of these enzymes can occasionally perform the chemistry associated with the other systems. Thus catalase can cause *N*-demethylation (Kadlubar *et al.* 1974), a reaction typical of cytochromes P_{450}, while P_{450} can function as an oxidase and cause the four-electron reduction of dioxygen to water (Gorsky *et al.* 1984).

This is a contribution from the Bioinorganic Group and was supported by the Natural Science and Engineering Research Council of Canada and the U.S. National Institutes of Health (AM 17989 and GM 29198). It is a special pleasure to acknowledge the continued and ongoing collaboration with Professor B. R. James, with whom much of this work was done.

REFERENCES

Barley, M. H., Dolphin, D. & James, B. R. 1984 Reversible intramolecular electron transfer within a ruthenium(III) porphyrin–ruthenium(II) porphyrin π-cation radical system induced by changes in axial ligation. *J. chem. Soc. chem. Commum.*, 1499–1500.

Basolo, F. & Pearson, R. G. 1958 *Mechanisms of inorganic reactions*. New York: John Wiley.

Blair, D. F., Martin, C. T., Gelles, J., Wang, H., Bradvig, G. W., Stevens, T. H. & Chan, S. I. 1983 The metal centers of cytochrome-c oxidase: structures and interactions. *Chem. Scr.* **21**, 43–53.

Boschke, F. L. (ed). 1983 Topics in current biochemistry, vol. 108 (radicals in biochemistry). Berlin: Springer-Verlag.

Chang, C. K. & Dolphin, D. 1975 Ferrous porphyrin–mercaptide complexes. Models for reduced cytochrome P_{450}. *J. Am. chem. Soc.* **97**, 5948–5950.

Collman, J. P. & Sorrell, T. N. 1975 A model for the carbonyl adduct of ferrous cytochrome P_{450}. *J. Am. chem. Soc.* **97**, 4133–4134.

Coon, M. J. & White, R. E. 1980 Cytochrome P_{450}, a versatile catalyst in mono-oxygenation reacti[illegible]. *Metal ion activation of dioxygen* (ed. T. G. Spiro), pp. 73–123. New York: John Wiley and Sons.

Dolphin, D. & Felton, R. H. 1974 The biochemical significance of porphyrin π-cation radicals. *Acct. chem. Res.* **7**, 26–32.

Dolphin, D., Forman, A., Borg, D. C., Fajer, J. & Felton, R. H. 1971 Compounds I of catalase and horse-radish peroxidase: π-cation radicals. *Proc. natn. Acad. Sci. U.S.A.* **68**, 614–618.

Dolphin, D., James, B. R. & Welborn, H. C. 1980 Oxygenation and carbonylation, of a reduced P_{450} (cam) enzyme and derivatives reconstituted with meso-, deutero-, dibromodeutero-, and diacetyldeuteroheme. *J. molec. Catal.* **7**, 201–213.

Dolphin, D., James, B. R. & Leung, T. 1983 Model and enzymatic studies with cytochrome P_{450}. *Inorg. chim. Acta* **79**, 25–27.

Fajer, J., Borg, D. C., Forman, A., Dolphin, D. & Felton, R. H. 1970 π-cation radicals and dications of metalloporphyrins. *J. Am. chem. Soc.* **92**, 3451–3459.

Fajer, J., Borg, D. C., Forman, A., Felton, R. H., Vegh, L. & Dolphin, D. 1973 E.S.R. studies of porphyrin π-cations: the $^2A_{1u}$ and $^2A_{2u}$ states. *Ann. N.Y. Acad. Sci.* **206**, 349–364.

Frommer, U., Ullrich, V. & Staudinger, H. 1970 Hydroxylation of aliphatic compounds by liver microsomes. I. Distribution pattern of isomeric alcohols. *Hoppe-Seyler's Z. physiol. Chem.* **351**, 903–912.

Gorsky, L. D., Koop, D. R. & Coon, M. J. 1984 On the stoichiometry of the oxidase and mono-oxygenase reactions catalyzed by liver microsomal cytochrome P_{450}. *J. biol. Chem.* **259**, 6812–6817.

Groves, J. T. 1980 Mechanisms of metal-catalyzed oxygen insertion. In *Metal ion activation of dioxygen* (ed. T. G. Spiro), pp. 125–162. New York: John Wiley and Sons.

Groves, J. T., Haushalter, R. C., Nakamura, M., Nemo, T. E. & Evans, B. J. 1981 High-valent iron–porphyrin complexes related to peroxidase and cytochrome P_{450}. *J. Am. chem. Soc.* **103**, 2884–2886.

Groves, J. T. & Kruper, W. J. 1979 Preparation and characterization of an oxoporphinatochromium(V) complex. *J. Am. chem. Soc.* **101**, 7613–7615.

Groves, J. T. & McClusky, G. A. 1976 Aliphatic hydroxylation via oxygen rebound oxygen transfer catalyzed by iron. *J. Am. chem. Soc.* **98**, 859–861.

Groves, J. T., McClusky, G. A., White, R. E. & Coon, M. J. 1978 Aliphatic hydroxylation by highly purified liver microsomal cytochrome P_{450}. Evidence for a carbon radical intermediate. *Biochem. biophys. Res. Commun.* **81**, 154.

Groves, J. T. & Nemo, T. E. 1983 Aliphatic hydroxylation catalyzed by iron porphyrin complexes. *J. Am. chem. Soc.* **105**, 6243–6248.

Groves, J. T., Nemo, T. E. & Meyers, R. S. 1979 Hydroxylation and epoxidation catalyzed by iron–porphyrin complexes. Oxygen transfer from iodosylbenzene. *J. Am. chem. Soc.* **101**, 1032–1033.

Groves, J. T. & Van Der Puy, M. 1976 Stereospecific aliphatic hydroxylation by iron–hydrogen peroxide. Evidence for a stepwise process. *J. Am. chem. Soc.* **98**, 5290–5297.

Guengerich, F. P. & Macdonald, T. L. 1984 Chemical mechanisms of catalysis by cytochromes P_{450}: a unified view. *Acct. chem. Res.* **17**, 9–16.

Hager, L. P., Doubek, D. L., Silverstein, R. M., Hargis, J. H. & Martin, J. C. 1972 Chloroperoxidase. IX. Structure of compound I. *J. Am. chem. Soc.* **94**, 4364–4366.

Hanson, L. K., Eaton, W. A., Sligar, S. G., Gunsalus, I. C., Gouterman, M. & Connel, C. R. 1976 Origin of the anomalous Soret spectra of carboxycytochrome P_{450}. *J. Am. chem. Soc.* **98**, 2672–2674.

Hewson, W. D. & Hager, L. P. 1979 Peroxidases, catalases and chloroperoxidases. In *The porphyrins* (ed. D. Dolphin), vol. 6 (biochemistry, part B), pp. 295–332. New York: Academic Press.

Hill, C. L. & Schardt, B. C. 1980 Alkane activation and functionalization under mild conditions by a homogeneous manganese(III) porphyrin–iodosylbenzene oxidizing system. *J. Am. chem. Soc.* **102**, 6374–6375.

Kadlubar, F. F., Morton, K. C. & Ziegler, D. M. 1974 Microsomal catalyzed hydroperoxide-dependent C-oxidation of amines. *Biochem. biophys. Res. Commun.* **54**, 1255–1261.

Kirmse, W. (ed.) 1964 *Carbene chemistry*, New York: Academic Press. Liebler, D. C. & Guengerich, F. P. 1983 Olefin oxidation by cytochrome P_{450}: Evidence for group migration in catalytic intermediates formed with vinylidene chloride and *trans*-1-phenyl-1-butene. *Biochemistry, Wash.* **22**, 5482–5489.

Lwowski, W. (ed). 1970 *Nitrenes*. New York: Interscience.

Mashiko, T., Dolphin, D., Nakano, T. & Traylor, T. G. 1985 *N*-alkylporphyrin formation during the reactions of cytochrome P_{450} model systems. *J. Am. chem. Soc.* **107**, 3735–3736.

McCandlish, E., Miksztal, A. R., Nappa, M., Sprenger, A. Q., Valentine, J. S., Stong, J. D. & Spiro, T. G. 1980 Reactions of superoxide with iron porphyrins in aprotic solvents. A high spin ferric porphyrin peroxo complex. *J. Am. chem. Soc.* **102**, 4268–4271.

Moss, T., Ehrenberg, A. & Bearden, A. J. 1969 Moessbauer spectroscopic evidence for the electronic configuration of iron in horse radish peroxidase and its peroxide derivatives. *Biochemistry, Wash.* **8**, 4159–4162.

Omura, T. & Sato, R. 1962 A new cytochrome in liver microsomes. *J. biol. Chem.* **237**, PC1375.

Ortiz de Montellano, P. R., Kunze, K. L., Beilan, H. S. & Wheeler, C. 1982 Destruction of cytochrome P_{450} by vinyl fluoride, fluoroxene, and acetylene. Evidence for a radical intermediate in olefin oxidation. *Biochemistry, Wash.* **21**, 1331–1339.

Poulos, T. L. 1984 The crystal structure of cytochrome P_{450} (cam). Paper presented at the 36th Southeastern Regional Meeting of the A.C.S., Raleigh, N.C.

Roberts, J. E., Hoffman, B. M., Rutter, R. & Hager, L. P. 1981 ^{17}O ENDOR of horseradish peroxidase compound I. *J. Am. chem. Soc.* **103**, 7654–7656.

Rutter, R., Valentine, M., Hendrick, M. D., Hager, L. D. & Debrunner, P. 1983 Chemical nature of the porphyrin π-cation radical in horseradish peroxidase compound I. *Biochemistry, Wash.* **22**, 4769–4774.

Schultz, C. E., Devang, P. W., Winkler, H., Debrunner, P. G., Doan, N., Chiang, R., Rutter, R. & Hager, L. P. 1979 Horseradish peroxidase compound I: Evidence for spin coupling between the haem iron and a 'free' radical. *FEBS Lett.* **103**, 102–105.

Sligar, S. G., Kennedy, K. A. & Pearson, D. C. 1980 Chemical mechanisms for cytochrome P_{450} hydroxylation: evidence for acylation of haem-bound dioxygen. *Proc. natn. Acad. Sci. U.S.A.* **77**, 1240–1244.

Traylor, P. S., Dolphin, D. & Traylor, T. G. 1984 Sterically protected haemins with electronegative substituents: Efficient catalysts for hydroxylation and epoxidation. *J. chem. Soc. chem. Commun.*, 279–280.

Welborn, H. C., Dolphin, D. & James, B. R. 1981 One-electron electrochemical reduction of a ferrous porphyrin dioxygen complex. *J. Am. chem. Soc.* **103**, 2869–2871.

Phil. Trans. R. Soc. Lond. B **311**, 593–603 (1985)
Printed in Great Britain

The necessary and the desirable production of radicals in biology

By R. J. P. Williams, F.R.S.

Inorganic Chemistry Laboratory, South Parks Road, Oxford OX1 3QR, U.K.

The production of radicals within biological systems can be necessary, as in the coupling of light capture or oxygen–hydride reactions to the generation of proton gradients in thylakoid and mitochondrial membranes; or it can be extremely desirable, to produce an effective catalyst centre in, for example, ribonucleotide reductases. To fit their function the radicals are usually of controlled reactivity, either through their redox properties or through confinement to a limited volume. The more indiscriminate generation of radicals can be of great use too in protective devices, for example, in the production of cross-linked polymers, or in the generation of oxygen radicals. The different distribution and release of initiating enzymes for all these radical processes ensures the spatial and temporal control of their reactions.

Introduction

This article is written to correct an impression that is growing in the biochemical community. Free radicals in biological systems are said to be dangerous and associated almost exclusively with the onset of diseased states such as cancer. To give a more balanced picture I shall deliberately look for those situations in which free radicals are essential for a particular biological system. There are two points of departure. First, free radicals are not necessarily extremely reactive. Oxygen itself, superoxide and flavin radicals *per se* are relatively stable and have either low thermodynamic reactivity or have considerable chemical kinetic barriers to reaction. The chemistry of free radicals is open to the same controls as that of non-free-radical chemistry. Secondly, even highly reactive free radicals can be generated in physically restricted parts of a biological system, where they may be extremely valuable. The spatial restrictions are dependent on the sites of production and thereafter on the prevention or limitation of diffusion of the radicals. By placing enzymes in special compartments, for example, extracellular rather than intracellular, by confining radical production to lipid bilayers through solubility restrictions, or by using radical intermediates in closed forms of hinged enzymes, the much discussed dangers inherent in radical reaction with DNA can be minimized and radicals can then be put to very effective use. If radicals had been so dangerous, surely during the process of evolution they would have been avoided; in fact they are used by all cells. The risks have always to be measured against the advantages in evolution, or elsewhere.

The redox potentials and reactivity of biological radicals

The simplest control over radical reactivity rests in the redox potential. Radicals in organic molecules are stabilized by conjugation so that in the reversible steps

$$X \rightleftharpoons X^{+\cdot} \rightleftharpoons X^{2+}$$

the radical may not be of such a high energy that disproportionation is grossly favoured. We see this in flavin and quinone chemistry where the two one-electron redox potentials of (1) in table 1 are not greatly different. The reasons for the stability of such radicals can be discussed in terms of the nature of their π-electron systems, and radicals stabilized by such orbitals usually have relatively low redox potentials. On the whole, radicals in σ-antibonding or non-bonding (lone-pair) orbitals are much less stable, for example, in $RCH_2^{\cdot}$. Such radicals are inherently dangerous if they can diffuse freely. They usually have high redox potentials.

Table 1. Redox potentials of some radicals[a]

oxidizing reagent	successive one-electron redox potentials (volts)	
O_2	-0.2 (O_2^-),	-0.6 ($OH^{\cdot}$), then $+2.0$ (H_2O)
		$+0.8$ (H_2O_2), then $+1.2$ (H_2O)
quinone	-0.2 ($Q^{\cdot}$),	$+0.3$ (QH_2)
flavin$^+$	-0.4 ($F^{\cdot}$),	$+0.2$ (FH)
$[Fe^{IV}O]^{2+}$	*ca.* $+1.00$ (Fe^{III}),	-0.3 (Fe^{II}) in peroxidases
$[Mo^{VI}O_2]^{2+}$	-0.3 (Mo^{V}),	-0.3 (Mo^{IV})
dehydro ascorbate	-0.2 ($A^{\cdot}$),	$+0.3$ (AH_2)
disulphides (glutathione)	-0.3 ($RS^{\cdot}$),	$+0.1$ (RSH)
simple phenol radicals		*ca.* $+1.00$ (phenol)
simple indole radicals		*ca.* $+1.00$ (indole)

[a] The potentials are the one-electron redox potentials required to go from the species on the left to the species on the right in each case, and a positive sign indicates a powerful oxidizing action. Measured at pH 7 and 25 °C. The redox potentials of all the coenzymes are adjusted within their enzymes to meet requirements of specific reactions.

Inorganic free radicals are not so readily described because we must distinguish light-atom free radicals, such as $NO^{\cdot}$, from metal-ion radicals. It is not conventional to call Cu^{II}, Fe^{III}, Mn^{II} or Mo^{V} free radicals although they do have unpaired electrons in d-shells. This is unfortunate in the context of this article, because much of the organic free-radical production in biology starts from reactions of these (free-radical) transition metal ions. Metal ions can be made into extremely reactive free radicals by exposing or elevating the energies of the unpaired d-electrons. Thus low-spin Fe^{III}, a 3d free radical, is often very reactive as compared with the high-spin Fe^{III} ion. Passing to elements of the second or third transition metal series the unpaired electrons become relatively more reactive as, for example, in Mo^{V}. We know that this is not a matter of redox potential but of the *exposure* of the electron, which in all cases is in screened d-orbitals, but is more exposed the higher the d-shell number. The extreme case of the hiding of unpaired electrons is shown in the 4f-core of the lanthanide cations, which do not behave as free radicals. Biological molecules can hide organic free electrons within complex molecules, proteins, as successfully as this hiding of free d-electrons in the core of metal ions. We see that it is the mixture of thermodynamic stability, local kinetic reactivity, and physical protection that allows both one-electron reactions of inorganic and organic chemistry to be controlled. Biology has learned to combine the inorganic and organic free-radical chemistry within such control structures to make extremely valuable devices. Sir Derek Barton (this symposium) introduces many of these ideas with respect to the not-so-obvious case of the B-class elements of the Periodic Table.

Radicals in extracellular space

When remote from the cell nucleus a radical is no more dangerous than a nucleophile or an electrophile. It is just a potential intermediate in a reaction. Some reactions need radicals and in such cases biology must use them. A clear example is free-radical polymerization, which gives a plastic. Thin layers of plastics are common devices in the surface structures of plants and animals, especially insects, and they are found in the skin of mammals in the form of melanin pigments. The initial radical polymerization is induced in response to growth hormones, as in insects, or in response to damage by sunlight or physical injury as when a potato or apple is cut. The reactions use a vesicular store of phenols or indoles related to tyrosine and tryptophan as the source of monomer. The exposure of these aromatic molecules to one-electron oxidation by oxygen or hydrogen peroxide in association with metallo-enzymes is the initiation step, and because many of these molecules are bifunctional they allow cross-linking as well as engaging in simple chain propagation. A firm plastic material such as insect cuticle, is generated. Thus the skeleton of insects and of many other animals, for example, the crab, is thus effectively a result of controlled free-radical reactions rather than mineralization. Many pigments are made this way too. We can look at the controls which make this possible.

The oxidizing agents O_2 and H_2O_2 diffuse freely across membrane barriers, but both reagents need metal catalysts to induce reaction. It is then the positioning of these catalysts and organic molecules that is the essential feature of radical production. The enzymes (metallo-enzymes) and the organic monomer substrates must be placed in separate compartments, only to be brought together when required. Some of these compartments are known (table 2). As far as the catalysts are concerned, most are copper proteins that are confined to extracellular fluids by their synthesis. It is a feature of biology that copper is very rarely found intracellularly.

Table 2. Examples of compartments of free-radical generation and reaction

reagent	catalyst
O_2 freely diffusing	extracellular phenol oxidases (Cu)
	extracellular ascorbate oxidase (Cu)
	extracellular caeruloplasmin (Cu)
H_2O_2 freely diffusing	peroxisomes, peroxidases (Fe)
light	components of melanin granule
oxygen and the superoxide	phagosomes (flavin enzymes)

(Superoxide dismutases are a very special case and are very well designed enzymes for a very particular purpose in the cytoplasm of cells as we shall see.) The monomers are then released into this extracellular space. The extracellar cross-linking activity extends from phenol oxidases and laccases to lysine oxidases and is important in cross-linking collagens for the production of bone as well as for the synthesis of radical polymers (Williams 1982).

Another type of catalyst is the iron enzymes, which are rarely if ever found free in extracellular fluids. Most iron enzymes are intracellular but the ones of interest here are held in vesicles, for example, peroxidases in peroxisomes and are only released from the vesicles in response to cell damage. The monomers of the reaction are stored in such bodies as melanocytes quite separately from these peroxidases. Once again, it is the release from special compartments into a particular common space that makes for radical polymerization in a controlled way.

There is another possible use for free radicals in extracellular space, and that is in the scavenging of other free radicals that have escaped or are potentially dangerous. A dangerous free radical in biology, given that O_2 is available, is Fe^{II}, the simple ferrous ion, because together iron and oxygen may generate $OH^{\cdot}$ radicals from oxygen reactions. Ferrous iron must be removed. It has often been suggested that the role of the enzyme caeruloplasmin in blood is for just this purpose: to remove Fe^{II} by converting it to Fe^{III}, which is virtually completely bound to transferrin and does not then undergo redox reactions. Another scavenger system is the ascorbate radical which is generated through the ascorbate oxidase reaction of ascorbate. This radical is relatively inert (it is a π free radical), but can react with other, more dangerous, radicals to give disproportionated products. Thus it removes other free radicals, mainly organic in character. The role of ascorbate oxidase could be to maintain a pool of the ascorbate free radical (note the small difference in the ascorbate redox potentials in table 1), which means that the standing potential of ascorbate radical is high. Note also that it is not an aggressive radical kinetically. Both caeroloplasmin and ascorbic acid oxidases are copper proteins, and it would appear again that copper enzymes generally have a special role outside cells, controlling through oxidation a variety of special free radicals and their reactions (Williams 1982).

Free radicals in vesicles and organelles

Higher cells have membrane-enclosed compartments which are not to be confused with the cytoplasm. They are of two kinds: (i) organelles related to primitive prokaryotes; and (ii) vesicles that contain no DNA or RNA. The membranes of both these compartments often have inwardly directed pumps for both proteins and small molecules. It is especially interesting that these pumps remove iron and manganese cations from the cell cytoplasm. Thus the cytoplasm is low in the free radicals Mn^{2+} and Fe^{2+}, presumably to protect eukaryotic DNA from mutation. The DNA of prokaryotes and of mitochondria and chloroplasts can readily allow a higher mutation rate, and the inside of these cells and organelles is relatively unprotected. The pumping of certain metals and certain proteins into these and other organelles generates in them special metallo-enzymes, such as the manganese and iron superoxide dismutases, aconitase, the manganese O_2-generating enzymes (see, for example, Bannister 1982). The other compartments, empty of DNA or RNA, also have Mn^{2+} pumped into them, but they are, in addition, the sites of storage of several haem iron enzymes, for example, catalase in peroxisomes as mentioned above. The protection of the cytoplasm of prokaryotes is therefore from accidental reactions of manganese, iron and copper (all the common free-radical metal ions), but eukaryotes are not so protected. Very different devices are used to remove the metal ions including ion, small chelate-molecule, and protein pumps.

Finally, there is the generation of superoxide and other oxygen radicals in phagosomes, which is part of the general protective activity of leucocytes. Here the reaction is the leakage of reducing equivalents from NADH to flavins and then to oxygen, across a vesicle membrane such that the anion superoxide is trapped inside the vesicle. Its destructive value is then harnessed.

Radicals in membranes

The most obvious parts of a biological system which use free radicals are those concerned with energy capture. Treating the capture of energy from light, there is an initial step, after the transfer of absorbed energy to a reaction centre, which can be written

$$h\nu + P \rightarrow P^* \rightarrow P^{+\cdot} + (A)^{-\cdot}.$$

Here, P^* is the excited state of a special reaction-centre complex of two chlorophyll molecules, and $P^{+\cdot}$ is a radical cation at this centre after the escape of an electron to a series of acceptors, which become $A^{-\cdot}$ and which may include pheophytins before quinones. The radicals of the quinones have been recognized by their e.p.r. spectra (Rich 1984), and the structure of the reaction complex from certain unicellular organisms is known. Reducing electrons are then given to the $P^{+\cdot}$ state from more remote donors that are usually iron-containing, i.e. free-radical containing proteins; although there is a suggestion that an organic pigment, Z, is involved too. In effect, a free-radical gradient of one-electron redox potential and a gradient of charge is created within the light-harvesting reaction centre and across the containing membrane. Both organic and inorganic centres trapped in membrane proteins are used. These initial reactions are extremely rapid (10^{-12} to 10^{-9} s), but the subsequent transfer of electrons to other acceptors or from donors back to the original centre (all in one-electron steps):

$$\text{donors} \underset{e}{\overset{(2)}{\rightarrow}} \text{P} \underset{e}{\overset{(1)}{\rightarrow}} \text{acceptors};$$

and where repeated excitation and steps (1) are followed successively by steps (2), all occur in the millisecond time range. While the first molecules involved in the radical reactions are bound, the molecules of Q in later steps are part of a membrane-contained pool in which they diffuse laterally. It is here that the peculiar protective nature of the medium is so valuable because such radicals as flavin$^{\cdot}$ and quinone$^{\cdot}$ or, for that matter, chlorophyll$^{\cdot}$, are normally readily attacked by oxygen. Inside a protein, especially a metalloprotein, this protection may be managed by steric constraints, but the differing entities in the membrane are at least somewhat exposed. However, in the lipid media it is probable that reaction with oxygen is not so thermodynamically favourable because the product O_2^-, an anion, is not lipid soluble. In fact the radicals of the electron-transfer chain are observed to be relatively stable, reacting only with selected partners within the membrane: a necessity if efficiency of energy capture is to be attained. The very hydrophobic nature of the molecules, such as the Q_{10} quinones, reflects the fact that they are forced to remain in membranes. The switch in the electron-transfer chain from metal to metal (free-radical) electron hops to organic free radicals is also essential to couple the chemistry of electrons to that of protons. Of course it is these dark reactions which lead from the initial charge separation to the generation of proton gradients across and in membranes, because it is the change in redox states of the organic acceptors and donors, unlike those of metal complexes, that lead to cyclic changes in pK_a values on oxidation–reduction:

$$(\text{H-donor}) \rightarrow \text{H}^+ + \text{donor}^{\cdot} + \text{e};$$

or, on reduction:

$$\text{H}^+ + \text{acceptor} + \text{e} \rightarrow (\text{H-acceptor})^{\cdot}.$$

Ultimately these donors and acceptors are involved in two-electron redox reactions and both the oxidation and reduction of the radicals may involve protonation–deprotonation. We must be concerned at this meeting with the coupled acid–base chemistry of radical species (see, for example Rich (1984)) apart from their redox potentials and their disposition in space, because these are seen to be an essential part of energy-capture devices. The dark reactions that occur during light capture also occur in the mitochondrial oxidative reactions: energy capture using O_2–organic molecule reactions. In fact the pathways of redox equivalents, electrons and protons, in mitochondria, chloroplast and bacterial membranes (dark reactions) are remarkably similar. They all have the same functional significance in that they produce initial charge gradients (metals), free-radical gradients, proton gradients, and then ATP or some other chemically trapped energy form (Mitchell 1968; Williams 1961).

Not all radicals or oxidized molecules in membranes are without hazard and particular risk appears to be associated with organic peroxides which cannot be removed by either catalase or the glutathione-linked systems of the cytoplasm, because these control systems are not lipid soluble. In fact they are removed by vitamin E which is only soluble in lipid phases, and is capable of forming rather innocuous free radicals itself from dangerously reactive ones, just as ascorbate does outside cells and gluthathione may do inside cells.

Radicals in the cytoplasm

A large part of the discussion of free radicals in biology is devoted to the superoxide anion (see other contributions to this symposium). The scavenger enzymes are the Cu–Zn proteins of the eukaryote cytoplasm and the Fe and Mn proteins of prokaryotes, mitochondria and chloroplasts. Elsewhere I have discussed the peculiarities and reasons for this different distribution of enzymes which perform apparently identical tasks but contain different metal ions. The overriding point is the danger from free cytoplasmic Fe^{II} and Mn^{II} to eukaryotes (Williams 1982). I shall not describe superoxide reactions further here. I turn instead to the reactions more often associated with hydrogen peroxide – the intracellular peroxidases – which may well scavenge radicals of a different kind. It must be remembered throughout this section that the production of a lethal mixture of H_2O_2 and O_2^- is an accepted risk, so that its protective value overrides its danger so long as it is produced in the right place. The escaping radicals and peroxides have, however, to be eliminated.

Peroxides are handled by haem enzymes (usually in vesicles), glutathione peroxidase (in the cytoplasm of most cells) and vitamin E (in membranes). As with the multiplicity of superoxide handling proteins there are reasons for the different locations and chemical components of the three classes of reactions (see table 3). The haem enzymes only handle peroxides, mainly H_2O_2,

Table 3. Enzymes destructive to superoxide or peroxides

enzyme	reaction	metal or heavy atom
superoxide dismutase	$O_2^- \rightarrow H_2O_2 + O_2$	Cu(Zn), Fe, Mn
catalase	$H_2O_2 \rightarrow O_2 + H_2O$	Fe(haem)
peroxidase		
phenols	$H_2O_2 + RH \rightarrow ROH + H_2O_2$	Fe(haem)
glutathione	$H_2O_2 + RH \rightarrow ROH + H_2O$ (2RSH → RSSR)	Se
vitamin E (in membranes)	$RO_2H \rightarrow ROH + H_2O$	none

in free aqueous solution in vesicles, and the reaction follows an enzyme one-electron pathway which would have clear dangers in the cytoplasm. The selenium-containing glutathione peroxidases generally react via an atom-transfer route without giving radicals and can handle all peroxides, even tertiary substituted peroxides, that are soluble in water. They are particularly valuable in the cytoplasm. Neither of these two water-soluble enzymes can control free radicals or peroxides that are organic soluble, so a third, non-enzymic, system is required: vitamin E. The account does not finish here, however, because there is a further protective device in the cytoplasm which links to the glutathione peroxidase. This is the glutathione couple itself, which is the equivalent of a very mild peroxide couple in which oxygen atoms are replaced by sulphur:

$$R_2O_2 \rightarrow 2ROH \quad \text{and} \quad R_2S_2 \rightarrow 2RSH.$$

The sulphur couple is of much lower redox potential (see table 1). In addition, the sulphur reaction is readily reversible in biology, so the glutathione couple acts as a redox buffer (Flohe 1979). It has the advantage that it can pick up dangerous oxidizing or reducing free radicals while its own radical is not very reactive, so purging the cell cytoplasm. The interesting chemical feature of all these systems is that as we descend Group VI of the Periodic Table from O to S to Se (peroxides, glutathione, glutathione peroxidase), so the chemistry becomes milder (table 1). Organic chemicals open to attack by hydrogen peroxide and radicals are protected by the rapid conversion to the relatively innocuous oxidized derivatives of –SH and –SeH (or to conjugated free radicals such as vitamin E and ascorbate). These two centres are such good scavengers that they, like ascorbate outside cells and vitamin E compounds within membranes, are not dangerous even though their reactive atoms are very exposed. Without free metal ions they do not activate O_2 reduction. The contrast here is with metals (free-radical metals), which can be used to scavenge only when hidden (as for O_2^- or H_2O_2 in catalase and superoxide dismutases), because they cannot resist O_2 reactions and would activate O_2 reduction powerfully if exposed to it.

Radicals in special enzymes

With hindsight it is possible to see that some reactions are extremely difficult to effect in two-electron steps. A typical example is the attack on a carbon atom in a fully saturated chain, $-CH_2-$. Probably the easiest route for this reaction is that of H-atom removal with homolytic radical bond break. Table 4 notes some of the cases which appear to require this pathway and

Table 4. Some radical intermediates in enzymes

enzymic reaction	radicals used
diol rearrangements	Co^{II} B_{12} and adenosine
ribonucleotide reduction	Co^{II} B_{12} and adenosine
	Fe^{III} dimer and tyrosine
haem oxygenases	FeO and porphyrin
cytochrome P_{450}	FeO and $CH^{\bullet}$(substrate)
cytochrome *c* peroxidation	FeO and tryptophan radical or Fe^{III} haem
methanogenesis	Ni^{II}
pyruvate decarboxylation	thiamin and ferredoxin

also lists some of the associated catalysts. The use of metal ions is notable, but it is equally clear that a particular metal ion has been selected for a particular reaction. Cobalt is used in many rearrangements that involve vitamin B_{12}; haem iron is used in many high-potential reactions and polynuclear iron, $Fe_n S_n$, and nickel are used in low-potential reactions. The choice of metal is interesting and must relate to the special nature of these metal ions opposite special organic reaction steps.

Cobalt can be made to yield a very reactive radical with a 3d unpaired electron that is very exposed. This situation is created in low-spin cobalt (II) in a square pyramid five-coordinate geometry. The single unpaired electron is forced into the σ-orbital, d_{z^2} and protrudes from the plane of the complex. This radical, though exposed, is not thermodynamically aggressive and several studies showed it to have a low redox potential. Another interesting feature of the radical is that it is generated in the enzymes, together with a $R-CH_2^{\bullet}$ radical, only when a substrate is bound (Cockle *et al.* 1972; Finke *et al.* 1984). There is a preferred order of reaction: inert system–substrate binding–activated system (radicals). The unavailability of the cobalt at first prevents attack of oxygen on the cobalt(II). In other words, when the radicals are formed they are highly restricted in their access to reagents within a locked-off enzyme–substrate complex.

One of the reactions which B_{12} enzymes can participate in is ribonucleotide reduction to deoxyribonucleotide. In some organisms this reduction is performed by an alternative (FeOFe)-dependent enzyme. Here the necessary radicals are observed to be present in the enzyme before substrate addition. Apart from the (FeOFe) centre, which contains ten unpaired electrons, there is a tyrosine free radical that is not attacked by oxygen (Sjoberg & Graslund 1978). This remarkable system is not understood. A tyrosine free radical is of high potential and is very reactive, and we must ask how and why it was formed in the enzyme and how it is prevented from reacting indiscriminately.

A rather different scheme can be given for another high redox potential radical: the $Fe^{IV}O$ and the formal $Fe^{V}O$ species that occur in several haem-containing oxygen or hydrogen peroxide using enzymes. The reactions of the enzymes range from phenol or tryptophan oxidation (peroxidases) to hydroxylation of a range of very inert hydrocarbons (P_{450} cytochromes). The initial state of the '$Fe^{V}O$' is really the complex $Fe^{IV}O$ bound to a porphyrin free radical, *or* an $Fe^{IV}O$ complex with an attendant tryptophan free radical, *or* an $Fe^{IV}O$ complex with a second haem unit (free radical), which undergoes only one-electron oxidation (Dolphin & Fenton 1974). These centres undergo either step-by-step one-electron reduction from a substrate *ca.* 1 nm away (peroxidase; Burns *et al.* 1975), or hydrogen-atom uptake and OH loss in an overall two-electron (O-transfer) reaction of the kind $FeO+RH\rightarrow FeOH+R^{\cdot}\rightarrow ROH$. The last reaction requires the RH group to lie immediately on top of the FeO unit. The value of hydroxylation is not just in the destruction of damaging molecules, but in the synthesis of hydroxylated species such as tyrosine and the prostaglandins.

Once again the radicals in these steps, both inorganic and organic, are hidden throughout the reaction cycle. In the case of cytochrome P_{450} the iron atom is not available (low-spin, bound axially to two protein ligands; compare Co^{III} in vitamin B_{12}) until the substrate binds, converting it to an open-sided high-spin iron. The FeO_2 and FeO complexes that are formed subsequently then see the bound substrate, RH, but not any side chain of the protein or any other molecules in solution when they cannot attack the wrong groups. The reaction proceeds in an isolated pocket until the products are generated and it is of great interest that the Fe=O haem unit is activated by a coordinated thiolate (Hill *et al.* 1970). A final example concerns nitrite or sulphite reductase, where the haem iron binds a unit NO_2^- or SO_3^{2-}, which is open to six-electron reduction. The one-electron (radical) steps are assisted by a bound Fe–S centre that, with the haem, makes a two-one electron unit (E. Munck 1984, personal communication).

In table 4 we have included a recent example of a radical pathway in an isolated reaction site which has been uncovered by Kerschev *et al.* (1982). The reaction is that of pyruvate oxidoreduction (figure 1), in which oxidative decarboxylation requires the presence of ferredoxins and the thiamin diphosphate free radical. The ability of enzymes to generate such isolated reaction or carrier pockets is here an elaboration of the haemoglobin story, but is of course common to many acid–base centres such as those of kinase, ATP-synthetase and so on. This parallel stresses the related ways in which enzymes protect reactions, whether they go by radical or non-radical paths.

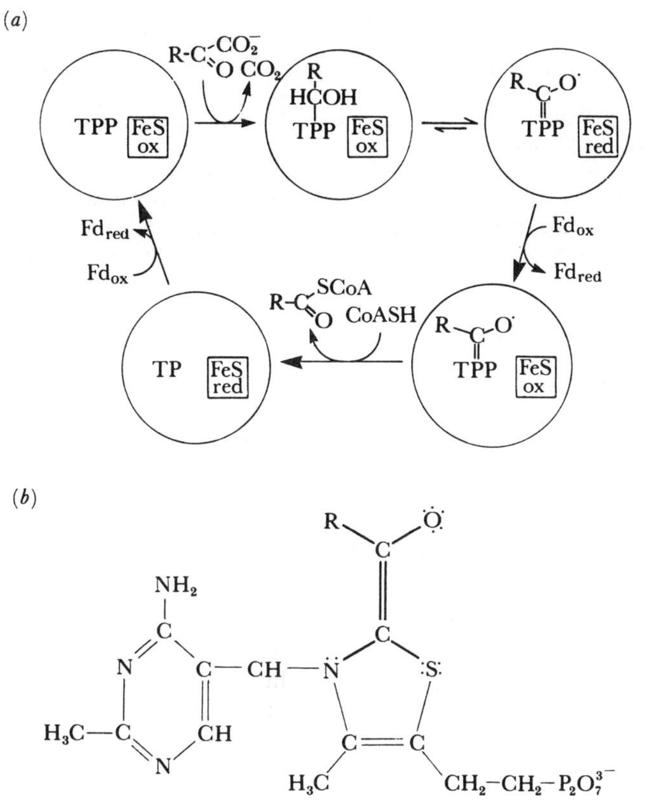

(a)

(b)

FIGURE 1. (a) The proposed reaction cycle of pyruvate oxidoreductase and (b) the thiamin radical involved (after Kerscher & Osterhelt 1982).

CONCLUSIONS

Biological systems have many essential or required ways of using free radicals which include:
(i) mechanistically required radical intermediates in different enzymes;
(ii) radicals for free-radical polymerization;
(iii) radicals for energy capture and transduction;
(iv) aggressive radicals for the destruction of invading organisms;
(v) radicals for removing more dangerous radicals.

These reactions carry risks, so all parts of biology need protection from escaping free radicals, as will be discussed in other contributions to this symposium. However, I trust that it is clear that radicals are extremely valuable intermediates in reactions and that their use can be very well controlled. Some free radicals are dangerous, but the whole group of reactions should not be classified by the errant few.

REFERENCES

Bannister, J. V. 1980 *Chemical and biochemical aspects of superoxide dismutase.* North Holland, New York: Elsevier.

Burns, P. S., Williams, R. J. P. & Wright, P. E. 1975 Conformational studies of peroxidase–substrate complexes. *J. chem. Soc. chem. Commun.*, pp. 758–759.

Cockle, S. A., Hill, H. A. O., Williams, J. P., Davies, S. P. & Foster, M. A. 1972 The detection of intermediates during a B_{12}-dependent enzyme reaction. *J. Am. chem. Soc.* **94**, 275–277.

Dolphin, D. & Fenton, R. H. 1974 The biochemical significance of porphyrin π-cation radicals. *Acct. chem. Res.* **7**, 26–36.

Finke, R. G., Schiraldi, D. A. & Mayer, B. J. 1984 Towards the unification of coenzyme B_{12}-dependent diol dehydratase studies: the bound radical mechanism. *Coord. Chem. Rev.* **54**, 1–22.

Flohe, L. 1979 Glutathione peroxidase: fact and fiction. *Ciba Foundation Symp.* **65**, pp. 95–112. Amsterdam: Excerpta Medica.

Hill, H. A. O., Roder, A. & Williams, R. J. P. 1970 Cytochrome P_{450}. Suggestions to the structure and mode of action. *Naturwissenschaften* **57**, 69–78.

Kerschev, L., Nowitzki, S. & Oesterholt, D. 1982 Thermoacidophilic Archaebacteria contain bacterial type ferredoxin acting as electron acceptors of 2-oxoacid-ferredoxin oxidoreductases. *Eur. J. Biochem.* **128**, 223–230.

Kerschev, L. & Oesterholt, D. 1982 Pyruvate: ferredoxin oxidoreduction. *Trends biol. Sci.* **7**, 371–374.

Mitchell, P. 1968 *Chemiosmotic coupling and energy transduction.* Bodmin Glynn Research, Bodmin, England.

Rich, P. R. 1984 Electron and proton transfer through quinones and cytochrome bc complexes. *Biochim. biophys. Acta* **768**, 53–79.

Sjoberg, B.-M. & Graslund, A. 1978 The free radical in ribonucleotide reductase from *E. coli. Ciba Foundation Symp.* *60*, pp. 187–193.

Williams, R. J. P. 1961 The function of chains of catalysts. *J. theoret. Biol.* **1**, 1–13.

Williams, R. J. P. 1982 Inorganic elements in biological space and time. *Pure appl. Chem.* **55**, 1089–1100.

Discussion

E. C. Baughan (27 *Canonbury Place*, *London N*1). A simple point that is perhaps worth making: ordinary electron-pair reactions differ from free-radical reactions by having a much larger activation energy; for an activation energy of 84 kJ mol^{-1} only one collision in 10^{14} leads to reaction at room temperature. Hence, free radicals in very small concentrations can often dominate the chemical scene. But this relative advantage diminishes with rising temperature, so that a more complex biochemistry becomes possible with organisms that keep themselves warm.

R. J. P. Williams. There is a danger in assuming that biological reactions are like solution reactions. While some molecules are very mobile in cells, some are more or less rigidly held in matrices. We must then look at solid-state reactions as much as at solution reactions. For example, the side chains of proteins such as valine and phenylalanine may have an activation energy for flipping of *ca.* 100 kJ. Thus solvent relaxation could control any reaction rate. The distinctive differences between electron, atom free-radical, or ion transfers seen in solution is not necessarily applicable to the complex matrices of biology. The reason for the tight temperature control of warm-blooded animals could be connected to phase changes of some kind and unrelated to conventional energies of solution kinetics. The question is interesting; the answer unknown.

R. L. Willson (*Department of Biochemistry*, *Brunel University*). I must challenge Professor Williams's statement that damage to proteins is not important but only coincidental. As a generalization I think this could be highly misleading. Clearly many proteins are present in high concentration, and any damage as such may be relatively small and therefore unimportant. Some proteins, however, may be only present in low concentration at some vital period in the life of the cell or tissue, and any damage could have serious repercussions. Polymerase enzymes known to be sensitive to oxidation and vital for the repair DNA damage, chemotactic factors, some of which contain methionine residues sensitive to oxidation, or protease inhibitors which in some cases are also sensitive to oxidation and which are thought to be involved in tumour promotion, are just some examples.

R. J. P. WILLIAMS. I accept the general statement that damage to any essential molecule that is present as a single or as very few copies in a cell is likely to be serious. DNA is *the* example. It is large and unique and most of its sequence must be faithfully retained. Radiation damage to DNA is undoubtedly very serious. Damage to proteins by radiation is far less likely to be serious for several reasons: most of such proteins exist as multiple copies; individually they are small molecules; a reasonable fraction of their amino acids can be damaged without affecting activity; and finally they can be reproduced. A quantitative examination of the probability of causing any damage to polymerase enzymes in a cell based on target size must give a value less than 0.1 % of the likelihood of causing damage to DNA. At the same time there will be some thousand-fold greater coincidental damage to proteins, taking all the proteins of the cell. With this proviso my statement stands and I remain unconvinced of the value of the study of radiation damage to proteins.

Phil. Trans. R. Soc. Lond. B **311**, 605–615 (1985)
Printed in Great Britain

Dioxygen-derived radicals in biological systems

By H. A. O. Hill
Inorganic Chemistry Laboratory, South Parks Road, Oxford OX1 3QR, U.K.

Three instances of the involvement of dioxygen-derived radicals in biological systems are considered. The first concerns the formation of radicals in the haemolytic reactions induced by treatment of erythrocytes by phenylhydrazine, as an example of the so-called 'oxidant drugs'. The evidence for the formation of phenyl radicals is considered and their origin in the oxidation of phenylhydrazine by a ferryl derivative of haemoglobin postulated. The relevance to the formation of phenylated iron and porphyrin species is described. It is suspected that many instances of oxidative damage to cellular systems result from the coincidence of unsequestered redox-active metal ions (particularly those of iron and copper), reductants, and dioxygen. As an example, the damage to hepatocytes, grown in a culture medium containing cysteine, is described. The formation of radical species derived from dioxygen during the respiratory burst associated with phagocytosis is discussed. A new electrochemical method of detecting the superoxide ion produced during the respiratory burst is described. Particular emphasis is placed on the relation between the production of radical species such as the hydroxyl radical and the superoxide ion, and the extent of phagocytosis.

Introduction

A reasonable assumption about the use of dioxygen in biological systems is that unless there is a loss of control in, or damage to, the cell, most reactions involving this molecule and capable of eliciting much free energy proceed without the production of the relevant radical derivatives, i.e. the superoxide ion and the hydroxyl radical (Hill 1981; Greenwood & Hill 1982). Things can go wrong (Halliwell & Gutteridge 1985) when that control is absent: when there is a surfeit of reductant; when the spatial or temporal mislocation of dioxygen, electrons (reductant) and redox-active metal ions such as iron and copper lead to a ready interchange of electrons between the components; or when the components of the defence are deficient, impaired or in disarray. Many of the same features apply to hydrogen peroxide as a thermodynamically vigorous, but kinetically restrained reactant.

A major problem in unravelling the reactions has been the internecine nature of the relations between the species. Thus the following series of reactions are known to occur:

$$O_2 + e\ \text{(a variety of reductants)} \rightarrow O_2^- ; \quad (1)$$

$$2O_2 + 2H^+ \rightarrow H_2O_2 + O_2 ; \quad (2)$$

$$H_2O_2 + Fe^{II}\ (\text{or } Cu^{I}) \rightarrow OH^{\cdot} + OH^- + Fe^{III} (\text{or } Cu^{II}) ; \quad (3)$$

$$O_2^- + OH^{\cdot} \rightarrow O_2 + OH^- ; \quad (4)$$

$$OH^{\cdot} + H_2O_2 \rightarrow OH^- + HO_2^{\cdot} . \quad (5)$$

In addition, most components of a cell will react with the hydroxyl radical leading to the propagation of chain reactions. There is little doubt that the hydroxyl radical exhibits a ferocious reactivity, little remaining unscathed in its neighbourhood. Its properties as a very weak acid are often overlooked but are relevant to the ability of its conjugate base, the oxene radical anion, $O^{-\cdot}$, to act as a ligand to metal ions. As D. Dolphin (this symposium) has described, there is very good evidence for such ferryl complexes as intermediates in reactions of cytochrome oxidase, mono-oxygenases and peroxidases:

$$Fe^{III} + H_2O_2 \rightarrow (Fe^{V}O^{2-} \leftrightarrow Fe^{IV}O^{-} \leftrightarrow Fe^{III}O) + H_2O. \quad (6)$$

The reactivity of the superoxide ion is pallid by comparison with that of the hydroxyl radical. Indeed, doubt has been cast (Fee *et al.* 1979; Fee 1982) on the so-called 'toxicity' of superoxide and its malignant role is simply assigned as a precursor of hydrogen peroxide, itself the presumed precursor of the hydroxyl radical. However, an agent that is the ready precursor of a toxic material is also a toxin. Those who dismiss the superoxide as a noxious agent overlook its unusual combination of properties: it is a good, and rapid, one-electron reductant; its acidity is such that in biological media it exists both as a strongly solvated anion, $O_2^-(aq)$, and, in its conjugate acid, $O_2H^{\cdot}$, a more reactive and neutral species, which, though solvated, might be able to partition more effectively into non-aqueous phases. Thus, once formed it could be effectively distributed throughout a cell. In addition, if the superoxide ion is formed in a non-aqueous environment, it will have the properties of a strong base. Few species combine these properties; indeed the closest analogues might be considered to be quinones and their reduction products. We can presume that the coincidence in time or space of dioxygen, reductants and metal ions, when one or more of these is uncontrolled, poses a threat to any cell. We shall be concerned with three examples; two involve damage to the cell while in the third the production of these dioxygen-derived species *by* the cell is used to *cause* damage.

Phenylhydrazine-induced haemolysis

There are (Beutler 1978) a number of agents (table 1) that cause haemolysis, especially in those (table 2) with deficiences in, or an impairment of, the cellular apparatus required to maintain the redox state of haemoglobin. The focal point of the reaction is oxyhaemoglobin and, for example, if the haemoglobin is in the carbon monoxy form, the reactions with these agents proceed quite differently or not at all. All the reagents react with oxyhaemoglobin freed from the cell with essentially the same dramatic consequences (table 3). A century ago, it was observed (Hoppe-Seyler 1885) that phenylhydrazine caused the haemolysis of erythrocytes and this reaction has served as a model for drug-induced haemolysis ever since. One of the results of treating erythrocytes with phenylhydrazine and like materials is the formation of methaemoglobin, which has given rise to the inappropriate description of such agents as 'oxidant drugs'; inappropriate because they are all reducing agents with respect to any form of haemoglobin other than deoxyhaemoglobin. A number of small molecular products arising from the reaction of phenylhydrazine with erythrocytes have been identified (Shetlar & Hill 1985) and others are suspected. Those identified include benzene, nitrogen, hydrogen peroxide, the superoxide anion and the phenyl radical; those suspected are phenyldiazene, and the phenylhydrazyl and phenyldiazenyl radicals.

We (Hill & Thornalley 1981, 1982, 1983) have been principally concerned with the identification of the phenyl radical. It is easily intercepted by the use of a number of spin traps. These, usually nitrones or nitroso derivatives, react (Janzen 1980) with radicals to give more

Table 1. Some agents that cause haemolysis

antimalarials such as primaquine and quinine
antipyretics and analgesics such as aspirin
sulphones such as 4,4′-diaminodiphenylsulphone
hydrazines such as acetylphenylhydrazine

Table 2. Conditions that give rise to enhanced sensitivity to oxidative damage to red cells

catalase deficiency
vitamin E deficiency
decreased NADPH production
abnormal glutathione metabolism
paroxysmal nocturnal haemoglobinuria
erythropoietic protoporphyria
thalassaemia syndromes
sickle-cell anaemia

Table 3. events associated with phenylhydrazine-induced haemolysis

blood assumes a brown–green coloration
haemoglobin forms 'green' haemoglobin in which the haem group is modified
destabilization of the globin portion leads to denaturation and precipitation
protein–protein cross-linking occurs within the supporting network
lipid peroxidation occurs within the membranes
oxyhaemoglobin forms methaemoglobin

stable nitroxides. The latter have characteristic electron paramagnetic resonance spectra which can be used both to detect and, in favourable cases, characterize the nitroxide and thus the reactive radical precursor. The production of the phenyl radical provides just such a favourable case; the product of the reaction with 5,5-dimethyl-1-pyrroline-*N*-oxide (DMPO) is stable such that it can be isolated and characterized by mass spectrometry. It gives rise to a six-line signal in the e.p.r. spectrum, the unpaired electron coupling to $[^{14}N]$ ($I = 1$; three lines) of the nitroxide group, each of which is also coupled to the neighbouring $C[^{1}H]$ ($I = \frac{1}{2}$; two lines). The superoxide anion had earlier been identified (Goldberg & Stern 1975) as a product. It was not detected in the spin-trapping experiments, but that is not surprising given its slow rate of reaction with most spin traps. Both water-soluble and lipid-soluble spin traps react with phenyl radicals though the latter seem to be more effective (figure 1). Certainly, they are better able to *protect* the erythrocytes from haemolysis. In a sense, it appears that there is an *endogenous* trap for the phenyl radical within the erythrocyte; the *iron* of haemoglobin! There is now (Kunze & Ortiz de Montellano 1983) convincing evidence that complexes containing the phenyl radical directly attached to the iron of haemoglobin are formed during the reaction. The η_1-$C_6H_5Fe^{III}$ myoglobin has been isolated (Ringe *et al.* 1984) and its structure determined by X-ray diffraction. Recent reports on analogous complexes of iron porphyrins have provided support (Mansuy *et al.* 1982; Battioni *et al.* 1983) for the proposal that this organometallic derivative is the precursor of the *N*-phenylated and *C*-phenylated porphyrins that are the cause of the green and blue compounds formed in the reaction of phenylhydrazine with haemoglobin or erythrocytes. It is therefore possible that the phenyl radical formed within the haem pocket can react with the iron or escape; if the former, the subsequent interaction with oxygen can lead to reaction at the nearby pyrrole nitrogen or, perhaps, at the *meso* position of the porphyrin.

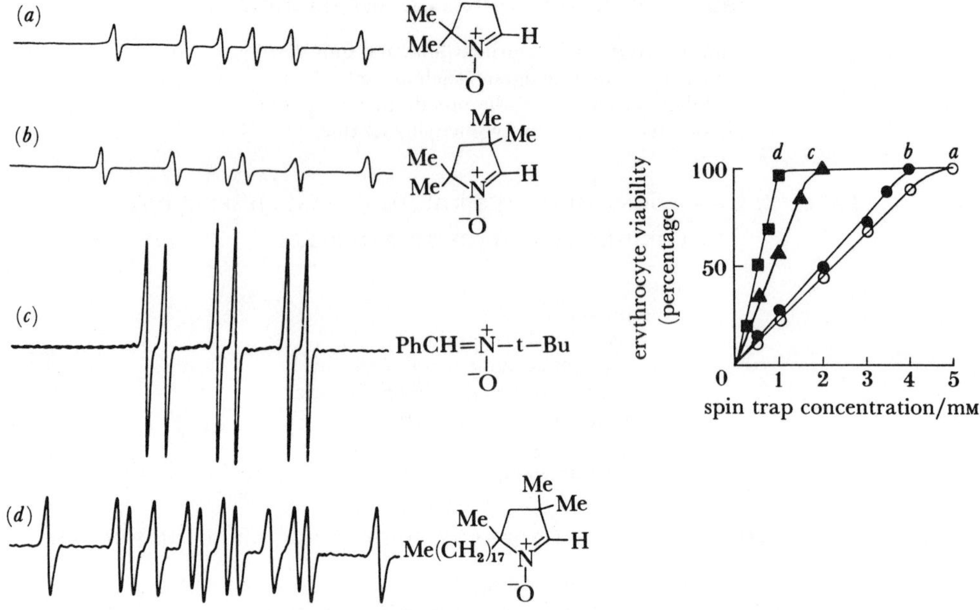

FIGURE 1. (*a*)–(*d*) shows the e.p.r. spectra of the spin adducts formed in the presence of erythrocytes, phenylhydrazine, and the indicated spin traps. The erythrocyte viability in the presence of the spin trap and hydrazine is shown in the inset. (Adapted from Hill & Thornalley 1983.)

If it escapes, it can abstract a hydrogen from a multitude of sources. However, it is a relatively long-lived radical and its mean free path might allow it to reach the cell membrane, there to initiate radical chain reactions. In the discussion of radical damage in biological systems, much emphasis has reasonably been placed on the most reactive radicals, such as the hydroxyl radical. However, it is possible that, because the latter will react with compounds very close to its site of formation, it might lead to damage that is inconsequential or easily repaired. Thc longer-lived radicals may cause more harm by reaching parts of the cell where damage is not easily repaired or where other reactants, like dioxygen, are available to amplify the damage.

How is the phenyl radical formed? It has been proposed (Goldberg *et al.* 1976) that ferrylhaemoglobin is formed during the reaction of haemoglobin with hydrogen peroxide formed in the erythrocyte, or by reduction of dioxyhaemoglobin by phenylhydrazine, in a manner analogous to the formation of ferryl intermediates in reactions of peroxidases or cytochrome P_{450}:

$$\mathrm{HbFe^{III} + H_2O_2 = HbFe^{III}OOH + H^+;} \tag{7}$$

$$\mathrm{HbFe^{III}OOH = HbFe^{V}O + OH^-.} \tag{8}$$

The ferryl intermediate, $\mathrm{HbFe^{V}O}$, can then form the $\mathrm{HbFe^{IV}O}$–porphyrin radical cation which may, by subsequent reaction with phenylhydrazine, give the phenylated haem derivatives, or it may react directly with phenylhydrazine to give a phenyldiazene complex:

$$\mathrm{HbFe^{V}O + PhNHNH_2 = HbFe^{III}{:}PhN{=}NH + H_2O.} \tag{9}$$

The latter could rearrange to give the phenyldiazenyl radical adduct as a precursor to the formation of the phenyl radical:

$$\mathrm{HbFe^{III}{:}PhN{=}NH = HbFe^{II}{:}PhN{=}N^{\cdot} + H^+.} \tag{10}$$

Loss of dinitrogen leads to the formation of Hb^{II} and the phenyl radical,

$$HbFe^{II}{:}PhN{=}N^{\cdot} = HbFe^{II} + N_2 + Ph^{\cdot}. \quad (11)$$

The latter may escape from the haem pocket or it may be trapped by the iron to form the phenylated derivative:

$$HbFe^{II} + Ph^{\cdot} = HbFe{-}Ph. \quad (12)$$

Subsequent decomposition of this organometallic derivative in the presence of dioxygen would lead to the formation of the phenylated products of the haem. The identification of the latter (Ortiz de Montellano & Kunze 1981; Saito & Itano 1981; Augusto *et al.* 1982), together with the characterization of the iron–phenyl derivatives, have provided much greater insight into the mechanism of this complex process.

There are two features that may be unique to the reactions of phenylhydrazine: its ability to act as both a multi-electron *and* multi-proton donor. In addition, its oxidation product, phenyldiazene, can be considered to be isoelectronic with dioxygen, which perhaps aids its entry to the haem site or even allows bonding to the iron. It should not therefore be concluded that insight derived from a study of the mechanism of phenylhydrazine-induced haemolysis may necessarily be relevant to the reactions of the other 'oxidant drugs'.

Radical damage to hepatocytes grown in culture

The coincidence of reducing agents, redox-active metal ions and dioxygen is a far from infrequent occurrence. Many media capable of supporting the growth of organisms in culture provide just such an opportunity. Common constituents of culture media are thiols; these are ready reductants and metal-sequestrants. In the presence of dioxygen, the autoxidation of, for example, cysteine yields hydrogen peroxide and various oxidized derivatives, including cysteine sulphinic and sulphonic acids and cystine (Benesch & Benesch 1955; Cavallini *et al.* 1969; Jocelyn 1972; Zwant *et al.* 1981). With the use of spin traps, it has been shown (Saez *et al.* 1982) that thiyl radicals derived from cysteine are formed under conditions appropriate to the growth of hepatocytes. This, perhaps, is not surprising. However, the spin adduct derived from the hydroxyl radical is also present and this is confirmed by the identification of the adduct formed by trapping the 1-hydroxylethyl radical when the autoxidation is effected in the presence of ethanol:

$$OH^{\cdot} + CH_3CH_2OH = H_2O + CH_3CH^{\cdot}(OH). \quad (13)$$

The relative amounts of the radical adducts depend on the cysteine concentration, on chelating agents or iron(III) or copper(II) salts added to the system, and on the concentration of the spin trap. The addition of superoxide dismutase alone stimulates the production of the two radicals, while catalase inhibits the formation of the radicals. This suggests that hydrogen peroxide is formed by the disproportionation of the superoxide ion, the latter remaining undetected because of its low rate of reaction with the spin trap. It is suggested that thiyl radicals ($RS^{\cdot}$), formed initially by the oxidation of cysteine by iron(III) or copper(II), react with dioxygen to give oxidized cysteine derivatives or with more cysteine to give reduced cystine:

$$RS^{\cdot} + RSH = RSSR^{-\cdot} + H^{+}. \quad (14)$$

This species may react with dioxygen to give the superoxide ion:

$$RSSR^{-\cdot} + O_2 = RSSR + O_2^{-}. \tag{15}$$

The latter may then form hydrogen peroxide and, if the Fenton reaction ensues, the hydroxyl radical. In addition, it is possible that the hydroxyl radical may be formed by the direct reaction of $RSSR^{-\cdot}$ with hydrogen peroxide:

$$RSSR^{-\cdot} + H_2O_2 = RSSR + OH^{-} + OH^{\cdot}. \tag{16}$$

The effect of autoxidizing cysteine on the hepatocytes is to cause the loss of most of their ATP and glutathione with an increase leakage of lactate dehydrogenase, presumably a result of oxidative damage to membrane lipids. The conditions used in the experiments were far from unusual, and thus care is advised in the culture of cells in media prone to autoxidation.

Production of oxygen-derived species by leukocytes

Leukocytes (in particular neutrophils and macrophages) show (Baehner *et al.* 1982) a marked increase in their use of oxygen when engaging in phagocytosis. This respiratory burst is elicited by a number of agents (table 4). The oxygen thus used leads to the formation of a variety of products. There is good evidence that the first product is superoxide, derived ultimately from the oxidation of NADPH. It was first identified as a product through the reduction of exogenous cytochrome *c*. We were concerned to find more direct methods of detecting this product. To that end, we first used (Green *et al.* 1979) the technique of spin trapping, though this is not a particularly effective method for superoxide because most spin traps react slowly with it, at least compared with their rates of reaction with other radicals. The superoxide was, however, detected under the following circumstances (Green *et al.* 1979; Hill & Okolow-Zubkowska 1981; Okolow-Zubkowska & Hill 1981): activation of the respiratory burst with the use of phorbol myristate actate; in the presence of azide ion; and at low temperatures (22 °C). We have recently found (Green *et al.* 1984) that it is possible to exploit the electrochemical oxidation of superoxide to detect the production of this species. An 'opsonized' electrode was prepared (figure 2) that had IgG adsorbed on the electrode surface. With the potential poised at a value that would oxidize any superoxide produced, neutrophils were placed in solution. The current that resulted could be inhibited by superoxide dismutase, and passed for a time consistent with

Table 4. Some agents capable of stimulating the respiratory burst in polymorphonuclear leukocytes

complement
phospholipase *c*
phorbol myristate acetate
formalin-treated red blood cells
C_3- or IgG-coated micro-organisms or latex particles
lipopolysaccharide C_3-coated paraffin oil droplets
C_3-coated zymosan (yeast cell wall)
digitonin and saponin (detergents)
ionophores such as A23187
antineutrophil antibodies
concanavalin A (lectin)
cytochalsin E
endotoxin

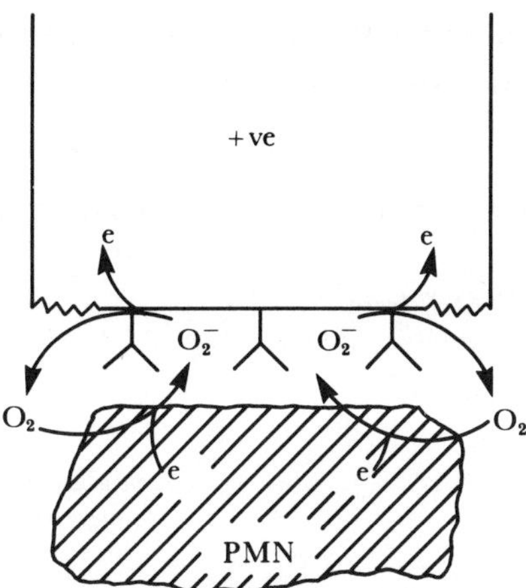

FIGURE 2. A scheme showing the formation of the superoxide ion by neutrophils (PMN) at an electrode surface modified with IgG. On formation, the superoxide is re-oxidized at the electrode.

it being associated with the respiratory burst. It was not elicited by, for example, bovine serum albumin, nor was it affected by the presence of catalase. It was straightforward to show that it was being produced *at* the electrode surface. Approximately 10% of the oxygen taken up gave rise to superoxide thus detected. Given the rapid disproportionation of superoxide, this was considered a reasonable level of detection. Recent results (H. A. O. Hill *et al.* 1985) have shown that, by using an electrode with dimensions not dissimilar to those of a neutrophil, it is possible to detect the superoxide produced by a single cell.

The superoxide anion is formed; is its formation associated with damage to the target organism? Its toxicity is now thought to reside in its role as a precursor of more lethal species, and thus interest centres on other oxygen-derived species, especially hydrogen peroxide and the hydroxyl radical. Of the presence of the former there is no doubt; it presumably arises through the disproportionation of the hydroperoxyl radical:

$$HO_2 + HO_2 = O_2 + H_2O_2. \tag{17}$$

Whence the hydroxyl radical? It has been detected (Clifford & Repine 1984) by a number of methods, including spin trapping. None of these methods are completely secure, though the evidence (table 5) is more than circumstantial. (The use of spin traps readily leads to the detection of the product associated with the formation of $OH^\cdot$. It is possible (Finkelstein *et al.* 1980) that the same product can be formed, albeit more slowly, by the decomposition of the product formed from the adduct of DMPO with HO_2. It has been stated (Halliwell 1984) that the absence of inhibition by hydroxyl radical scavengers such as formate and ethanol indicated that hydroxyl radicals are not formed during the respiratory burst. However, these scavengers were *not used* (Green *et al.* 1979), because they inhibit the respiratory burst. Mannitol, which at the concentrations used does not inhibit the respiratory burst, reduced the yield of the hydroxyl radical adduct of DMPO by 50%. The purpose of using spin traps is to allow the detection of radicals under conditions where the normal functions of the cell *are not perturbed.*) The hydroxyl radical can be formed via the Fenton reaction

$$H_2O_2 + Fe^{II} = OH^\cdot + OH^- + Fe^{III}; \tag{18}$$

TABLE 5. A SUMMARY OF EVIDENCE FOR THE FORMATION OF SUPEROXIDE AND HYDROXYL RADICALS DURING THE RESPIRATORY BURST ASSOCIATED WITH PHAGOCYTOSIS

observations	inferences
superoxide dismutase (SOD inhibits the production of ethylene from methional	hydroxyl radicals arise from an O_2^--dependent reaction
SOD inhibits the production of DMPO(OH) from DMPO	
SOD inhibits the reduction of cytochrome *c* by stimulated neutrophils	O_2^- is produced by stimulated cells
catalase only slightly reduces the yield of DMPO(OH)	catalase is not as effective an inhibitor as SOD, suggesting that H_2O_2 is not the immediate precursor of $OH^{\cdot}$
catalase has only a slightly inhibitory effect on the generation of ethylene from methional	
hydroxyl radical scavengers, e.g. mannitol, inhibit the formation of DMPO(OH)	hydroxyl radicals are formed during the respiratory burst
hydroxyl radical scavengers, e.g. benzoic acid, ethanol, or mannitol, inhibit the formation of ethylene from methional	
phorbol myristate acetate (PMA)-stimulated cells yield DMPO(OOH) from DMPO	PMA-soluble stimulus results in the release of specific granules only; not the azurophic granules which contain MPO
both DMPO(OH) and DMPO(OOH) are formed from DMPO in the presence of azide	azide inhibits myeloperoxidases, thereby preventing formation of hypochlorite

but what is the source of the iron? Lactoferrin, present in the neutrophils and released during the respiratory burst, may be a source of iron, but the ability of this protein to release iron within the phagocytic vacuole, which is crucially dependent on its state of repletion, has not yet been confirmed. The formation of the superoxide adduct of DMPO, is increased relative to that of the hydroxyl radical in the presence of azide. The latter inhibits myeloperoxidase, which catalyses the reaction:

$$H_2O_2+Cl^- = OCl^-+H_2O. \qquad (19)$$

This leads to the suggestion (Okolow-Zubkowska & Hill 1981) that the following reaction might, in part, also play a role in the formation of $OH^{\cdot}$:

$$O_2^-+HOCl = OH^{\cdot}+O_2+Cl^-. \qquad (20)$$

One difficulty in designating a single species as the toxic entity is the superfluity of reactions between the oxygen-derived products. Thus,

$$HOCl+H_2O_2 = {}^1O_2+H_2O+HCl \qquad (21)$$

and

$$Cl^-+OH^{\cdot} = Cl^{\cdot}+OH^- \qquad (22)$$

add to the complexity of the situation. Of course, it is possible that this complexity is exploited in the sense that the reduction of dioxygen in the presence of chloride, iron and the appropriate catalysts leads to a range of highly reactive species, of which at least one will prove lethal to any given cell. The reduction of dioxygen in the respiratory burst is an essential part of the host defence, as witnessed by the consequences of the absence of this facility in those suffering from chronic granulomatous disease. A deficit of myeloperoxidase does not appear to lead to such devastating effects.

The exploitation of such reactive chemistry cannot be without risk. In 'normal' circumstances the dioxygen-derived products are presumably directed towards the target cell; some of the consequences are described in table 6. If some other physiological function of the host is

TABLE 6. CONSEQUENCES OF PRODUCTION OF DIOXYGEN-DERIVED SPECIES DURING THE RESPIRATORY BURST

cell lysis
synthesis of *N*-chloramines
transformation of prostaglandins
methaemoglobin formation in erythrocytes
degradation of hyaluronic acid
bacterial mutagenesis

impaired, it is possible that this array of reagents may be misplaced. It may therefore be important that the leukocyte membrane and the cell wall of the target are contiguous and even that the target is fully contained within the phagocytic vacuole. If that does not happen, the radical species will be able to reach the host tissues. We have found (M. J. Green, H. A. O. Hill & D. G. Tew, unpublished observations) that the ability to detect the superoxide anion, by using the reduction of cytochrome *c* as a monitor, does not correlate with oxygen utilization by neutrophils at low target:cell ratios. It is possible that cytochrome *c* is able to intercept only that fraction of superoxide that escapes from the neutrophil–target boundary interface. At high target:cell ratios, imperfectly formed interfaces may be more numerous and therefore the amount of superoxide detected increases. Many experiments using, for example, IgG-coated latex particles are performed at such high ratios; they might therefore have more relevance to situations *in vivo* where the neutrophils are similarly 'overloaded'. It will be interesting to observe if the quantity of radicals that can so 'escape' is related, not only to the target:cell ratio but also to the ease with which the two surfaces can be made contiguous. This may depend on the surface topography and target shape.

CONCLUSIONS

The accidental coincidence of dioxygen, electrons and redox-active metal ions must be avoided if deleterious effects are not to ensue. The intrusion of exogenous materials, not subject to enzymatic controls, can lead to the production of noxious species that require action by an elaborate defensive system to restrict damage. If this defensive apparatus is absent, attenuated or overcome, then cellular death will follow. In phagocytosis by neutrophils or macrophages, the positions are reversed and the attack, exploiting dioxygen-derived species, is mounted by bringing together the precursors: dioxygen, NADPH, and as-yet uncharacterized metal ions. Oxidation products derived from chloride are added for good measure. It is possible that there is an element of control of the disposition of these attacking forces that is at present unidentified, such that damage to the launching cell is minimized. Also, it is not yet known which of the reactive species is responsible for the death of the target cell, nor are the events that lead to this terminal event recognized. What is certain is that of all reagents exploited by living systems, dioxygen carries with it the greatest risk, yet bestows the greatest benefit.

I am grateful to Dr J. V. Bannister, Dr M. J. Green, Dr P. J. Thornalley and Mr D. G. Tew for permission to describe their work. I thank Professor M. D. Shetlar for much profitable discussion. I am glad to acknowledge the support of the British Heart Foundation.

References

Augusto, O., Kunze, K. L. & Ortiz de Montellano, P. R. 1982 *N*-Phenylprotoporphyrin IX formation in the hemoglobin–phenylhydrazine reaction. *J. biol. Chem.* **257**, 6231–6241.

Baehner, R. L., Boxer, L. A. & Ingraham, L. M. 1982 Reduced oxygen by-products and white blood cells. In *Free radicals in biology*, vol. 5 (ed. W. A. Pryor), pp. 91–114. New York: Academic Press.

Battioni, P., Mahy, J.-P., Gillet, G. & Mansuy, D. 1983 Iron porphyrin dependent oxidation of methyl- and phenylhydrazine: isolation of iron(II)-diazene and alkyliron(III) (or aryliron(III)) complexes. Relevance to the reactions of hemoproteins with hydrazines. *J. Am. chem. Soc.* **105**, 1399–1401.

Benesch, R. E. & Benesch, R. 1955 The acid strength of the –SH group in cysteine and related compounds *J. Am. chem. Soc.* **77**, 5877–5881.

Beutler, E. 1978 *Hemolytic anemia in disorders of red blood cell metabolism.* New York: Plenum Press.

Cavallini, D., DeMarco, C., Dupre, S. & Rotilio, G. 1969 The copper catalysed oxidation of cysteine to cystine. *Archs Biochem. Biophys.* **130**, 354–361.

Clifford, D.P. & Repine, J. E. 1984 Measurement of oxidizing radicals by polymorphonuclear leukocytes. In *Methods in enzymology*, vol. 105 (ed. L. Packer), pp. 393–399. New York: Academic Press.

Fee, J. A. 1982 On the question of superoxide toxicity and the biological function of superoxide dismutases. In *Oxidase and related redox systems* (ed. T. E. King, H. S. Mason & M. Morrison), pp. 101–112. Oxford: Pergamon Press.

Fee, J. A., Lees, A. C., Bloch, P. L. & Neidhardt, F. C. 1979 The role of superoxide in the induction of superoxide dismutase and oxygen toxicity. In *Biochemical and clinical aspects of oxygen* (ed. W. S. Caughey), pp. 635–646. New York: Academic Press.

Finkelstein, E., Rosen, G. M. & Rauckman, E. J. 1980 Spin-trapping of superoxide and hydroxyl radicals. Practical aspects. *Archs Biochem. Biophys.* **200**, 1–16.

Goldberg, B. & Stern, A. 1975 The generation of O_2^- by the interaction of the hemolytic agent, phenylhydrazine, with human hemoglobin. *J. biol. Chem.* **250**, 2401–2403.

Goldberg, B., Stern, A. & Peisach, J. 1976 The mechanism of superoxide anion generation by the interaction of phenylhydrazine with hemoglobin. *J. biol. Chem.* **251**, 3045–3051.

Green, M. J., Hill, H. A. O., Tew, D. G. & Walton, N. J. 1984 An opsonised electrode. The direct electrochemical detection of superoxide generated by human neutrophils. *FEBS Lett.* **170**, 69–74.

Green, M. J. & Hill, H. A. O. 1984 Chemistry of dioxygen. In *Methods in enzymology* (ed. L. Packer), pp. 3–22. New York: Academic Press.

Green, M. R., Hill, H. A. O., Okolow-Zubkowska, M. J. & Segal, A. W. 1979 The production of hydroxyl and superoxide radicals by stimulated human neutrophils – measurements by e.p.r. spectroscopy. *FEBS Lett.* **100**, 23–29.

Greenwood, C. & Hill, H. A. O. 1982 Oxygen in biological systems. *Chemy Br.*, pp. 194–198.

Halliwell, B. 1984 Superoxide dismutase and the superoxide theory of oxygen toxicity. A critical appraisal. In *Copper proteins and copper enzymes* (ed. R. Lontie), pp. 63–102. Florida: C.R.C. Press.

Halliwell, B. & Gutteridge, J. M. C. 1985 *Free radicals in biology and medicine.* Oxford University Press.

Hill, H. A. O. 1981 Oxygen, oxidases and the essential trace metals. *Phils. Trans. R. Soc. Lond.* B**294**, 119–126.

Hill, H. A. O. & Okolow-Zubkowska, M. J. 1981 The exploitation of molecular oxygen by human neutrophils: spin trapping of radicals produced during the respiratory burst. In *Oxygen and life: Proc. Second Priestley conf.*, pp. 98–106. London: Royal Society of Chemistry.

Hill, H. A. O. & Thornalley, P. J. 1981 Phenyl radical production during the oxidation of phenylhydrazine and in phenylhydrazine-induced haemolysis. *FEBS Lett.* **125**, 235–240.

Hill, H. A. O. & Thornalley, P. J. 1982 Free radical production during phenylhydrazine-induced hemolysis. *Can. J. Chem.* **60**, 1528–1534.

Hill, H. A. O. & Thornalley, P. J. 1983 The effect of spin traps on phenylhydrazine-induced haemolysis. *Biochim. biophys. Acta* **762**, 44–52.

Hoppe-Seyler, G. 1885 Uber die wirkung des phenylhydrazins auf den organisms. *Hoppe-Seyler's Z. physiol. Chem.* **9**, 34–39.

Janzen, E. G. 1980 A critical review of spin trapping in biological systems. In *Free radicals in biology*, vol. 4 (ed. W. A. Pryor), pp. 116–154. New York: Academic Press.

Jocelyn, P. C. 1972 *Biochemistry of the –SH group*, pp. 94–115. New York: Academic Press.

Kunze, K. L. & Ortiz de Montellano, P. R. 1983 Formation of a σ-bonded aryliron complex in the reaction of arylhydrazines with hemoglobin and myoglobin. *J. Am. chem. Soc.* **105**, 1380–1381.

Mansuy, D., Battioni, J.-P., Dupre, D., Sartori, E. & Chottard, G. 1982 Reversible iron-nitrogen migration of alkyl, aryl or vinyl groups in iron poryphyrins: a possible passage between Fe(III)(porphyrin)(R) and Fe(II)(N–R)(porphyrin) complexes. *J. Am. chem. Soc.* **104**, 6159–6161.
Okolow-Zubkowska, M. J. & Hill, H. A. O. 1981 An alternative mechanism for the production of the hydroxyl radical by neutrophils. In *Biochemistry and function of neutrophils: Proc. Second European Conf. on Phagocytosis*, pp. 423–428. New York: Plenum Press.
Ortiz de Montellano, P. R. & Kunze, K. L. 1981 Formation of *N*-phenylheme in the hemolytic reaction of phenylhydrazine with hemoglobin. *J. Am. chem. Soc.* **103**, 6534–6536.
Ringe, D., Petsko, G. A., Kerr, D. E. & Ortiz de Montellano, P. R. 1984 Reaction of myoglobin with phenylhydrazine: a molecular doorstop. *Biochemistry* **23**, 2–4.
Saez, G., Thornalley, P. J., Hill, H. A. O., Hems, R. & Bannister, J. V. 1982 The production of free radicals during autoxidation of cysteine and their effect on isolated rat hepatocytes. *Biochim. biophys. Acta* **719**, 24–31.
Saito, S. & Itano, H. A. 1981 *Meso*- phenylbiliverdin IX and *N*-phenylprotoporphyrin IX, products of the reaction of phenylhydrazine with oxyhemoglobin. *Proc. natn Acad. Sci. U.S.A.* **78**, 5508–5512.
Shetlar, M. D. & Hill, H. A. O. 1985 The reactions of hemoglobin with phenylhydrazine: a review of selected aspects. In *Environmental health perspectives* (ed. R. P. Mason), National Institute of Environmental Health Sciences. (In the press.)
Zwant, J., Van Wolput, J. H. M. C. & Koningsberger, D. C. 1981 The mechanism of the copper ion catalyzed autoxidation of cysteine in alkaline medium. *J. molec. Catal.* **12**, 85–101.

Discussion

Catherine Rice-Evans (*Department of Biochemistry, Royal Free Hospital School of Medicine*). Does the phenylhydrazyl radical interact with the iron ion in the haem group in the same way as Dr Hill described for the phenyl radical? Is there any evidence that radicals derived from unsubstituted hydrazine, for example, the hydrazyl or diazyl radical, interact in the same way with the haem iron?

H. A. O. Hill. There is no *direct* evidence that nitrogen-centred radicals react in the same way with the iron in the haem group. It is possible, however, that such interactions play an important part in the mechanism (Shetlar & Hill 1985).

Phil. Trans. R. Soc. Lond. B **311**, 617–631 (1985)
Printed in Great Britain

Oxidative stress: damage to intact cells and organs

H. Sies and E. Cadenas

Institut für Physiologische Chemie I, Universität Düsseldorf, Moorenstrasse 5, *D*-4000 *Düsseldorf* 1, *F.R.G.*

Oxidative cell damage can be monitored by detection of (*a*) photoemission of singlet molecular oxygen formed from radical interactions (so-called low-chemical chemiluminescence), (*b*) end products of lipid peroxidation, such as ethane, and (*c*) glutathione disulphide release. These methods, preferably used in a complementary fashion, provide insight into the pro-oxidant–antioxidant balance in the intact cell or organ.

Recent work from this laboratory on the metabolism of hydroperoxides and aldehydes as well as on redox cycling of the quinone menadione is presented. The comparison of GSSG transport systems in liver and heart reveals a limitation of capacity in the latter, thus making GSSG export potentially critical in the heart.

As part of an inter-organ feedback system between extrahepatic tissues and liver, the newly described hormone stimulation of GSH release from liver is also presented.

The exposure of cells to oxidative conditions of a diverse nature can be accompanied by an elevated production of free radicals, which in turn is expressed as an enhanced generation of electronically excited states leading to the production of low-level chemiluminescence. Among others, lipid peroxidation reactions may occur, leading to products that include volatile hydrocarbons. Changes in the cellular glutathione status are linked to the reduction of hydroperoxides which can result from free-radical intermediates.

We give here a survey of recent work on these topics in our laboratory. Several reviews of more general coverage are available (see, for example, Chance *et al*; 1979; Sies 1985). Techniques of monitoring cell damage, for example, the release of enzymes to the extracellular space or the uptake of dyes such as trypan blue, are not within the scope of the present article.

Low-level chemiluminescence

The detection of light emission from biological samples is a useful method for studying oxidative reactions in intact systems (Boveris *et al.* 1979; Cadenas & Sies 1984). The generation of electronically excited states during intracellular oxidative conditions can result from free-radical interactions that may or may not be associated with the peroxidation of membrane fatty acids. Alternatively, the direct generation of excited states can occur in enzyme-catalysed reactions (Cilento 1982), as demonstrated in model systems using peroxidases. The excited states discussed are (*a*) singlet molecular oxygen as measured by photoemission during its decay to the triplet ground state in the monomol (1) or dimol (2) reactions;

$$^{1}O_2 \rightarrow {}^{3}O_2 + h\nu \ (1270 \text{ nm}), \tag{1}$$

$$2\,{}^{1}O_2 \rightarrow 2\,{}^{3}O_2 + h\nu \ (634 \text{ and } 703 \text{ nm}); \tag{2}$$

and (*b*) excited triplet carbonyls (RO*) exhibiting a weak emission in the blue–green region of the spectrum (reaction (3)) or an indirect emission after energy transfer to a suitable acceptor

A (reaction (4*a*)), thus eliciting sensitized emission (reaction (4*b*));

$$RO^* \rightarrow RO + h\nu, \quad (3)$$

$$RO^* + A \rightarrow RO + A^*, \quad (4a)$$

$$A^* \rightarrow A + h\nu. \quad (4b)$$

Interactions of lipid peroxy radicals can be a source of excited state(s) as in reactions (5*a*,*b*) (Russell 1957; Howard & Ingold 1968). Because peroxy radicals are produced at the final stages of lipid peroxidation, they might be considered as the common mechanism for chemiluminescence (reactions (5*a*, *b*)), shared by different oxidative conditions which promote lipid peroxidation: CCl_4 poisoning, iron overload, oxidative breakdown of hydroperoxides, hyperoxia, etc.

$$ROO^{\bullet} + ROO^{\bullet} \rightarrow ROH + RO + {}^1O_2, \quad (5a)$$

$$ROO^{\bullet} + ROO^{\bullet} \rightarrow ROH + RO^* + O_2. \quad (5b)$$

Free-radical interactions supporting redox cycling, however, elicit chemiluminescence that is not associated with lipid peroxidation. Although O_2^- and $HO^{\bullet}$ are generally produced during the activation of different xenobiotics by redox cycling, a direct link between them and the electronically excited state(s) generated remains to be determined. The fact that redox-cycling-supported photoemission might be inhibited by superoxide dismutase indicates that the O_2^- generated is required at some stage for the generation of photoemission.

The molecular mechanism for the production of singlet oxygen during several peroxidase or peroxidase-like reactions seems to follow the dismutation of hydroperoxides without involving free radical intermediates (Cadenas *et al.* 1983*a*). The production of triplet excited carbonyl compounds during the peroxidase-catalysed oxidation of aliphatic aldehydes to their lower analogues (Cilento 1982) is thought to proceed via the formation of a dioxetane intermediate.

Alkane production

Volatile hydrocarbons such as ethane or pentane (and others) are liberated from membranes after peroxidation and fragmentation of lipids. Possible alternative sources for these alkanes have not yet been characterized. The general route for formation of ethane and pentane includes the formation of free-radical products from hydroperoxide decomposition (Tappel 1980). A pentyl free radical is formed on cleavage of an ω-6 fatty-acid chain on the side of the peroxy group closest to the methyl end (reaction (6)). Hydrogen abstraction by the resulting free radical forms pentane (reaction (7)). These mechanisms suggest that the most likely precursor for pentane and ethane formation is an alkoxy radical ($RO^{\bullet}$).

$$CH_3{-}(CH_2)_3{-}CH_2{-}\underset{\displaystyle O^{\bullet}}{\underset{|}{CH}}{-}CH_2{-}R \rightarrow CH_3{-}(CH_2)_3{-}CH_2^{\bullet} + HOC{-}CH_2{-}R, \quad (6)$$

$$CH_3{-}(CH_2)_3{-}CH_2^{\bullet} + H^{\bullet} \rightarrow CH_3{-}(CH_2)_3{-}CH_3 \quad (7)$$

The methodology for the assay of alkanes with intact animals or with isolated organs has been described by Müller & Sies (1984). The monitoring of alkanes as products of lipid peroxidation was introduced by Riely *et al.* (1974) into the field of biochemical toxicology. Alkane measurements from experimental models such as isolated liver microsomes, isolated hepatocytes, and the isolated perfused rat liver afford the possibility for the direct study of metabolic influences and mechanistic aspects with respect to lipid peroxidation.

Glutathione and oxidative stress

Glutathione and reactive oxygen species: GSSG *release and oxidative stress*

The well described function of glutathione peroxidases (GSH Px) is that of reducing hydroperoxides at the expense of GSH thiol equivalents, thus leading to the formation of glutathione disulphide, GSSG (see figure 1). GSSG is then reduced back to GSH by GSSG reductase at the expense of NADPH, thus allowing the maintenance of flux through GSH Px.

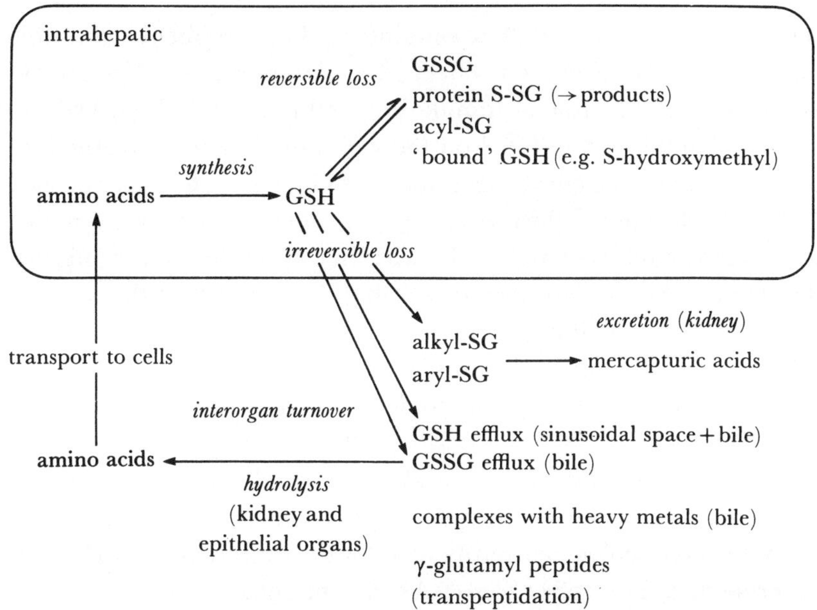

FIGURE 1. Processes (chemical and translocation) affecting intrahepatic and extrahepatic glutathione status. Modified from Sies *et al.* (1983*a*).

Interestingly, about 3% of the flux through GSH Px is not going back to GSH but rather is dealt with by an efflux system. Thus, for every 1000 nanomoles of hydroperoxide reduced per gram of liver per minute about 30 nanomoles of hydroperoxide reduced per gram of liver per minute about 30 nanomoles of GSSG are released per minute per gram of liver (Sies & Summer 1975). This apparently is a means of keeping the GSSG level from rising, and so GSSG efflux may serve a protective purpose. The efflux of GSSG occurs selectively into bile (Sies *et al.* 1978), and the correlation between the rise in intracellular (cytosolic) GSSG with the rate of GSSG efflux across the canalicular membrane (Akerboom *et al.* 1982*a*) suggest the existence of a transport system in the plasma membrane of the canalicular area of the hepatocyte. Carrier-mediated transport of GSSG has recently been detected in canalicular membrane vesicles (Akerboom *et al.* 1984*b*).

The export of GSSG can be competitively inhibited by other glutathione derivatives, such as glutathione S-conjugates (Akerboom *et al.* 1982*b*). It was found that when

1-chloro-2,4-dinitrobenzene was infused into the portal vein to form S-dinitrophenyl glutathione in the hepatocytes (Wahlländer & Sies 1979), the release of GSSG into bile was inhibited. This was accompanied by an increase in intracellular GSSG, as would be expected if the rate of GSSG formulation remained unchanged during the formation of the S-conjugate. However, we detected an intracellular accumulation of GSSG even above that of the value expected by the inhibition of transport across the canalicular membrane, suggesting that the partition ratio of 0.03 for efflux: GSSG reduction was altered. It was found that the accumulation of S-conjugate in the cytosol led to an inhibition of GSSG reductase (Bilzer *et al.* 1984). This was verified with the isolated GSSG reductase, the K_i being about 30 μM for the enzymes from yeast or from human erythrocytes. Further, in an X-ray crystallographic analysis of the binding of S-dinitrophenyl glutathione it was found that the conjugate binds to the enzyme in the active site, largely but not completely overlapping with the binding arrangement for GSSG (Bilzer *et al.* 1984).

More recently, it was also found that binding of S-dinitrophenyl glutathione occurs to another enzyme, and by this can exert an effect of regulatory interest. Kondo *et al.* (1984) have shown that S-conjugate plays a role in stimulating erythrocyte GSH synthesis, presumably by releasing the feedback inhibition of GSH on the γ-glutamyl cysteine synthetase.

In the state of selenium deficiency, the amount of GSH Px selenium decreases to very low levels (2–5% of control values). Burk *et al.* (1978) observed that while hydrogen peroxide infusion did not lead to an increase in GSSG release in this condition, the infusion of an organic hydroperoxide such as t-butyl hydroperoxide (t-BuOOH) still did; the latter hydroperoxide is a substrate for the non-selenium GSH Px activity of glutathione S-transferases, notably isozyme B. It is interesting that in Se deficiency the plasma concentrations of GSH were found to be double that of controls fed a selenium-adequate diet (Hill & Burk 1982; see also next section).

A listing of the response of difference cells and tissues to cellular oxidative conditions coupled to a release of glutathione disulphide into the extracellular space was given, together with a critical survey on the methodology currently used for the detection of GSH, GSSG, and mixed disulphides (Akerboom & Sies 1981; Sies & Akerboom 1984).

Glutathione release into plasma: GSH *release and hormonal responses*

Hepatic efflux of GSH, initially described by Bartoli & Sies (1978), substantially accounts for the turnover of the tripeptide in this organ (for a review, see, for example, Sies 1983; Meister & Anderson 1983). A system of transport of GSH across hepatocyte sinusoidal plasma membrane vesicles has been described (Inoue *et al.* 1984). Factors that govern GSH release under physiological conditions that are not well known. Recently, we observed that GSH efflux across the sinusoidal plasma membrane in isolated perfused rat liver was stimulated by the addition of hormones such as vasopressin, phenylephrine and adrenaline, whereas glucagon or dibutylryl cyclic AMP were without effect (Sies & Graf 1985). As shown in figure 2, the addition of adrenaline (epinephrine) leads to an increase in thiol release. Such hormonal response of the GSH transport system may explain the known loss of GSH during conditions of experimental shock (traumatic or endotoxin) and stress. For example, hepatic GSH content decreases to half that of controls in peripheral inflammation (Bragt & Bonta 1980). Because reactive oxygen species are known to be involved in the process of inflammation (for a review, see for example, Flohé *et al.* 1985), the newly discovered hormone dependence of GSH release from the liver (Sies & Graf 1985) may be part of a servo system to maintain the thiol redox

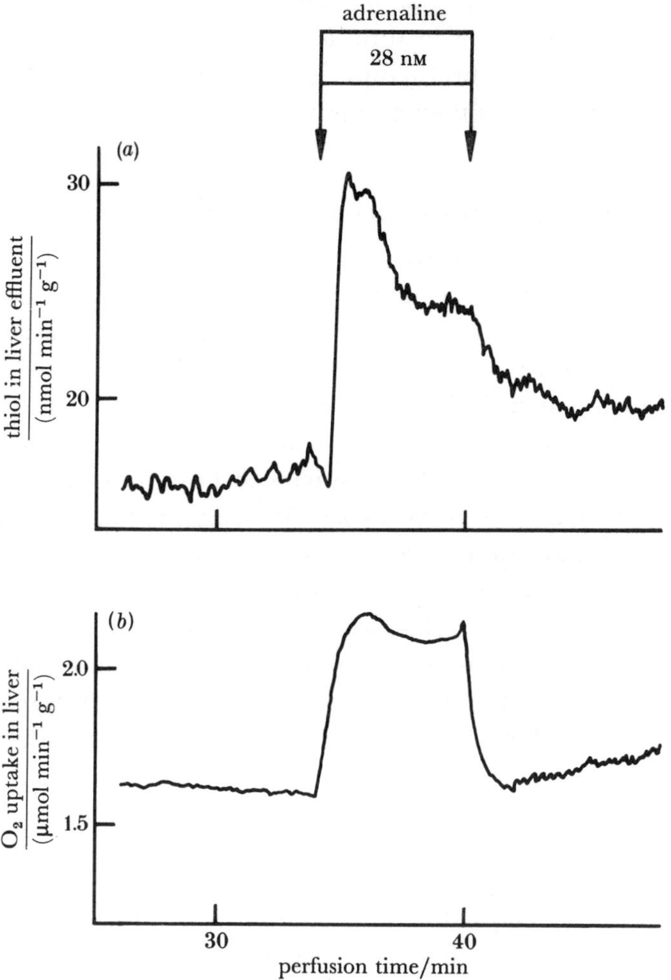

FIGURE 2. Epinephrine (adrenaline) stimulation of thiol efflux from isolated perfused rat liver (*a*) and stimulation of O_2 uptake (*b*). Modified from Sies & Graf (1985).

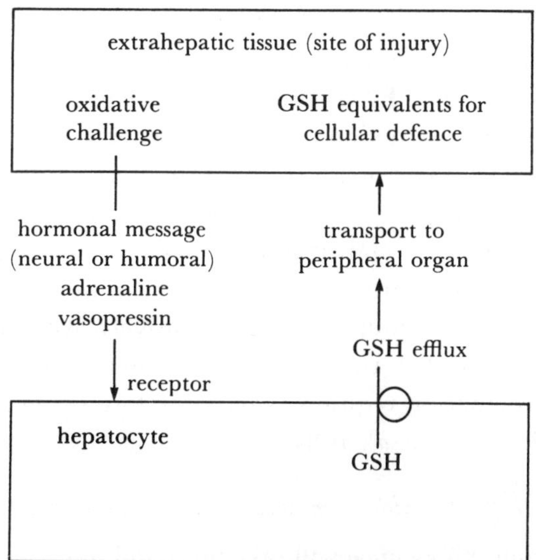

FIGURE 3. Simplified scheme or interorgan feedback system of hormone-stimulatable GSH efflux between extrahepatic tissue and liver. Interrelations of hormone receptor and GSH transport system are yet unknown. Uptake of GSH equivalents by peripheral (extrahepatic) cells may involve prior hydrolysis and subsequent *de novo* synthesis in cells (see figure 1).

balance. The peripheral site (for example, an inflammatory response, or tissue injury) would generate reactive oxygen species which, in turn, lead to a transportable signal of hormone nature. This can be of humoral or neural nature. The stimulation of GSH release form the liver then would counteract the loss of thiol groups in the periphery (see figure 3).

Metabolism of hydroperoxides

Glutathione-dependent hydroperoxide reduction

Reduction of model hydroperoxides (t-butyl hydroperoxide, hydrogen peroxide, cumene hydroperoxide) by perfused rat liver is accompanied by an intracellular decrease of GSH, an intracellular increase of GSSG, GSSG efflux into the bile (see Sies *et al.* 1983*a*). The glutathione content in heart is only about one-fifth of that in liver. The existence of a transport system for GSSG in rat heart were recently shown (Ishikawa & Sies 1984*a*), its capacity also being substantially lower than in liver. Thus, the sensitivity of heart to oxidative stress may be explained, in part, by a low capacity of GSSG export. It is suggested that this transport system takes part in detoxication in cooperation with cardiac glutathione S-transferase. The release by perfused heart of substantial amounts of glutathione S-conjugate (S(2,4-dinitrophenyl) glutathione) in response to infusion of 1-chloro-2,4-dinitrobenzene (Ishikawa & Sies 1984*b*) indicates the existence of cardiac glutathione S-transferases and of an export system for the glutathione S-conjugate formed. The isozyme pattern of glutathione S-transferases in heart differs from that found in liver (Ishikawa & Sies 1984*b*); the cardiac S-transferases consist predominantly of acidic proteins, many of which are absent in the liver (as also found in the testis (Guthenberg *et. al* 1983)). These acidic S-transferases may take a biologically significant share which differs from that for the basic enzymes in liver.

The oxidation of NADPH during either the glutathione-dependent organic hydroperoxide reduction or the cytochrome P_{450}-dependent drug oxidations is linked to calcium movements as shown perfused rat liver (Sies *et al.* 1981) and in isolated hepatocytes (Bellomo *et al.* 1982; Moore *et al.* 1983).

Glutathione peroxidase like and antioxidant activity of the seleno-organic compounds PZ-51

A new aspect on cellular hydroperoxide metabolism was introduced by the use of a seleno-organic compound that exhibits glutathione peroxidase activity, 2-phenyl-1,2-benzoisoselenazol 3(2H)-one, called PZ-51 or Ebselen. The function of PZ-51 can be described within the frame of a glutathione-peroxidase-like activity and a free-radical quenching activity. The former activity was demonstrated *in vitro* by the reduction of t-butyl hydroperoxide by PZ-51 in contrast to its sulphur analogue, PZ-25 (Müller *et al.* 1984; Wendel *et al.* 1984). The temporary protection of microsomes against ascorbate–Fe^{3+}-induced lipid peroxidation did not require GSH and therefore is thought to be a result of the antioxidant capacity of PZ-51. This effect is also exhibited by compounds that are relatively well soluble in lipids, such as diethyldithiocarbamate (Bartoli *et al.* 1983). In contrast, in isolated hepatocytes the protection of PZ-51 against lipid peroxidation requires GSH, therefore being attributed to the GSH Px activity of the compound (Müller *et al.* 1985).

Lipid peroxide formation and chemiluminescence

The occurrence of chemiluminescence will take place in intact hepatocytes or perfused liver when the glutathione-dependent enzymatic detoxication of hydroperoxides is saturated.

Glutathione-depleted hepatocytes (after treatment with phorone (2,6-dimethyl(1)-2,5-heptadiene-4-one) yield low-level chemiluminescence with levels of t-butyl hydroperoxide much lower than those with intact hepatocytes (Sies *et al.* 1983*b*). Likewise, O_2-induced (lipid peroxidation mediated) chemiluminescence of glutathione-depleted hepatocytes is about three times higher in GSH-depleted hepatocytes than in controls (Cadenas *et al.* 1981). Chemiluminescence traces in different cellular glutathione states are shown in figure 4.

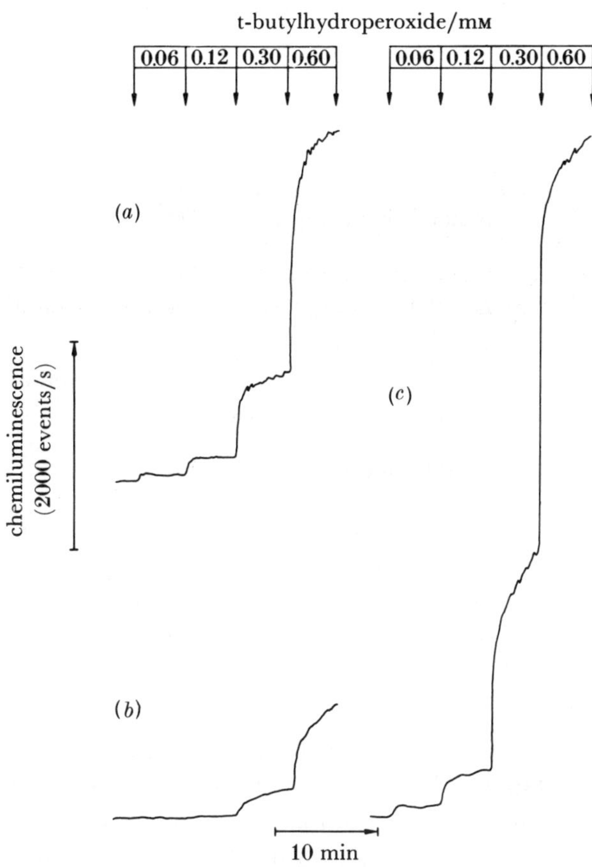

FIGURE 4. Low-level chemiluminescence recorded from surface of perfused rat liver in control (*b*), selenium deficiency (*a*), and selenium deficiency plus GSH-depleted (phorone) state (*c*), in response to increasing concentrations of t-butyl hydroperoxide. Three separate livers were exposed to the same sequence of hydroperoxide concentrations. Modified from Sies *et al.* (1983*b*).

Hydroperoxide metabolism in endoplasmic reticulum

In vitro studies with isolated cytochrome P_{450} and microsomal fractions led to proposed molecular mechanisms for the hydroperoxide-supported mono-oxygenations in terms of homolytic (White & Coon 1980; McCarthy & White 1983) or heterolytic (Ullrich 1977) scission. The homolytic cleavage of the O—O bond of the hydroperoxide is understood as a 'quasi-Fenton' pathway, whereas the heterolytic cleavage can be called an 'oxenoid' pathway. Figure 5 illustrates these types of mechanism for the generation of excited states and, hence, chemiluminescence during the metabolism of hydroperoxides by cytochrome P_{450}.

During the heterolytic cleavage the hydroperoxide coordinates to the haem iron followed by extrusion of alcohol to generate a transient $(FeO)^{3+}$ complex similar to compound I. This

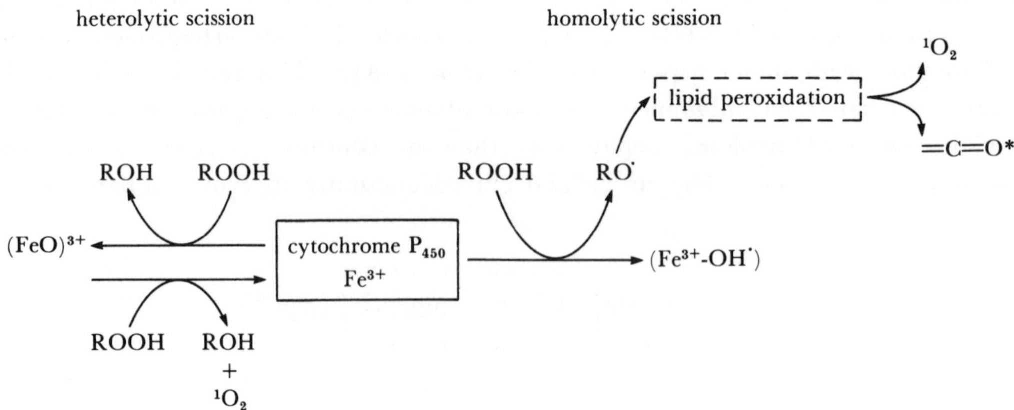

FIGURE 5. Scheme of heterolytic and homolytic scission of hydroperoxides by haemproteins. Iron protoporphyrin is denoted as Fe^{3+}.

$(FeO)^{3+}$ species could either hydroxylate or epoxidize a mono-oxygenation substrate or react with a second molecule of hydroperoxide and yield singlet oxygen (Cadenas *et al.* 1983*c*). During the heterolytic cleavage the generation of singlet oxygen can be viewed as a disproportionation of hydroperoxides (Cadenas *et al.* 1983*a*) (reaction (8)),

$$ROOH + ROOH \rightarrow 2ROH + {}^1O_2 \tag{8}$$

which formally operates in a manner similar to that described for the generation of singlet oxygen during the enzymatic reduction of prostaglandin G_2 to prostaglandin H_2 (Cadenas *et al.* 1983*d*) and during the disproportionation of hydrogen peroxide catalysed by lacto-, myelo-, or chloroperoxidase–halide systems (Khan 1984; Kanofsky 1984). The products of the homolytic pathway are alkoxy radical (RO˙) and a ferric hydroxide haem. The formation of RO˙ in the presence of large amounts of hydroperoxide and absence of a hydroxylatable substrate can account for the destruction of the haem observed under these conditions and the accompanying lipid peroxidation. The formation of excited states by a homolytic mechanism results from the attack of such free radicals on microsomal unsaturated fatty acids; therefore, this photoemission is difficult to distinguish from the lipid peroxidation supported chemiluminescence (Cadenas & Sies 1982).

In addition to various reports suggesting the generation of singlet oxygen as a consequence of lipid peroxidation, it was recently found that excited triplet carbonyls are also formed (Cadenas *et al.* 1984*b*). The detection is based on energy transfer from a triplet carbonyl to a suitable acceptor such as chlorophyll *a* (see reactions 4*a*, *b*). This approach was based on previous reports of an efficient energy transfer from triplet carbonyls generated enzymatically to chlorophyll *a* as such, or incorporated in micelles or as constituents of chloroplasts (Nassi & Cilento 1984; Brunetti *et al.* 1983).

METABOLISM OF ALDEHYDES

The cytotoxic effects derived from aldehyde metabolism in isolated hepatocytes and perfused liver has been evaluated by detection of volatile hydrocarbons and chemiluminescence. Either exogenously added aldehydes, or enzymatically generated aldehydes, or lipid peroxidation derived aldehydes promote alkane release from the perfused liver (Müller & Sies 1982; 1983*a*, *b*). Aldehyde cytotoxicity might rely on (*a*) the ongoing production of radicals (as indicated by a protective effect of radical quenchers such as (+)-cyanidanol-3 and vitamin E (Müller & Sies 1982) and (*b*) the accompanying decreased cellular level of GSH, which might in part be a result of conjugation of the aldehyde with glutathione (Vina *et al.* 1981).

Enzymatically generated aldehydes

The hepatotoxicity elicited by ethanol requires its oxidation to acetaldehyde as indicated by a decreased rate of production of volatile hydrocarbons during alcohol dehydrogenase inhibition (by methyl- or propylpyrazole) and by a hundred-fold higher efficiency of acetaldehyde than ethanol in promoting alkane release by the perfused liver. Moreover, a decreased rate of ethane formation is observed during the inhibition of acetaldehyde oxidation by pargyline, disulphiram, or cyanamide. This indicates that in the reaction chain alcohol–aldehyde–acid, the last step is responsible for ethane production. Neither acetate nor NADH have an effect on ethane release (Müller & Sies 1983*a*, *b*).

The oxidation of monoamines via monoamine oxidase represents a metabolic source of aldehydes. The ethane release observed during the oxidation of benzylamine or phenylethylamine is inhibited in the presence of monoamine oxidase inhibitors, such as pargyline or the more specific tranylcypromine (Müller & Sies 1983*b*). Another aspect of monoamine oxidation cytotoxicity is brought forward by the increased GSSG efflux from liver caused during benzylamine infusion into the perfused liver (Akerboom *et al.* 1982*a*). This, along with the substantially increased cellular content of GSSG, is interpreted as an increased rate of hydrogen peroxide production formed during the oxidation of benzylamine or tryptamine. Reduction of hydrogen peroxide by the glutathione peroxidase system is indicated by an increased cellular content of GSSG (Akerboom *et al.* 1982*a*) at the expense of GSH. If 1-chloro-2,4-dinitrobenzene (CDNB) is given in addition to the oxidative conditions caused by benzylamine, a further increase in GSSG content is observed. This is partly explained by the action of glutathione S-transferase converting CDNB to a thioether of glutathione, which depresses GSSG release, as described above (Akerboom *et al.* 1982*a*).

Lipid peroxidation derived aldehydes

Aldehyde products from the free-radical peroxidative breakdown of polyunsaturated fatty acids include, among others, 4-hydroxyalkenals (Esterbauer *et al.* 1982). The aldehydes 4-hydroxynonenal, 4-hydroxyoctenal, and 4-hydroxyundecenal exert powerful cytotoxic effects, which have been evaluated by detection of low-level chemiluminescence and alkane formation. Cell damage elicited by 4-hydroxynonenal is reflected by an enhanced chemiluminescence intensity and alkane production, in addition to changes in the glutathione status of hepatocytes involving a rapid loss of GSH with formation of a glutathion S-conjugate with 4-hydroxynonenal (Cadenas *et al.* 1983*b*). Several aldehydes, when added to the perfused liver, elicit an extra ethane release that mimics the effect produced by endogenously generated aldehydes during ethanol- or monoamine oxidation. Examples of this are the ethane release promoted by acetaldehyde, benzaldehyde, crotonaldehyde, propionaldehyde, etc. (Müller *et al.* 1983*a*, *b*). As stated above for the endogenously generated aldehydes, the oxidation of the carbonyl is required to observe ethane release, a process which probably involves generation of free-radical intermediates leading to membrane lipid peroxidation.

Redox cycling

The molecular mechanism for activation of certain xenobiotics (often of quinone structure) involves the univalent reduction of the compound with formation of the superoxide anion radical and, subsequently, other oxygen radical derived species (Kappus & Sies 1981; Doroshow & Hochstein 1982). This reduction is accomplished by different enzymatic activities present in the endoplasmic reticulum, mitochondria and nuclei of different tissues. O_2 is required for cytotoxicity; O_2^-, H_2O_2, $HO^{\bullet}$, and $^1O_2$ were thought to be the species responsible for cell damage, whereas the radical form of the xenobiotic appeared less likely to be responsible for cytotoxicity. The relative contribution by the quinone itself remains to be evaluated. The one-electron reduction of the redox cyclers that leads to O_2^- and subsequent hydrogen peroxide generation (via superoxide dismutase reaction) provides substrate for GSH peroxidase, hence an increase in cellular GSSG and (probably catalysed by thioltransferase) an increase in mixed disulphides. These observations suggest that substantial losses of GSH caused by intracellular redox cycling are linked to an increase in intracellular GSSG and, concomitantly, to an increase in mixed disulphides.

Changes in the cellular GSH status may have regulatory significance for several metabolic processes, including transport, by altering the membrane thiol status and the amount of protein mixed disulphides. Redox cycling of menadione affects the hepatic disposition of taurocholate (Akerboom *et al.* 1984). Biliary taurocholate efflux is almost completely inhibited, whereas the hepatic uptake of taurocholate remains unaltered. The intracellular GSSG content rises from 18 (controls) to 540 nmol g^{-1} liver. GSH decreased from 5.5 to 3.5 μmol g^{-1} liver because of the formation and biliary excretion of GSSG. In livers from Se-deficient rats the inhibition of taurocholate transport amounted to only 25% of that in Se-adequate controls, thus suggesting that taurocholate transport is related to the flux through Se-dependent glutathione peroxidase. Adriamycin, a known anticancer drug that was shown *in vitro* to be activated by a redox cycling mechanism (Doroshow & Hochstein 1982), promoted no GSSG efflux from either perfused rat liver or heart (T. Ishikawa, T. P. M. Akerboom & H. Sies, unpublished results).

In addition to these changes in the glutathione status and an intracellular increase of $NADP^+$, menadione redox cycling is associated with the production of low-level chemiluminescence when infused into the perfused liver or during its metabolism by isolated hepatocytes or microsomal fractions (Wefers & Sies 1983*a*). Light emission is increased during conditions which support univalent reduction of the quinone as indicated by (*a*) the enhanced emission during inhibition of the two-electron transfer by dicoumarol or (*b*) the inhibition of chemiluminescence during an enhanced NAD(P)H:quinone reductase activity promoted by pretreatment of the animals with butylated hydroxyanisole (Wefers *et al.* 1984). The cellular decrease of GSH is partly explained by the formation of a glutathione S-conjugate with menadione. This conjugation reaction is not in itself of protective nature and does not abolish semiquinone formation. The glutathione conjugate of menadione can also redox cycle with formation of excited states which lead to chemiluminescence. This could explain the weaker intensity of menadione-induced photo-emission in glutathione-depleted liver, at variance with the intensity observed under similar conditions during hydroperoxide-induced photoemission (Wefers & Sies 1983*a*).

It is interesting that, in a model system *in vitro* with the use of xanthine or acetaldehyde as substrate and xanthine oxidase for O_2^- generation, GSH was shown to increase low-level chemiluminescence in the red spectral region (above 620 nm) and, as a stable end product of GSH, GSSG, and GSO_2^-, the glutathione sulphonate (Wefers & Sies 1983*b*). The mechanism proposed involves the formation of thiyl radical and, further, of the peroxysulphenyl radical (reactions (9) and (10)).

$$GSH + HO_2^{\bullet} \rightarrow GS^{\bullet} + H_2O_2, \tag{9}$$

$$GS^{\bullet} + GS^{\bullet} \rightarrow GSSG, \tag{10a}$$

$$GS^{\bullet} + O_2 \rightarrow GSOO^{\bullet}. \tag{10b}$$

The reactions of thiyl radicals in biological tissue are of interest not only in radiation biochemistry but also in metabolism. For example, the biological significance of the observation by Glatt *et al.* 1983 of a mutagenicity of glutathione and cysteine in the Ames test remains to be evaluated.

This work was supported by Deutsche Forschungsgemeinschaft, Schwerpunktsprogramm 'Mechanismen toxischer Wirkungen von Fremdstoffen' and by the National Foundation for Cancer Research, Washington.

References

Akerboom, T. P. M., Bilzer, M. & Sies, H. 1982*a* The biliary glutathione disulfide efflux and intracellular GSSG content in perfused rat liver. *J. biol. Chem.* **257**, 4248–4252.

Akerboom, T. P. M., Bilzer, M. & Sies, H. 1982*b* Competition between transport of glutathione disulfide (GSSG) and glutathione S-conjugates from perfused rat liver into bile. *FEBS Lett.* **140**, 73–76.

Akerboom, T. P. M. & Sies, H. 1981 Assay for GSH, GSSG and glutathione mixed disulfides in biological samples. *Meth. Enzym.* **77**, 373–382.

Akerboom, T. P. M., Inoue, M., Sies, H., Kinne, R. & Arias, I. M. 1984 Biliary transport of glutathione disulfide studied with isolated rat liver canalicular membrane vesicles. *Eur. J. Biochem.* **141**, 211–215.

Bartoli, G. M., Müller, A., Cadenas, E. & Sies, H. 1983 Antioxidant effect of diethyldithiocarbamate on microsomal lipid peroxidation assessed by low-level chemiluminescence and alkane production. *FEBS Lett.* **164**, 371–374.

Bartoli, G. M. & Sies, H. 1978 Reduced and oxidized glutathione efflux from liver. *FEBS Lett.* **86**, 89–91.

Bellomo, G., Jewell, S. A., Thor, H. & Orrenius, S. 1982 Regulation of intracellular calcium compartmentation: Studies with isolated hepatocytes and t-butyl hydroperoxide. *Proc. natn Acad. Sci. U.S.A.* **79**, 6842–6846.

Bilzer, M., Krauth-Siegel, R. L., Schirmer, R. H., Akerboom, T. P. M., Sies, H. & Schulz, G. E. 1984 Interaction of a glutathione S-conjugate with glutathione reductase. Kinetic and X-ray crystallographic studies. *Eur. J. Biochem.* **138**, 373–378.

Boveris, A., Cadenas, E., Reiter, R., Filipkowski, M., Nakase, Y. & Chance, B. 1979 Organ chemiluminescence: noninvasive assay for oxidative radical reactions. *Proc. natn Acad. Sci. U.S.A.* **77**, 347–351.

Bragt, P. C. & Bonta, I. L. 1980 *Agents, Actions* **10**, 536–539.

Brunetti, I. L., Cilento, G. & Nassi, L. 1983 Energy transfer from enzymatically generated triplet species to acceptors in micelles. *Photochem. Photobiol.* **38**, 511–519.

Burk, R. F., Nishiki, K., Lawrence, R. A. & Chance, B. 1978 Peroxide removal by selenium-dependent and selenium-independent glutathione peroxidases in hemoglobin-free perfused rat liver. *J. biol. Chem.* **253**, 43–46.

Cadenas, E., Brigelius, R., Akerboom, T. P. M. & Sies, H. 1983*a* Oxygen radicals and hydroperoxides in mammalian organs: aspects of redox cycling and hydrogen peroxide metabolism. In *Biological oxidations* (ed. H. Sund & V. Ullrich), pp. 288–310. Berlin: Springer-Verlag.

Cadenas, E., Müller, A., Brigelius, R., Esterbauer, H. & Sies, H. 1983*b* Effects of 4-hydroxynonenal on isolated hepatocytes. Studies on chemiluminescence response, alkane production, and glutathione status. *Biochem. J.* **214**, 479–487.

Cadenas, E. & Sies, H. 1982 Low-level chemiluminescence of microsomal fractions initiated by t-butyl hydroperoxide. Relation to microsomal hemoproteins, oxygen consumption, and lipid peroxidation. *Eur. J. Biochem.* **124**, 349–356.

Cadenas, E. & Sies, H. 1984 Low-level chemiluminescence as an indicator of singlet molecular oxygen in biological systems. *Meth. Enzym.* **105**, 221–231.

Cadenas, E., Sies, H., Campa, A. & Cilento, G. 1984*b* Electronically excited states in microsomal membranes: the use of chlorophyll *a* as an indicator of triplet carbonyls. *Photochem. Photobiol.* **40**, 661–666.

Cadenas, E., Sies, H., Graf, H. & Ullrich, V. 1983*c* Oxene donors yield low-level chemiluminescence with microsomes and isolated cytochrome P_{450}. *Eur. J. Biochem.* **130**, 117–121.

Cadenas, E., Sies, H., Nastainczyk, W. & Ullrich, V. 1983*d* Singlet oxygen formation detected by low-level chemiluminescence during enzymatic reduction of prostaglandin G_2 to H_2. *Hoppe-Seyler's Z. physiol. Chem.* **364**, 519–528.

Cadenas, E., Wefers, H. & Sies, H. 1981 Low-level chemiluminescence of isolated hepatocytes. *Eur. J. Biochem.* **119**, 531–536.

Chance, B., Sies, H. & Boveris, A. 1979 Hydroperoxide metabolism in mammalian organs. *Physiol. Rev.* **59**, 527–605.

Cilento, G. 1982 Electronic excitation in dark biological processes. In *Chemical and biochemical generation of excited states* (ed. W. Adam & G. Cilento). pp. 277–307. New York: Academic Press.

Doroshow, J. & Hochstein, P. 1982 Redox cycling and the mechanism of action of antibiotics in neoplastic diseases. In *Pathology of oxygen* (ed. A. P. Autor), pp. 245–260. New York: Academic Press.

Esterbauer, H., Cheeseman, K. H., Dianzani, M. U., Poli, G. & Slater, T. F. 1982 Separation and characterization of the aldehydic products of lipid peroxidation stimulated by ADP–Fe^{2+} in rat liver microsomes. *Biochem. J.* **208**, 129–140.

Flohé, L., Beckmann, R., Giertz, H. & Loschen, G. 1985 Oxygen-centred radical as mediators of inflammation. In *Oxidative stress* (ed. H. Sies). London: Academic Press. (In the press.)

Glatt, H., Protic-Sabljic, M. & Oesch, F. 1983 Mutagenicity of glutathione and cysteine in the Ames test. *Science, Wash.* **220**, 961–962.

Guthenberg, C., Astrand, F., Alin, P. & Mannervik, B. 1983 Glutathione transferases in rat testis. *Acta chem. scand.* **37**, 261–262.

Hill, K. E. & Burk, R. F. 1982 Effect of selenium deficiency and vitamin E deficieny on glutathion metabolism in isolated rat hepatocytes. *J. biol. Chem.* **257**, 10668–10672.

Howard, J. A. & Ingold, K. U. 1968 Rate constants for the self-reactions of *n*- and t-butyl peroxy radicals and cyclohexylperoxy radicals. The deuterium isotope effects in the termination of secondary peroxy radicals. *J. Am. chem. Soc.* **90**, 1058–1059.

Inoue, M., Kinne, R., Tran, T. & Arias, I. M. 1984 Glutathione transport across hepatocyte plasma membranes. Analysis using isolated rat liver sinusoidal-membrane vesicles. *Eur. J. Biochem.* **138**, 491–495.

Ishikawa, T. & Sies, H. 1984*a* Cardiac transport of glutathione disulfide and S-conjugate. *J. biol. Chem.* **259**, 3838–3843.

Ishikawa, T. & Sies, H. 1984*b* The isozyme pattern of glutathione S-transferases in rat heart. *FEBS Lett.* **169**, 155–160.

Kanofsky, J. R. 1984 Singlet oxygen production by lactoperoxidase: halide dependence and quantification of yield. *J. Photochem.* **25**, 105–113.

Kappus, H. & Sies, H. 1981 Toxic drug effects associated with oxygen metabolism: redox cycling and lipid peroxidation. *Experientia* **37**, 1233–1241.

Khan, A. U. 1984 Discovery of enzyme generation of singlet molecular oxygen: spectra of $(0, 0)^1\Delta_g$–$^3\Sigma_g$ i.r. transition. *J. Photochem.* **25**, 327–334.

Kondo, T., Taniguchi, N. & Kawakami, Y. 1984 Significance of glutathione S-conjugate for glutathione metabolism in human erythrocytes. *Eur. J. Biochem.* **145**, 131–136.

McCarthy, M. B. & White, R. E. 1983 Functional differences between peroxidase compound I and the cytochrome P_{450} reactive oxygen intermediate. *J. biol. Chem.* **258**, 9153–9158.

Meister, A. & Anderson, M. E. 1983 Glutathione. *A. Rev. Biochem.* **42**, 711–760.

Moore, G. A., Jewell, S. A., Bellomo, G. & Orrenius, S. 1983 On the relationship between calcium efflux and membrane damage during t-butyl hydroperoxide metabolism by liver mitochondria. *FEBS Lett.* **153**, 289–292.

Müller, A., Cadenas, E., Graf, P. & Sies, H. 1984 A novel biologically active seleno-organic compound I. Glutathione-peroxidase-like activity in vitro and antioxidant capacity of PZ-51 (Ebselen). *Biochem. Pharmacol.* **33**, 3235–3239.

Müller, A., Gabriel, H. & Sies, H. 1985 A novel biologically active seleno-organic compound. IV. Protective glutathione-dependent effect of PZ-51 (Ebselen) against ADP–Fe-induced lipid peroxidation in isolated hepatocytes. *Biochem. Pharmacol.* **34**, 1185–1189.

Müller, A. & Sies, H. 1982 Role of alcohol dehydrogenase activity and aldehyde on ethanol-induced ethane and pentane production by isolated perfused liver. *Biochem J.* **206** 153–156.

Müller, A. & Sies, H. 1983*a* Inhibition of ethanol and acetaldehyde-induced release of ethane from isolated perfused liver by pargyline and disulfiram. *Pharmacol. biochem. Behav.* **18** (suppl. 1), 423–432.

Muller, A. & Sies, H. 1983*b* Ethane release during metabolism of aldehydes and monoamines in perfused rat liver. *Eur. J. Biochem.* **134**, 599–602.

Müller, A. & Sies, H. 1984 Assay of ethane and pentane from isolated organs and cells. *Meth. Enzym.* **105**, 311–319.

Nassi, L. & Cilento, G. 1984 Excitation of micelle-solubilized chlorophyll *a* during the peroxidase-catalyzed aerobic oxidation of isonicotinic acid hydrazide. *Archs Biochem. Biophys.* **229**, 220–225.

Riely, C. A., Cohen, G. & Lieberman, M. 1974 Ethane evolution: a new index of lipid peroxidation. *Science, Wash.* **183**, 208–210.

Russell, G. A. 1957 Deuterium-isotope effects in the autoxidation of aralkyl hydrocarbons. Mechanism of interaction of peroxy radicals *J. Am. chem. Soc.* **79**, 3871–3877.

Sies, H. 1983 Reduced and oxidized glutathione efflux from liver. In *Glutathione: storage, transport, and turnover in mammals* (ed. Y. Sakamoto, T. Higashi & N. Tateishi), pp. 63–88. Tokyo: Japan Scientific Society Press.

Sies, H. (ed.) 1985 *Oxidative stress*. Orlando: Academic Press.

Sies, H. & Akerboom, T. P. M. 1984 Glutathione disulfide (GSSG) efflux from cells and tissues. *Meth. Enzym.* **105**, 445–451.

Sies, H., Brigelius, R. & Akerboom, T. P. M. 1983*a* Intrahepatic glutathione status In *Functions of glutathione: biochemical, physiological, toxicological, and clinical aspects* (ed. A. Larsson, S. Orrenius, A. Holmgren & B. Mannervik), pp. 51–64. New York: Raven Press.

Sies, H., Brigelius, R., Wefers, H., Müller, A. & Cadenas, E. 1983*b* Cellular redox changes and response to drugs and toxic agents. *Fundam. Appl. Toxicol.* **3**, 200–208.

Sies, H. & Graf, P. 1985 Hepatic thiol and glutathione efflux under the influence of vasopresin, phenylephrine, and adrenaline. *Biochem. J.* **226**, 545–549.

Sies, H., Graf, P. & Cadenas, E. 1984 Hydroperoxide reduction in liver cells: calcium release and relationship to arachidonate metabolism. In *Prostaglandins and membrane ion transport* (ed. P. Braquet, S. Nicosia, J. C. Frölich & R. Garay), pp. 119–128. New York: Raven Press.

Sies, H., Graf, P. & Estrela, J. M. 1981 Hepatic calcium efflux during cytochrome P_{450}-dependent drug oxidation at the endoplasmic reticulum in intact liver. *Proc. Natn Acad. Sci. U.S.A.* **78**, 3358–3362.

Sies, H. & Summer, K. H. 1975 Hydroperoxide-metabolizing systems in rat liver. *Eur. J. Biochem.* **57**, 503–512.

Sies, H., Wahlländer, A. & Waydhas, C. 1978 Properties of glutathione disulfide (GSSG) and glutathione S-conjugate release from perfused rat liver. In *Functions of glutathione in liver and kidney* (ed. H. Sies & A. Wendel), pp. 120–126. Berlin: Springer-Verlag.

Tappel, A. L 1980 Measurements of and protection from *in vivo* lipid peroxidation. In *Free radicals in biology* vol. 4 (ed. W. A. Pryor), pp. 2–48. New York: Academic Press.

Ullrich, V. 1977 The mechanisms of cytochrome P_{450} action. In *Microsomes and drug oxidations* (ed. V. Ullrich, I. Roots, A. Hildebrandt, R. O. Estabrook & A. H. Conney), pp. 192–210. New York: Plenum.

Vina, J., Estrela, J. M., Guerri, C. & Romero, F. J. 1980 Effect of ethanol on GSH concentration in isolated hepatocytes. *Biochem. J.* **188**, 549–552.

Wahlländer, A. & Sies, H. 1979 Glutathione S-conjugate formation from 1-chloro-2,4-dinitrobenzene and biliary S-conjugate excretion in the perfused rat liver. *Eur. J. Biochem.* **96**, 441–446.

Wefers, H., Komai, T., Talalay, P. & Sies, H. 1984 Protection against oxidative stress by NAD(P)H:quinone induced by the dietary antioxidant butylated hydroxyanisole (BHA). *FEBS Lett.* **169**, 63–66.

Wefers, H. & Sies, H. 1983*a* Hepatic low-level chemiluminescence during redox cycling of menadione and the mandione glutathione conjugate. Relation to glutathione and NAD(P)H:quinone reductase (DT diaphorase) activity. *Arch. Biochem, Biophys.* **224**, 568–578.

Wefers, H. & Sies, H. 1983*b* Oxidation of glutathione by the superoxide radical to the disulfide and the sulfonate yielding singlet oxygen. *Eur. J. Biochem* **137**, 29–36.

Wendel, A., Fausel, M., Safayi, H. & Otter, R. 1984 A novel biologically active selenoorganic compound. II. Activity of PZ-51 (Ebselen) in relation to glutathione peroxidase. *Biochem. Pharmacol.* **33**, 3241–3245.

White, R. E. & Coon, M. J. 1980 Oxygen activation of cytochrome P_{450}. *Am. Rev. Biochem.* **49**, 315–336.

Discussion

M. C. R. Symons (*Department of Chemistry, Leicester University*). I accept that it is conventional wisdom that the two visible emission bands associated with $^1\Delta$ oxygen are a result of emission from the dimer $^1\Delta O_2 \cdots {}^1\Delta O_2$, and that the near-i.r. band is simply due to $^1\Delta O_2$ itself. However, I find this hard to understand, and wonder if there might be some other explanation. This is because the intensity of the visible emission seems to be greater than the i.r. emission. Because $^1\Delta O_2$ has only a very limited lifetime, the probability of any sort of 'collision' involving orbital overlap must surely be small. If one argues that the i.r. band is weak because of thermal pathways, this will also serve to reduce the visible bands because the concentration of dimer will also be reduced. Has Professor Sies any explanation for the apparently extraordinary efficiency of dimer emission? Is it possible that one fails to 'see' much of the i.r. emission because of absorption by water, which has vibrational overtone features in this region? (If this were so, then the i.r. emission would be enhanced by using D_2O, but not the visible emission.)

H. Sies. This problem has not yet been solved in simple physical–chemical systems. In our biological models, as in intact cells, we have a multi-phase system, so that there are differences in localized concentrations of $^1\Delta O_2$; this can contribute to the preference of dimol over monomol emission. In a purely aqueous phase with H_2O_2–NaOCl *in vitro*, we found that DABCO enhanced both monomol and dimol emission, but enhancement was more pronounced for the latter.

Reference

Lengfelder, E., Cadenas, E. & Sies, H. 1983 *FEBS Lett.* **164**, 366–370.

M. C. R. Symons. In Professor Sies's scheme outlining the reactions of GSH, he included reaction (1) below as a route to GSSG. In my view, an important alternative mechanism is given in steps (2)–(4), which leaves open the route involving electron loss from $GSSG^-$.

$$2GS^{\cdot} \longrightarrow GS{-}SG, \tag{1}$$

$$GS^{\cdot} + GSH \longrightarrow GS\overset{\cdot}{-}S\begin{smallmatrix}\diagup H \\ \diagdown G\end{smallmatrix}, \tag{2}$$

$$GS\overset{\cdot}{-}S\begin{smallmatrix}\diagup H \\ \diagdown G\end{smallmatrix} + H_2O \rightleftharpoons GS\overset{\cdot}{-}SG^- + H_3O^+, \tag{3}$$

$$GS\overset{\cdot}{-}SG^- \longrightarrow GSSG + [e^-]. \tag{4}$$

We think that we have e.s.r. evidence for intermediates of the type $RS\overset{\cdot}{-}SHR$, (Nelson *et al.* 1977, 1978) although this remains a matter of controversy (Hadley & Gordy 1974; Symons 1985). The anions $RS\overset{\cdot}{-}SR^-$ are, of course, relatively stable, and at neutral pH, proton loss from $RS\overset{\cdot}{-}SHR$ should be favoured.

I would also like to support Professor Sies's suggestion that $GS^{\cdot}$ radicals can react with oxygen to give the 'peroxide' radical $GS{-}OO^{\cdot}$. We find that in the presence of oxygen at low temperatures, $RS\overset{\cdot}{-}SHR$–$RSSR^-$ is suppressed in favour of a species having e.s.r. properties commensurate with the formation of a bent species, **1**, which is the structure expected for this radical. Such a 'peroxide', however, may be much less reactive than alkyl peroxide radicals or $HO_2^{\cdot}$. Furthermore, at room temperature, the equilibrium may not favour $RSOO^{\cdot}$ formation as much as it does at low temperatures.

$$R{-}S{-}\dot{O}{\diagdown}O$$

1

References

Hadley, J. H. & Gordy, W. 1974 *Proc. natn Acad. Sci. U.S.A.* **71**, 3106–3110.
Nelson, D. J., Petersen, R. L. & Symons, M. C. R. 1977 *J. chem. Soc. Perkin Trans.* II, 2005–2015.
Nelson, D. J., Petersen, R. L. & Symons, M. C. R. 1978 *J. chem. Soc. Perkin Trans.* II, 225–231.
Symons, M. C. R. 1985 *Int. J. Radiat. Oncology.* (in the press.)

H. Sies. As I said, there are several potential reactions involving $GS^{\bullet}$, and reaction (2) is certainly one of them; alternatively, $GS^{\bullet}$ may react with GS^{-} also directly. Thus, I agree fully with your comment. One interesting electron acceptor for reaction (4) can be O_2, this way leading to formation of the superoxide anion radical.

G. Scott (*Department of Molecular Sciences, Aston University*). In his talk Professor Sies referred to the formation of glutathione sulphonate species, by which I understand him to mean that the contain two oxygens attached to sulphur. We know that *in vitro* these species are very effective antioxidants (Armstrong *et al.* 1979) by virtue of their ability to trap radicals and destroy hydroperoxides. Is there any evidence that the oxidized sulphur species he has observed are antioxidants *in vivo* or do you consider this a 'wasteful' reaction of thiyl radicals?

Reference

Armstrong, C., Husbands, M. & Scott, G. 1979 *Eur. Polym. J.* **15**, 241.

H. Sies. We have detected glutathione sulphonate, $\mathrm{G{-}S(=O)_2{-}O^-}$, in our *in vitro* system. So far, this species has not yet been detected *in vivo*. Regarding the sulphinate, $\mathrm{G{-}S(O){-}O^-}$, and the peroxy sulphenyl radical $GSOO^{\bullet}$, we have no information as to their potential pro-oxidant or antioxidant activity.

Phil. Trans. R. Soc. Lond. B **311**, 633–645 (1985)
Printed in Great Britain

Carbon tetrachloride toxicity as a model for studying free-radical mediated liver injury

By T. F. Slater[1], K. H. Cheeseman[1] and K. U. Ingold, F.R.S.[2]

[1] *Department of Biochemistry, Brunel University, Uxbridge, Middlesex UB8 3PH, U.K.*
[2] *Division of Chemistry, National Research Council of Canada, Ottawa, Canada, KIA 0R6*

A single dose of CCl_4 when administered to a rat produces centrilobular necrosis and fatty degeneration of the liver. These hepatotoxic effects of CCl_4 are dependent upon its metabolic activation in the liver endoplasmic reticulum to reactive intermediates, including the trichloromethyl free radical. Positive identification of the formation of this free radical *in vivo*, in isolated liver cells and in microsomal suspensions *in vitro* has been achieved by e.s.r. spin-trapping techniques. The trichloromethyl radical has been found to be relatively unreactive in comparison with the secondarily derived peroxy radical $CCl_3O_2^{\cdot}$, although each free radical species contributes significantly to the biological disturbances that occur. Major early perturbations produced to liver endoplasmic reticulum by exposure *in vivo* or *in vitro* to CCl_4 include covalent binding and lipid peroxidation; studies of these processes occurring during CCl_4 intoxication have uncovered a number of concepts of general relevance to free-radical mediated tissue injury. Lipid peroxidation produces a variety of substances that have high biological activities, including effects on cell division; many liver tumours have a much reduced rate of lipid peroxidation compared with normal liver. A discussion of this rather general feature of liver tumours is given in relation to the liver cell division that follows partial hepatectomy.

Introduction

It has been known for more than 100 years that carbon tetrachloride is a very toxic substance. However, although its damaging actions on the liver were recorded in the early literature, the detailed analysis of its hepatotoxic actions can be traced back mainly to the series of studies by Cameron and his colleagues, commencing in the 1930s (Cameron & Karunaratne 1936). When powerful new techniques of biochemistry became available in the period 1950–1960 there was a proliferation of biochemical and toxicological studies on CCl_4 because it was widely recognized to have features making it well suited as a model agent for studying hepatotoxicity. These features include its ready availability in pure form, the reproducibility of its effects on liver in different species, and the diversity of effects it can produce under appropriate conditions. Most studies have concentrated on the early acute effects of CCl_4 on rat liver: a single dose of CCl_4 (for example, 0.5 ml per kilogram of body mass) administered to a rat causes centrilobular necrosis and fatty degeneration of the liver. Repeated doses of CCl_4 can lead to the onset of cirrhosis and, under certain conditions, to liver tumours. In this paper we will concentrate mainly on the early acute effects of CCl_4 on the liver, but some remarks germane to liver tumours will be included at the end.

Butler (1961) and an associated commentary by Wirtschafter & Cronyn (1964) suggested that CCl_4 was converted to a trichloromethyl radical ($CCl_3^{\cdot}$) that was of significance to the ensuing lesion. In the data and discussions of those early papers there were no indications of any specific metabolic route for the production of $CCl_3^{\cdot}$ from CCl_4. The suggestion that CCl_4 was converted to $CCl_3^{\cdot}$ in liver was developed by Slater (1966) into the concept of the metabolic activation of CCl_4 by enzymes in liver endoplasmic reticulum with toxic consequences such as lipid peroxidation. Almost simultaneously, and quite independently, Recknagel and coworkers (Ghoshal & Recknagel 1965) considered the formation of $CCl_3^{\cdot}$ from CCl_4, and also stressed the importance of lipid peroxidation, initiated and stimulated by $CCl_3^{\cdot}$, in the liver damage. Since the publications of Slater (1966) and of Ghoshal & Recknagel (1965) many other studies have confirmed the metabolic activation of CCl_4, its relevance to liver injury, and its association with the NADPH–cytochrome-P_{450} electron-transport chain. There is no intention here to review basic features of CCl_4-mediated liver injury; full reviews of the general background can be obtained in Recknagel (1967) and Slater (1972).

In this paper we will consider some key aspects of the liver injury caused by CCl_4 that are still in doubt or controversial; we will outline a number of concepts of general importance in relation to biochemical studies on tissue injury, and which have been derived largely from work with CCl_4; the toxicological significance of lipid peroxidation with respect to CCl_4-induced liver injury will be briefly discussed; and some of our recent studies on lipid peroxidation in relation to cell division and cancer will also be described.

Metabolic activation and free-radical products

Although many early biochemical studies were based upon the working hypothesis that CCl_4 was metabolically activated to $CCl_3^{\cdot}$ in the endoplasmic reticulum of liver, it took a relatively long time to obtain unequivocal evidence that this was indeed so. Experiments (Ingall *et al.* 1978) to demonstrate the formation of $CCl_3^{\cdot}$ in whole liver or in liver microsomes by direct electron spin resonance (e.s.r.) spectroscopy were not successful, probably because the concentration of $CCl_3^{\cdot}$ was too low for the sensitivity of the method, and because of the well known fact that liver samples examined by e.s.r. show an envelope of overlapping signals close to $g = 2$, where the absorption of $CCl_3^{\cdot}$ would occur. Attempts were then made to establish the occurrence of $CCl_3^{\cdot}$ by the less direct method of e.s.r. spin trapping, but no clear evidence was obtained by using *N*-methyl-nitrosopropane (Ingall *et al.* 1978). McCay's group introduced the spin trap phenylbutyl nitrone (PBN) for biological use and obtained good evidence for the production of $CCl_3^{\cdot}$ from CCl_4 in liver microsomes and in liver *in vivo* (Poyer *et al.* 1978, 1980). Consistent and complementary studies (Albano *et al.* 1982) have provided unequivocal evidence for the formation of $CCl_3^{\cdot}$ by liver microsomes plus NADPH, in isolated hepatocytes and *in vivo*. These spin-trapping studies, together with radioisotope labelling analysis of covalently bound products, and the formation of hexachloroethane and chloroform (for background see, for example, Slater 1972) establish beyond doubt that $CCl_3^{\cdot}$ is a 'normal' metabolite of CCl_4 in rat liver.

It is known that $CCl_3^{\cdot}$ reacts quickly with oxygen to yield the trichloromethyl peroxy radical $CCl_3O_2^{\cdot}$ (Packer *et al.* 1978) and that the $CCl_3O_2^{\cdot}$ radical is much more reactive chemically than $CCl_3^{\cdot}$. Attempts to trap $CCl_3O_2^{\cdot}$ in liver systems are hindered by the lability of peroxy adducts (Niki *et al.* 1983) and the high reactivity of $CCl_3O_2^{\cdot}$ with molecules in its immediate environment. Direct evidence for the interaction of $CCl_3^{\cdot}$ and O_2 has been obtained by low-temperature e.s.r. experiments in chemical model systems (see, for example, Symons *et al.* 1982), but a direct demonstration in liver samples has not yet been achieved. In recent work a secondarily derived peroxy species has been trapped (M. Davies, K. A. K. Lott & T. F. Slater 1985, unpublished results), but this is probably the lipid peroxy radical resulting from the interaction of $CCl_3O_2^{\cdot}$ with a polyunsaturated fatty acid PUFA:

$$CCl_3O_2^{\cdot} + PUFA \rightarrow PUFA^{\cdot} + CCl_3O_2H,$$

$$PUFA^{\cdot} + O_2 \rightarrow PUFAO_2^{\cdot}.$$

An apparently identical spectrum for this peroxy adduct has been found with Halothane ($CClBrH{-}CF_3$) and chloroform, although the relative yields are smaller.

Recent studies have demonstrated that spin-trap adducts can also be obtained from other halogenoalkanes: $CHCl_3$, $CHBr_3$, CHI_3, CH_2Br_2, Halothane, and dibromoethane (Tomasi *et al.* 1983*a*, *b*, 1984, 1985). In such experiments the concentration of O_2 is a critical feature (for Halothane see, for example, Tomasi *et al.* 1983*a*; for a discussion of this in relation to CCl_4 see, for example, Noll & de Groot 1984).

Reactivity of the free-radical intermediates

The chemical reactivity of $CCl_3^{\cdot}$ has been extensively studied in the liquid phase by the classical kinetic techniques of physical organic chemistry (Walling 1957), although most of the reactions so studied are not directly relevant to the biological situation. After the realization that CCl_4 has to undergo a metabolic activation to $CCl_3^{\cdot}$ to exert its full range of hepatotoxic effects, it was natural to consider that the early damaging reactions that occur were a result of the reactivity of the primary metabolite $CCl_3^{\cdot}$. To gain quantitative information about the chemical reactivity of $CCl_3^{\cdot}$ with important biomolecules, we decided in 1973 to approach that task by using the technique of pulse radiolysis (Willson & Slater 1975). During these early studies it was found that when the reactions were conducted under strictly anaerobic conditions the CCl_3 free radical is relatively unreactive; in fact, no detectable reaction could be observed with a range of compounds such as thiols, nucleotides, amino acids, etc. The introduction of small concentrations of O_2 greatly changed the kinetic features (Packer *et al.* 1978) and leads to the formation of the peroxy free radical $CCl_3O_2^{\cdot}$.

As already mentioned, $CCl_3O_2^{\cdot}$ has a much higher chemical reactivity with biomolecules in solution than $CCl_3^{\cdot}$. The rate constants for $CCl_3O_2^{\cdot}$ in such reactions fall generally in the range 10^6–10^9 $M^{-1}\,s^{-1}$ (Packer *et al.* 1978, 1981). In contrast, the reactions of $CCl_3^{\cdot}$ with such substances were undetectable with the pulse radiolysis system used and have rate constants less than 10^5 $M^{-1}\,s^{-1}$. This does not mean that the chemical reactivity of $CCl_3^{\cdot}$ is negligible in relation to cellular damage (far from it!), but that its reactivity is *relatively* much less than for $CCl_3O_2^{\cdot}$. Because $CCl_3O_2^{\cdot}$ reacts much faster with a PUFA (such as arachidonate) than does $CCl_3^{\cdot}$ it has been suggested that lipid peroxidation is preferentially initiated in the endoplasmic reticulum by $CCl_3O_2^{\cdot}$ rather than $CCl_3^{\cdot}$ (Slater 1982). Conversely, $CCl_3^{\cdot}$ may be responsible for most of the covalent binding detected in liver after exposure to $^{14}CCl_4$, as it is probable that any covalent binding of $CCl_3O_2^{\cdot}$ would be unstable during experimental work-up procedures (Slater 1982).

Finally, in this short account of the reactivity of $CCl_3^{\cdot}$ and $CCl_3O_2^{\cdot}$ it is interesting to note the study by Packer *et al.* (1981), which demonstrated the important influence of the chlorine substituents on chemical reactivity: the reactivity of the free radicals decreases in the order $CCl_3O_2^{\cdot} > CHCl_2O_2^{\cdot} > CH_2ClO_2^{\cdot} > CH_3O_2^{\cdot}$.

Locus of formation of $CCl_3^{\bullet}$ in liver endoplasmic reticulum

The metabolic activation of CCl_4 in liver endoplasmic reticulum probably occurs through a process of dissociative electron capture (Gregory 1966):

$$CCl_4 + e^- \rightarrow CCl_4^{-\bullet} \rightarrow CCl_3^{\bullet} + Cl^-.$$

In principle, the electron could be supplied directly by the NADPH–cytochrome-P_{450} system (for example, via the NADPH–flavoprotein or cytochrome P_{450}), or indirectly by donation from secondary reductants, such as the superoxide anion radical or an iron chelate, which are themselves reduced by the primary NADPH–cytochrome-P_{450} electron-transport chain. Whether direct or indirect, the activation is certainly closely associated with the NADPH–cytochrome-P_{450} system as many studies have demonstrated (for a review see, for example, Slater 1972, 1982).

A possible secondary route of activation that has already been mentioned is through an interaction of O_2^- and CCl_4 in the non-polar environment of the endoplasmic reticulum. It is known from the work of Sawyer and his colleagues that O_2^- can reduce CCl_4 in aprotic media (Roberts & Sawyer 1981). This pathway, if it occurs at all in the biological situation, would be aided by the destructive effect of CCl_4 activation on cytochrome P_{450} (Glende 1972), thereby encouraging electron outflow to O_2. In liver microsomal suspensions, however, the activation of lipid peroxidation by CCl_4 is not significantly diminished by the addition of superoxide dismutase (O. P. Sharma, K. H. Cheeseman & T. F. Slater 1985, unpublished results). The latter point, together with the effects of several free-radical scavengers on the microsomal lipid peroxidation which is stimulated by CCl_4, are illustrated in table 1.

Table 1. Effects of superoxide dismutase (SDM) and other substances on CCl_4-stimulated lipid peroxidation in rat-liver microsomes[a]

addition	concentration/μM or units	percentage inhibition	EC_{50}/μM[b]
SDM	1000 units	15	—
urate	200	2	—
	500	10	—
indomethacin	20	0	—
	50	12	—
propyl gallate	—	—	2.0
promethazine	—	—	0.5
metiazinic acid	—	—	33
nafazatrom	—	—	16

[a] Data from Sharma *et al.* (1985). [b] Concentration producing a 50% inhibition.

The discussion above has briefly considered the interactions of CCl_4 with the NADPH–cytochrome-P_{450} electron-transport chain; another feature of CCl_4 activation to consider, however, is the location of activation among the different regions of the liver lobules. Administration of CCl_4 *per os*, by inhalation, or by injection into the peritoneum produces necrosis that is essentially centrilobular in nature. It is reasonable to assume that a major contribution to this lobular location of injury is the distribution of the NADPH–cytochrome-P_{450} system itself. It is known (Gooding *et al.* 1978) that cytochrome P_{450} is more concentrated in the centrilobular regions of rat liver than in the periportal regions. However, other factors may contribute significantly: for example, the lobular gradient of O_2 (Ji *et al.* 1982), which may ensure the optimal conditions for lipid peroxidation (Noll & de Groot 1984); and the lobular distribution of protective mechanisms, about which little is known, although glutathione is preferentially distributed periportally (Smith *et al.* 1979).

In connection with the metabolic activation of CCl_4 in tissues other than liver, it is known that the NADPH–flavoprotein and cytochrome P_{450} are widely distributed, even if the amounts of P_{450} tissue may be small in many tissues (Benedetto *et al.* 1981). Covalent binding of $^{14}CCl_4$, and adduct formation with the spin trap phenylbutyl nitrone (PBN) *in vitro* generally were correlated with the tissue distribution of P_{450} (Benedetto *et al.* 1981). Studies *in vivo* with PBN (A. Tomasi & T. F. Slater, unpublished results) give broadly similar results. Because cytochrome P_{450} is known to be located in specific cell types in various non-hepatic tissues (such as lung), it is probable that the overall measures of covalent binding, spin-adduct formation, stimulation of lipid peroxidation, etc., expressed per gram wet mass of tissue, hide much greater extents of activation and damage in specific cell types.

Damage to the plasma membrane

Most studies on the damaging actions of CCl_4 on isolated liver cell membranes have been on microsomes, for obvious reasons. The earliest damage *in vivo* that is morphologically evident is to the endoplasmic reticulum (Oberling & Rouiller 1956) and the NADPH–cytochrome-P_{450} system is firmly associated with this intracellular membrane function. Studies with isolated hepatocytes, however, have demonstrated relatively early damage to the plasma membrane when exposed to CCl_4 or Halothane (Perrissoud *et al.* 1981; Jewell *et al.* 1982; Tomasi *et al.* 1983*a*); morphologically this damage is evident as substantial blebbing. It is thus of interest to consider the possibility that CCl_4 is activated by the plasma membrane.

We have isolated highly purified plasma membrane subfractions from rat liver, by using modifications of the method of Wisher & Evans (1975), and have studied the effects of CCl_4 in relation to covalent binding, spin trapping and stimulation of lipid peroxidation. No evidence was found (table 2; Le Page *et al.* 1985) for any significant activation of CCl_4 by the highly

Table 2. Rat liver plasma membranes and activation of CCl_4 compared with data obtained from rat-liver microsomal suspensions.[a] Also shown are some results obtained after exposing normal rat hepatocytes and Novikoff tumour cells to γ-radiation

	microsomes (percentage)	plasma membrane (percentage)
cytochrome P_{450}	100	12
NADPH–cytochrome *c* reductase	100	20
$C_{20:4}$ (percentage total fatty acids)	29	26
NADPH–CCl_4-stimulated lipid peroxidation	100	4
$^{14}CCl_4$ covalent binding	100	3
ascorbate–Fe^{2+}-stimulated lipid peroxidation	100	114
lipid peroxidation stimulated by γ-irradiation	100	73
Novikoff cells against isolated normal hepatocytes as 100%:	normal hepatocytes	Novikoff cells
lipid peroxidation stimulated by γ-irradiation	100	2

[a] Values from Le Page *et al.* (1985).

purified sinusoidal membrane subfraction; less extensive data for the lateral membrane and canalicular membrane subfractions also point to the same conclusion. In relation to the canalicular membrane data it is known from studies *in vivo* that bile flow is not substantially altered in the early stages of CCl_4 intoxication (Delaney & Slater 1971).

We can conclude from these results that the liver plasma membrane does not significantly metabolize CCl_4 to the $CCl_3^{\cdot}$ free radical. In consequence, the early changes seen in the plasma membrane of isolated hepatocytes may reflect secondary consequences of metabolic activation in the endoplasmic reticulum or, less likely in our view, artefactual perturbations in the plasma membrane of the isolated hepatocyte resulting from the proteolytic method of preparation.

Reactions of $CCl_3^{\cdot}$ and $CCl_3O_2^{\cdot}$

A reactive oxidizing species such as $CCl_3O_2^{\cdot}$ (and, to a lesser extent, $CCl_3^{\cdot}$) can be expected to interact in damaging ways with a variety of substances in the local environment around the locus of metabolic activation (Slater 1984*a*). For example, primary consequences of the metabolic activation of CCl_4 can be expected to be (i) oxidation of thiol groups which may be essential for enzyme activity; (ii) covalent binding to lipid protein, nucleotides, haem, etc., that may greatly change or even destroy biochemical activity. An example is the destruction of NADPH in the liver *in vitro*, during the early phase of CCl_4-induced liver injury (Slater *et al.* 1964); and (iii) by initiating lipid peroxidation which can result in membrane disturbances due to loss of PUFA, cross-linking and production of reactive products.

The destruction of NADPH referred to above can also be studied *in vitro* by using isolated hepatocytes; figure 1 gives some corresponding data with the powerful hepatotoxic agent

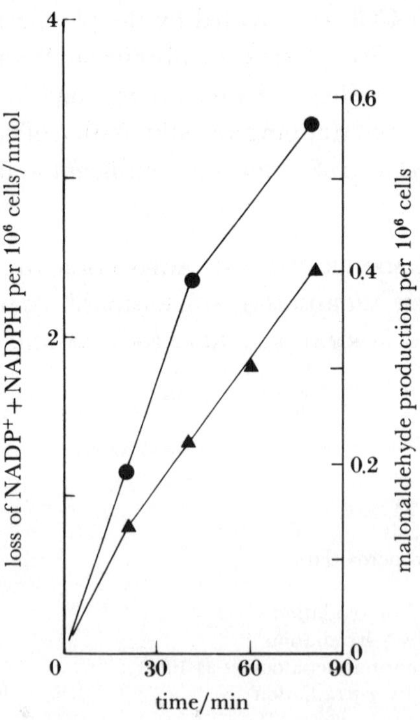

FIGURE 1. Effect of incubating normal isolated rat hepatocytes with CCl_3Br; ●, loss of $NADP^+$+NADPH; ▲, production of malonaldehyde (arbitrary absorbance units at A_{535}). Data from R. Scott & T. F. Slater, unpublished results.

CCl_3Br. It was found (R. Scott & T. F. Slater, unpublished results) that the loss of $NADP^+$ and NADPH during incubation of rat hepatocytes with CCl_3Br correlated with the amount of malonaldehyde-like material produced. CCl_3Br can also be used to study reactions of $CCl_3^{\bullet}$ in solution because CCl_3Br is photochemically degraded to $CCl_3^{\bullet}$ and other products. For example, exposure of a mixture of CCl_3Br and the spin trap PBN to a wide-spectrum mercury lamp results in the appearance of the PBN–CCl_3 adduct, which can be readily detected by e.s.r. (M. Davies, K. A. K. Lott & T. F. Slater 1985, unpublished results). An interesting interaction of $CCl_3^{\bullet}$ with membrane PUFA has recently been discussed by Link *et al.* (1984). In this interaction, the addition of $CCl_3^{\bullet}$ across a double bond is followed by cross-linking with neighbouring fatty-acid chains. If this were of appreciable extent then it could result in a decreased membrane fluidity, with associated consequences for membrane function. In this context it is worth noting that lipid peroxidation is known to decrease microsomal membrane fluidity (Dobretsov *et al.* 1977; Slater 1979). It can be seen from the discussion earlier that CCl_4 can produce a variety of damaging effects on liver cells. In consequence, we believe that the liver damage produced by CCl_4 is multicausal in origin (Slater 1982), and that the relative contributions of such important features as covalent binding and lipid peroxidation to the overall cellular perturbation will vary somewhat with the particular experimental system under study.

Lipid peroxidation

Changes consistent with a stimulation of lipid peroxidation can be detected in liver *in vivo* very shortly after administering CCl_4 to a rat (Rao & Recknagel 1969). With isolated hepatocytes a marked lipid peroxidation can be detected on adding CCl_4 to the incubating medium (Poli *et al.* 1979), without a significant lag phase and long before the appearance of signs of major cell injury, such as the loss of trypan blue staining and leakage of cytoplasmic enzymes. In cytochemical experiments with isolated hepatocytes incubated with CCl_4 and then stained for products of lipid peroxidation (G. Nöhammer, E. Schauenstein, G. Poli, M. U. Dianzani & T. F. Slater 1985, unpublished results), the increased amount of peroxidation products was evident in almost all cells examined, thereby eliminating the possibility that early increases in lipid peroxidation in isolated hepatocytes are confined to a small percentage of the total cell population and, moreover, a small percentage that is composed of 'dead cells'.

There are interesting species differences in the extents of CCl_4-stimulated lipid peroxidation in liver microsomes. Although it was claimed that mouse-liver microsomes did not readily peroxidize when incubated with CCl_4 (Toranzo *et al.* 1978), this result was contradicted by Lee *et al.* 1982). We also find (Proudfoot *et al.* 1985) good peroxidative activity in mouse (and guinea pig) microsomes in comparison with the rat, but rabbit microsomes are much less active (table 3).

Lipid peroxidation is known to produce a variety of products (Slater 1984*b*), including biologically reactive lipid hydroperoxides and aldehydes such as 4-hydroxy-alkenals (Esterbauer *et al.* 1982; Poli *et al.* 1985). Because these products have much longer half-lives in the biological environment in which they are formed than $CCl_3^{\bullet}$ and $CCl_3O_2^{\bullet}$, they can diffuse for much greater distances, even to extracellular regions. With isolated hepatocytes following incubation with ADP–iron or CCl_4, for example, the suspending medium contains significant amounts of aldehydic products of cellular lipid peroxidation (Poli *et al.* 1985). From such considerations we may understand how a precisely localized metabolic activation in the endoplasmic reticulum can result in metabolic disturbances at considerable distances, because of the biological

TABLE 3. CCl_4-STIMULATED LIPID PEROXIDATION IN LIVER MICROSOMAL SUSPENSIONS PREPARED FROM DIFFERENT ANIMALS

species	number of experiments	stimulation of lipid peroxidation[a]
rat	6	285 ± 27
mouse	5	277 ± 46
guinea pig	7	269 ± 23
rabbit	3	100 ± 18

[a] Values are picomoles of malonaldehyde per minute per milligram of protein (Proudfoot *et al.* 1985). Mean values are given ±s.e.m.

reactivity and diffusion of products of lipid peroxidation. Figure 2 illustrates this concept (Slater 1976).

Because lipid hydroperoxides can affect the activity of cyclo-oxygenase and of other enzymes of the prostaglandin cascade (Hemler *et al.* 1979), it is of interest to consider whether liver injury due to CCl_4 is associated with disturbances of liver eicosanoid metabolism. Our preliminary studies (S. Hewertson, R. G. McDonald-Gibson, J. Hurst, A. Morgan & T. F. Slater 1985, unpublished results) indicate that CCl_4 administration *in vivo* produces an increase in the thromboxanc content of liver (measured as TXB_2), but the cellular origin of this material remains to be investigated (Spolarics *et al.* 1984). The increase in TXB_2, if confirmed, is interesting in view of the suggestion (Kanzaki *et al.* 1979) that thromboxane is the trigger for liver-cell division after partial hepatectomy. It is well known that CCl_4-induced liver necrosis is followed by a regenerative phase.

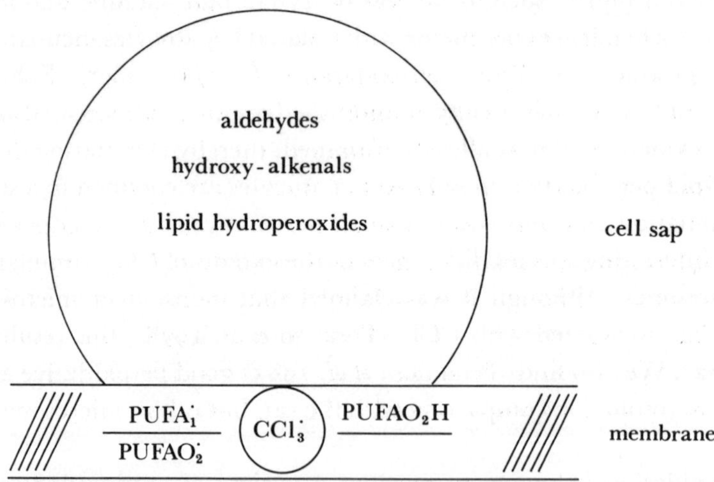

FIGURE 2. A diagrammatic impression of the metabolic activation of CCl_4 to the trichloromethyl radical in the membranes of the liver endoplasmic reticulum. The diffusion of $CCl_3^{\cdot}$ (and even more so, of $CCl_3O_2^{\cdot}$) is shown as restricted to the micro-environment of the site of activation owing to the chemical reactivity of the free-radical intermediate. Other products of the lipid peroxidation, which results from the attack of $CCl_3O_2^{\cdot}$ on membrane polyunsaturated fatty acids, are shown diffusing in the plane of the membrane, or escaping from the membrane into the cytosol. Modified from Slater (1976).

Protection

Protection against the type of cell injury produced by CCl_4 can be achieved, at least in part, in many ways (for a review see, for example, Slater 1978), including the efficient scavenging of the primary reactive free radicals.

Because of the restricted diffusion and short lifetimes of reactive free radicals such as $CCl_3O_2^{\cdot}$ (and $CCl_3^{\cdot}$), it is evident that effective scavenging of such species must satisfy a number of demanding criteria (Slater 1981); the scavenger must penetrate to the precise intracellular locus of metabolic activation; it must achieve a local concentration sufficient to compete successfully with neighbouring biomolecules that would otherwise be 'damaged' by the attack of $CCl_3^{\cdot}$ or $CCl_3O_2^{\cdot}$; it must reach the critical zones of metabolic activation in time to prevent seriously damaging secondary radical and non-radical processes from occurring; and its own cytotoxic effects must be acceptably low. However, because most free-radical scavengers have many other actions *in vivo* that may significantly affect the development of the injury under study, then, as pointed out elsewhere (Slater 1984*a*), even where a substance has an established scavenging activity, and is present *in vivo* in concentrations appropriate for effective scavenging, it cannot be assumed that this feature is the only or even the major mechanism by which it exerts protective functions *in vivo*.

Liver tumours

Repeated doses of CCl_4 produce cirrhosis and may produce tumours of the liver; CCl_4 is not especially effective as a liver carcinogen (for a literature review see, for example, World Health Organization 1979), and is essentially unreactive in modified Ames tests for mutagenicity (World Health Organization 1979).

Although CCl_4 stimulates lipid peroxidation in acute situations *in vivo* or during incubations *in vitro*, and can produce liver tumours under appropriate conditions, it is of interest that many liver tumours have a much reduced rate of lipid peroxidation (Slater *et al.* 1984; Burton *et al.*

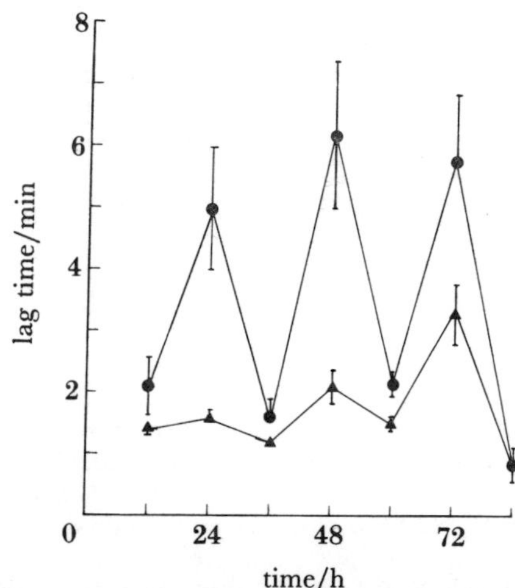

FIGURE 3. Changes in lipid peroxidation (as measured by the lag time in the NADPH–ADP–Fe^{2+} microsomal system) in regenerating rat liver after partial hepatectomy. The data are from Cheeseman *et al.* (1985). Sham-operated rat-liver samples (▲); regenerating liver samples (●).

1983). Novikoff cells even peroxidize very slowly when γ-irradiated by ^{60}Co (see table 2). A major reason for the slow rate of peroxidation is the increased content of α-tocopherol (Cheeseman *et al.* 1984).

Of course, it is possible that such changes are not directly related to malignant transformation but are perhaps reflections of an increased propensity for cell division. We have investigated this aspect by using regenerating liver after partial hepatectomy as a model of liver cells that are greatly stimulated to divide. In preliminary studies (Cheeseman *et al.* 1985) we have found that there are big changes in the rate of lipid peroxidation at the times of cell division (figure 3).

We are grateful for financial assistance supporting the studies described here from the National Foundation for Cancer Research, the Association for International Cancer Research, the Medical Research Council, the Cancer Research Campaign and Ciba-Geigy p.l.c. We also acknowledge the scientific assistance and stimulation generously given by our collaborating colleagues, notably Professor M. U. Dianzani, Professor G. Poli, Professor H. Esterbauer, Professor R. L. Willson, Professor J. E. Packer, Dr E. Albano, Dr A. Tomasi and Professor P. A. Riley.

References

Albano, E., Lott, K. A. K., Slater, T. F., Stier, A., Symons, M. R. C. & Tomasi, A. 1982 Spin trapping studies on the free radical products formed by metabolic activation of carbon tetrachloride in rat liver microsomal fractions, isolated hepatocytes and *in vivo* in the rat. *Biochem. J.* **204**, 593–603.

Benedetto, C., Dianzani, M. U., Ahmed, M., Cheeseman, K., Connelly, C. & Slater, T. F. 1981 NADPH–cytochrome c reductase, cytochrome P_{450}, other microsomal enzyme activities, and activation of CCl_4 in rat tissues. *Biochim. biophysis. Acta* **677**, 373–372.

Burton, G. W., Cheeseman, K. H., Ingold, K. U. & Slater, T. F. 1983 Lipid peroxidation and products of lipid peroxidation as potential tumour protective agents. *Biochem. Soc. Trans.* **11**, 261–262.

Butler, T. C. 1961 Reduction of carbon tetrachloride *in vivo* and reduction of carbon tetrachloride and chloroform *in vitro* by tissues and tissue constituents. *J. Pharmac. exp. Ther.* **134**, 311–319.

Cameron, G. R. & Karunaratne, W. A. E. 1936 Carbon tetrachloride cirrhosis in relation to liver regeneration. *J. Path. Bact.* **42**, 1–21.

Cheeseman, K. H., Benedetto, C., Burton, G. W., Collins, M., Ingold, K. U., Maddix, S., Milia, A., Proudfoot, K. & Slater, T. F. 1985 Lipid peroxidation in regenerating rat liver. (In preparation.)

Cheeseman, K. H., Burton, G. W., Ingold, K. U. & Slater, T. F. 1984 Lipid peroxidation and lipid antioxidants in normal and tumour cells. *Toxic. Path.* **12**, 235–239.

Delaney, V. B. & Slater, T. F. 1971 The effects of various drugs and toxic agents on bile flow rate and composition in the rat. *Toxic. appl. Pharmac.* **20**, 157–174.

Dobretsov, G. E., Borschevskaya, T. A., Petrov, V. A. & Vladimirov, Y. U. 1977 The increase of phospholipid bilayer rigidity after lipid peroxidation. *FEBS Lett.* **84**, 125–128.

Esterbauer, H. E., Cheeseman, K., Dianzani, M. U., Poli, G. & Slater, T. F. 1982 Separation and characterization of the aldehydic products of lipid peroxidation stimulated by ADP–Fe^{2+} in rat liver microsomes. *Biochem. J.* **208**, 129–140.

Ghoshal, A. K. & Recknagel, R. O. 1965 Positive evidence of acceleration of lipoperoxidation in rat liver by carbon tetrachloride: *in vitro* experiments. *Life Sci.* **4**, 1521–1530.

Glende, E. A. Jr 1972 Carbon tetrachloride-induced protection against carbon tetrachloride toxicity: the role of the liver microsomal drug-metabolizing system. *Biochem. Pharmac.* **21**, 1697–1702.

Gooding, P. E., Chayen, J., Sawyer, B. & Slater, T. F. 1978 Cytochrome P_{450} distribution in rat liver and the effect of sodium phenobarbitone administration. *Chem. Biol. Interact.* **20**, 299–310.

Gregory, N. L. 1966 Carbon tetrachloride toxicity and electron capture. *Nature, Lond.* **212**, 1460–1461.

Hemler, M. E., Cook, H. W. & Lands, W. E. M. 1979 Prostaglandin synthesis can be triggered by lipid peroxides. *Archs Biochem. Biophys.* **193**, 340–345.

Ingall, A., Lott, K. A. K., Slater, T. F., Finch, S. & Stier, A. 1978 Metabolic activation of carbon tetrachloride to a free radical product: studies using a spin trap. *Biochem. Soc. Trans.* **6**, 962–964.

Jewell, S. A., Bellomo, G., Thor, H., Orrenius, S. & Smith, M. T. 1982 Bleb formation in hepatocytes during drug metabolism is caused by disturbances in thiol and calcium ion homeostasis. *Science, Wash.* **217**, 1257–1259.

Ji, S., Le Masters, J. J., Christenson, V. & Thurman, R. G. 1982 Periportal and pericentral pyridine nucleotide fluorescence from the surface of the perfused liver: evaluation of the hypothesis that chronic treatment with ethanol produces pericentral hypoxia. *Proc. natn. Acad. Sci. U.S.A.* **79**, 5415–5419.

Kanzaki, Y., Mahmud, I., Asanagi, M., Fukui, N. & Miura, Y. 1979 Thromboxane as a possible trigger of liver regeneration. *Cell Molec. Biol.* **25**, 147–152.

Lee, P. Y., McCay, P. B. & Hornbrook, K. R. 1982 Evidence for carbon tetrachloride-induced lipid peroxidation in mouse liver. *Biochem. Pharmac.* **31**, 405–409.

Le Page, R. N., Cheeseman, K. H. & Slater, T. F. 1985 Lipid peroxidation in purified plasma membrane fractions prepared from rat liver in relation to the hepatotoxicity of carbon tetrachloride. (In preparation.)

Link, B., Dürk, H., Thiel, D. & Frank, H. 1984 Binding of trichloromethyl radicals to lipids of the hepatic endoplasmic reticulum during tetrachloromethane metabolism. *Biochem. J.* **223**, 577–586.

Niki, E., Yokoi, S., Tsuchiya, J. & Kamiya, Y. 1983 Spin trapping of peroxy radicals by phenyl-*N*-(tert-butyl)nitrone and methyl-*N*-duryl nitrone. *J. Am. chem. Soc.* **105**, 1498–1503.

Noll, T. & de Groot, H. 1984 The critical steady-state in hypoxic conditions in carbon tetrachloride-induced lipid peroxidation in rat liver microsomes. *Biochim. biophys. Acta* **795**, 356–362.

Oberling, Ch. & Rouiller, Ch. 1956 Les effets de l'intoxication aiguë au tétrachlorure de carbone sur le foie du rat. *Ann. anat. Path.* **1**, 401–407.

Packer, J. E., Mahood, J. S., Mora-Arellano, V. O., Slater, T. F., Willson, R. L. & Wolfenden, B. S. 1981 Free radicals and singlet oxygen scavengers: reactions of a peroxy-radical with β-carotene, diphenylfuran and 1,4-diazobicyclo(2,2,2) octane. *Biochem. biophys. Res. Commun.* **98**, 901–906.

Packer, J. E., Slater, T. F. & Willson, R. L. 1978 Reactions of the carbon tetrachloride-related peroxy free radical ($CCl_3O_2^{\cdot}$) with amino acids: pulse radiolysis evidence. *Life Sci.* **23**, 2617–2620.

Packer, J. E., Willson, R. L., Bahnemann, D. & Asmus, K.-D. 1980 Electron transfer reactions of halogenated aliphatic peroxyl radicals: measurement of absolute rate constants by pulse radiolysis. *J. chem. Soc. Perkin Trans.* II, pp. 296–299.

Perrissoud, D., Anderset, G., Reymond, O. & Maignan, M. F. 1981 The effect of carbon tetrachloride on isolated rat hepatocytes: early morphological alterations of the plasma membrane. *Virchows Arch. path. Anat. Physiol.* (*Cell Path.*) **35**, 83–91.

Poli, G., Dianzani, M. U., Cheeseman, K. H., Slater, T. F., Lang, J. & Esterbauer, H. 1985 Separation and characterisation of the aldehydic products of lipid peroxidation stimulated by carbon tetrachloride on ADP-iron in isolated rat hepatocytes and rat liver microsomal suspensions. *Biochem. J.* (In the press.)

Poli, G., Gravela, E., Albano, E. & Dianzani, M. I. 1979 Studies on fatty liver with isolated hepatocytes. II. The action of carbon tetrachloride on lipid peroxidation, protein, and triglyceride synthesis and secretion. *Expl Mol. Path.* **30**, 116–127.

Poyer, J. L., Floyd, R. A., McCay, P. B., Janzen, E. G. & Davis, E. R. 1978 Spin-trapping of the trichloromethyl radical produced during enzymic NADPH oxidation in the presence of carbon tetrachloride or bromotrichloromethane. *Biochim. biophys. Acta* **539**, 402–409.

Poyer, J. L., McCay, P. B., Lai, E. K., Janzen, E. G. & Davis, E. R. 1980 Confirmation of assignment of the trichloromethyl radical spin adduct detected by spin trapping during ^{13}C-carbon tetrachloride metabolism *in vitro* and *in vivo*. *Biochem. biophys. Res. Commun.* **94**, 1154–1160.

Proudfoot, K., Cheeseman, K. H. & Slater, T. F. 1985 Studies on lipid peroxidation in microsomal suspensions prepared from livers of rats, mice, guinea pigs and rabbits. (In preparation.)

Rao, K. S. & Recknagel, R. O. 1969 Early incorporation of carbon-labelled carbon tetrachloride into rat liver particulate lipids and proteins. *Expl Mol. Pathol.* **10**, 219–228.

Recknagel, R. O. 1967 Carbon tetrachloride toxicity. *Pharmac. Rev.* **19**, 145–208.

Roberts, J. L. Jr & Sawyer, D. T. 1981 Facile degradation by superoxide ion of carbon tetrachloride, chloroform, methylene chloride and p,p′-DDT in aprotic media. *J. Am. chem. Soc.* **103**, 712–714.

Slater, T. F. 1966 Necrogenic action of carbon tetrachloride in the rat: a speculative hypothesis based on activation. *Nature, Lond.* **209**, 36–40.

Slater, T. F. 1972 *Free radical mechanisms in tissue injury*, pp. 1–283. London: Pion Ltd.

Slater, T. F. 1976 Biochemical pathology in microtime. In *Recent advances in biochemical pathology: toxic liver injury* (ed. M. U. Dianzani, G. Ugazio & L. M. Sena), pp. 381–390. Torino: Minerva Medica.

Slater, T. F. 1978 Mechanisms of protection. In *Biochemical mechanisms of liver injury* (ed. T. F. Slater), pp. 745–801. London: Academic Press.

Slater, T. F. 1979 Biochemical studies of transient intermediates in relation to chemical carcinogenesis. In *Submolecular biology and cancer* (*Ciba Foundation Symposium no. 67*) (ed. G. E. W. Wolstenholme, D. W. FitzSimons & J. Whelan), pp. 301–328. Amsterdam: Excerpta Medica.

Slater, T. F. 1981 Free radical scavengers. In *International workshop on (+)-cyanidanol-3 in diseases of the liver* (ed. H. O. Conn), pp. 11–15. Royal Society of Medicine International Congress and Symposium Series no. 47. Academic Press and Grune & Stratton: Royal Society of Medicine.

Slater, T. F. 1982 Activation of carbon tetrachloride: chemical principles and biological significance. In *Free radicals, lipid peroxidation and cancer* (ed. D. C. H. McBrien & T. F. Slater), pp. 243–270. London: Academic Press.

Slater, T. F. 1984*a* Free radical mechanisms in tissue injury. *Biochem. J.* **222**, 1–15.

Slater, T. F. 1984*b* An overview of methods used for detecting lipid peroxidation. In *Oxygen radicals in biological systems. Methods in enzymology, vol. 105* (ed. L. Packer), pp. 283–293. New York: Academic Press.

Slater, T. F., Benedetto, C., Burton, G. W., Cheeseman, K. H., Ingold, K. U. & Nodes, J. T. 1984 Lipid peroxidation in animal tumours: a disturbance in the control of cell division? In *Icosanoids and cancer* (ed. H. Thaler-Dao, A. Crastes de Paulet & R. Paoletti), pp. 21–29. New York: Raven Press.

Slater, T. F., Sträuli, U. & Sawyer, B. 1964 Changes in liver nucleotide concentrations in experimental liver injury. 1. Carbon tetrachloride poisoning. *Biochem. J.* **93**, 260–266.

Smith, M. T., Loveridge, N., Wills, E. D. & Chayen, J. 1979 The distribution of glutathione in the rat liver lobule. *Biochem. J.* **182**, 103–108.

Spolarics, Z., Tauács, B., Garzo, T., Mandl, J., Mucha, I., Antoni, F., Machovich, R. & Horvath, I. 1984 Prostaglandin and thromboxane synthesizing activity in isolated murine hepatocytes and non-parenchymal liver cells. *Prostagland. Leuk. Med.* **16**, 379–388.

Symons, M. C. R., Albano, E., Slater, T. F. & Tomasi, A. 1982 Radiolysis of tetrachlormethane. *J. chem. Soc. Faraday Trans.* I **78**, 2205–2214.

Tomasi, A., Albano, E., Biasi, F., Slater, T. F., Vannini, V. & Dianzani, M. U. 1985 Activation of chloroform and related trihalomethanes to free radical intermediates in isolated hepatocytes as detected by the e.s.r. spin trapping technique. *Chem. Biol. Interact.* (Submitted.)

Tomasi, A., Albano, E., Bini, A., Botti, E., Slater, T. F. & Vannini, V. 1984 Free radical intermediates under hypoxic conditions in the metabolism of halogenated hydrocarbons. *Toxic. Path.* **12**, 240–246.

Tomasi, A., Albano, E., Dianzani, M. U., Slater, T. F. & Vannini, V. 1983*a* Metabolic activation of 1,2-dibromoethane to a free radical intermediate by rat liver microsomes and isolated hepatocytes. *FEBS Lett.* **160**, 191–194.

Tomasi, A., Billing, S., Garner, A., Slater, T. F. & Albano, E. 1983*a* The metabolism of halothane in hepatocytes: a comparison between free radical spin trapping and lipid peroxidation in relation to cell damage. *Chem. Biol. Interact.* **46**, 353–368.

Toranzo, E. G. D. de, Díaz Gómez, M. I. & Castro, J. A. 1978 Carbon tetrachloride activation, lipid peroxidation and liver necrosis in different strains of mice. *Res. Commum. Chem. Pathol. Pharmacol.* **19**, 347–352.

Walling, Ch. 1957 *Free radicals in solution.* New York and London: John Wiley.

Willson, R. L. & Slater, T. F. 1975 CCl_4 and biological damage: pulse radiolysis studies of associated free radical reactions. In *Fast processes in radiation chemistry and biology* (ed. G. E. Adams, E. M. Fielden & B. D. Michael), pp. 147–161. New York: John Wiley.

Wirtschafter, Z. T. & Cronyn, M. W. 1964 Free radical mechanism for solvent toxicity. *Environ. Health* **9**, 186–191.

Wisher, M. H. & Evans, W. H. 1975 Functional polarity of the rat hepatocyte surface membrane: isolation and characterization of plasma-membrane sub-fractions from the blood-sinusoidal, bile-canalicular and contiguous surfaces of the hepatocyte. *Biochem. J.* **146**, 375–388.

World Health Organization 1979 IARC Monographs on the evaluation of the carcinogenic risk of chemicals to humans, vol. 20 (some halogenated hydrocarbons). Lyon: International Agency for Research on Cancer.

Discussion

Catherine Rice-Evans (*Department of Biochemistry, Royal Free Hospital School of Medicine*). Could Professor Slater please explain his observation that hydroxy-alkenals inhibit the aggregation of platelets? Which stimulating agents are involved and what is the mechanism of action?

T. F. Slater. I am unable to give much detail in reply to this question as the results are still preliminary and unpublished. However, I can say that, together with Dr John Hurst in Brunel, we have found that 4-hydroxy-nonenal inhibits platelet aggregation by ADP or arachidonate but has little effect on aggregation stimulated by thrombin, collagen or calcium ionophore. We have also found that 4-hydroxy-nonenal is much more active in these respects than 4-hydroxy-pentenal.

H. Sies (*Institut für Physiologische Chemie I, Universität Düsseldorf, F.R.G.*). The liver seems to be particularly vulnerable at the perivenous end of the liver lobule, given the activation of CCl_4 and the low oxygen tension. Further, as Smith *et al.* (1979) have shown, there are lower contents of GSH in perivenous than periportal cells. Regarding other antioxidants, is the subcellular distribution of vitamin E known?

Reference

Smith, M. T., Loveridge, N., Wills, E. D. & Chayen, J. 1979 The distribution of glutathione in the rat liver lobule. *Biochem. J.* **182**, 103–108.

T. F. Slater. Not so far as I know. Quite clearly, the intralobular distribution of protective agents and enzymes is one important aspect to consider in relation to the intralobular location of the injury, in addition to the location and activity of associated activating systems (such as the NADPH–P_{450} pathway), and other necessary components such as O_2. These aspects have intrigued us for many years but specific data on vitamin E and related free-radical scavengers are lacking.

Phil. Trans. R. Soc. Lond. B **311**, 647–657 (1985)
Printed in Great Britain

Paraquat toxicity

By L. L. Smith

Imperial Chemical Industries p.l.c., Central Toxicology Laboratory, Alderley Park, Macclesfield, Cheshire SK10 4TJ, U.K.

Paraquat (1,1′-dimethyl-4,4′-bipyridylium dichloride) is marketed as a contact herbicide. Although it has proved safe in use there have been a number of cases of poisoning after the intentional swallowing of the commercial product. The most characteristic feature of poisoning is lung damage, which causes severe anoxia and may lead to death. The specific toxicity to the lung can be explained in part by the accumulation of paraquat into the alveolar type I and type II epithelial cells by a process that has been shown to accumulate endogenous diamines and polyamines. When accumulated, paraquat undergoes an NADPH-dependent, one-electron reduction to form its free radical, which then reacts avidly with molecular oxygen to reform the cation and produce superoxide anion, which in turn will dismutate to form H_2O_2. This may lead to the formation of more reactive (and hence toxic) radicals which have the potential to cause lipid peroxidation and lead to cell death.

Biochemical changes provoked by paraquat in the lung suggest that it causes a rapid, pronounced and prolonged oxidation of NADPH that initiates compensatory biochemical processes in the lung. NADPH may be further depleted as it is consumed in an attempt to detoxify H_2O_2 or lipid hydroperoxides. Thus it is possible that with toxic levels of paraquat in the cell, compensatory biochemical processes are insufficient to maintain levels of NADPH consistent either with cell survival or with the ability to detoxify H_2O_2 or prevent lipid peroxidation.

Introduction

Paraquat (1,1′-dimethyl-4,4′-bipyridylium dichloride) was first described in the literature in the latter part of the nineteenth century by Weidel & Rosso (1882). In 1933 its redox properties were discovered by Michaelis & Hill (1933) and since then it has been used as a redox indicator, known by its trivial name methyl viologen. Its herbicidal properties were discovered in the mid-1950s and it has been marketed as a herbicide for approximately 25 years. Although it has proved remarkably safe in use there have been a number of fatalities largely as a consequence of the intentional swallowing of the concentrated commercial product for suicidal purpose (Fletcher 1974).

When paraquat is swallowed the symptoms of poisoning depend largely on the amount consumed. Those who die from paraquat poisoning can be divided into two broad categories; (i) those who die within one to five days of its ingestion and (ii) those who die later than five days and as long as several weeks after ingestion. In the cases of death which occur within a few days of poisoning, extremely large amounts of paraquat have been ingested. Death, when it occurs, results from multi-organ failure associated with damage to the adrenal gland (Nagi 1970), liver (Fennelly *et al.* 1971), kidney (Oreopoulos *et al.* 1968), lung (Bronkhurst *et al.* 1968) and brain (Nienhaus & Ehrenfeld 1971). In these cases of poisoning the precise cause of death is difficult to diagnose because so many of the vital organs have been affected. In patients who survive five days or longer but eventually die from paraquat poisoning, the most characteristic features of poisoning are damage to the lung and kidney (Fairshter *et al.* 1976), with the cause of death usually attributed to anoxia as a consequence of extensive lung damage.

In this paper the mechanism of toxicity of paraquat in the lung will be described. While it is generally agreed that it is the cyclical reduction and reoxidation of paraquat that represents its primary mechanism of toxicity, consideration will also be given to the delivery of paraquat to the lung, the biochemical consequences of redox cycling and the pathology that results.

Pathology in the lung

The toxic effects of paraquat in experimental animals was first reported by Clark *et al.* (1966) who showed that the histological effects of paraquat in rats, mice and dogs are similar. The most extensive studies on the pathogenesis of paraquat-induced lung damage have been performed on rats. Within 24 h of the administration of an approximate LD_{50} dose of paraquat the alveolar type I and type II epithelial cells are damaged (Vijeyaratnam & Corrin 1971). The destruction of the alveolar epithelium continues such that by two to four days after dosing, areas of the alveolar epithelium are destroyed. During this time an alveolitis develops with concomitant infiltration of inflammatory cells into the lung (Vijeyaratnam & Corrin 1971). Considerable oedema is produced during this phase and many animals will die within the first few days of dosing (Smith & Rose 1977). A few rats that develop this severe alveolitis survive for many days or weeks after dosing but then develop an extensive hypercellular lesion characterized by the proliferation of fibroblasts (Smith & Heath 1976). This phase of the lesion, along with residual oedema and alveolar collapse, results in death due to anoxia. Thus it appears that the initial lesion in the lung is damage to the alveolar epithelium and as a consequence of this a proliferative fibrosis develops (probably as part of the repair process) that, if extensive enough, will cause death.

Delivery of paraquat to the lung

Most authors agree that paraquat is not metabolized in experimental animals. Sharp *et al.* (1972) first demonstrated that the lungs of rats intravenously dosed with paraquat had the highest concentration and selectively retained the compound in comparison with other organs. When rats are dosed orally with paraquat the concentration in the plasma remains relatively constant over a period of 30 h, whereas that in the lung increases with time such that by 30 h there is six to seven times more in the lung than in the plasma (figure 1). Of the organs studied, no other tissue shows a similar time-dependent increase in paraquat concentration (Rose *et al.* 1976). Thus the lung, which is the organ most severely damaged by paraquat, has the ability to accumulate the bipyridyl from the plasma after oral dosing (Smith *et al.* 1974) and retains paraquat independent of the fall in the plasma concentration (Sharp *et al.* 1972). This selective accumulation–retention of paraquat in the lung provides a convincing explanation why this organ is selectively damaged by paraquat.

Experiments using lung slices have shown that the lung accumulates paraquat in a time-dependent manner by a process that is energy dependent (Rose *et al.* 1974). As with the *in vivo* situation, slices of organs taken from control rats were unable to accumulate paraquat,

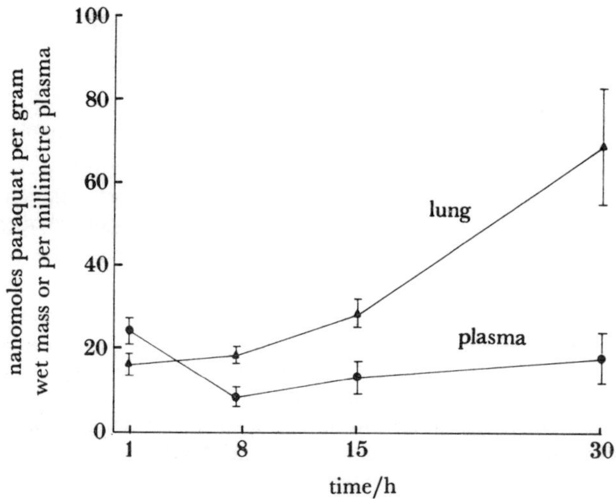

FIGURE 1. The level of paraquat in the lung and plasma of rats given 680 μmol of paraquat per kilogram of body mass orally. Points on the graph represent the mean ± s.e. (standard error) with at least five rats per time point. Rate of uptake: 4–6 nmol per gram wet mass per hour.

with the exception of brain slices, which could accumulate paraquat, but to a much lesser extent than lung slices (Rose *et al.* 1976).

The efflux of paraquat from lung slices has been shown to be biphasic, having a very fast component and a slow component (Smith *et al.* 1981). The fast component is thought to represent the diffusion of paraquat from the extracellular space or from cells damaged during slicing of the lung (Smith *et al.* 1981). The slow component is first order, being characterized by a $t_{\frac{1}{2}}$ of 17 h, which is similar to that found *in vivo* (Smith *et al.* 1981).

The discovery that paraquat was accumulated into the lung led to the suggestion that its uptake was the consequence of its ability to be mistaken for an endogenous compound (or compounds) that is normally accumulated by the lung. A number of compounds were examined for their ability to inhibit paraquat accumulation into the lung (Lock *et al.* 1976). Among these inhibitors was the diamine putrescine, which competitively blocks paraquat accumulation and is itself accumulated by an energy-dependent process in the lung (Smith & Wyatt 1981). The uptake of both paraquat and putrescine in the lung obeys saturation kinetics and the apparent K_m for the accumulation of paraquat is 70 μM with a V_{max} of 300 nmol per gram wet mass per hour (Rose *et al.* 1974) whereas that of putrescine is 7 μM with a V_{max} of 330 nmol per gram wet mass per hour (Smith 1982). The polyamines spermidine and spermine are also accumulated into the lung by a process that obeys saturation kinetics (Smith *et al.* 1982). The apparent K_m and V_{max} values for the accumulation of these oligoamines are similar (table 1). It seems reasonable to conclude that a single process is responsible for the accumulation of diamines and polyamines into the lung and that the reason paraquat is accumulated is that it is transported by this process in mistake for these or other endogenous substrates.

The most probable reason for the pulmonary uptake of paraquat is its structural similarity to the diamines and polyamines with respect to the separation of the quaternary nitrogen atoms of paraquat and the amino groups of the oligoamines. A characterization of the structural requirements of this process has been made by Gordonsmith *et al.* (1983).

Definitive evidence for the cellular compartments into which paraquat is accumulated is not

TABLE 1. THE KINETIC CONSTANTS FOR THE ACCUMULATION OF DIAMINES AND POLYAMINES BY SLICES OF RAT LUNG

compound	method of preparing lung slice	K_m/μM	V_{max}/(nmol g h^{-1})
spermine	tissue chopper	8.8	639 (527–811)
spermidine	tissue chopper	15.7	606 (452–917)
cadaverine	tissue chopper	15.9	832 (601–1351)
putrescine	tissue chopper	11.4	635 (399–1546)
putrescine	hand slice	7.0	330

Slices of rat lung were incubated at 37 °C in K.R.P. glucose medium containing 1, 3, 10, 30 or 100 μM of ^{14}C-labelled spermine, spermidine, cadaverine or putrescine as appropriate. The rate of accumulation of label into the slice was determined for each concentration of each compound by using a least-squares regression on four slices at three time points. By using the rate of accumulation for each concentration, estimates of the Lineweaver–Burke relation were determined with the use of unweighted least-square linear regression and the apparent K_m and V_{max} determined for each compound. The figures in parentheses are the 95% confidence limits.

yet available. Waddell & Marlowe (1980) concluded that paraquat is accumulated almost entirely into cells having the typical distribution of alveolar type II epithelial cells. There is also indirect evidence that paraquat and putrescine are accumulated into the same cell types (Smith & Wyatt 1981). In a series of as yet unreported studies with the use of autoradiographic techniques the distribution of paraquat, putrescine and the polyamines spermidine and spermine in lung slices has been found to be confined to the alveolar type I and type II cells and Clara cells of the lung (Soames & Smith, unpublished results). Thus, it appears from both direct and indirect evidence that the alveolar epithelium is at least partly a site for the accumulation of paraquat.

The realization that paraquat is accumulated into specific cell types in the lung is of considerable importance in attempting to understand its mechanism of toxicity. It is claimed that there are more than 40 cell types in the lung (Sorokin 1970) and therefore if paraquat is accumulated into only a small proportion of the total cell population then the intracellular concentration of paraquat in those cells it damages (alveolar type I and type II epithelial cells) could be seriously underestimated. This is because the data describing the amount of paraquat in the lung is usually expressed as a function of the wet weight of tissue. Therefore the concentration, when expressed in this way, may underestimate by as much as two orders of magnitude the intracellular concentration of paraquat in those cells into which it has accumulated.

PRIMARY MECHANISM OF TOXICITY

Paraquat has the ability to undergo a one-electron reduction from the cation to form a stable blue coloured free radical in the absence of oxygen (Michaelis & Hill 1933). In the presence of oxygen the radical will immediately reform the cation with the concomitant production of superoxide anion (O_2^-). This reaction between paraquat radical and oxygen is so rapid that it is diffusion-limited (Farrington *et al.* 1973). Thus, provided there is a continuous supply of electrons to paraquat, and oxygen is present, paraquat will rapidly cycle from its oxidized to reduced form with the continuous production of O_2^-.

Gage (1968) first reported that under anaerobic conditions, NADPH together with a flavoprotein could reduce paraquat from its cation to radical. Under aerobic conditions the radical is reoxidized and this redox cycling continues until the available NADPH is consumed.

The studies of Gage (1968) were extended by Baldwin *et al.* (1975) who demonstrated that microsomal preparations from liver, lung and kidney were all able to generate radicals of paraquat and eventually produce H_2O_2. The production of O_2^- and H_2O_2 may then lead to the formation of more reactive oxygen radicals, which in turn may be more toxic to the cell (Bus & Gibson 1984). Thus the cycling of paraquat from its reduced to reoxidized form provides a plausible primary mechanism of its toxicity that is entirely analogous with the proposed mechanism of the phytotoxicity of paraquat (Dodge 1971).

Biochemical changes associated with toxicity

The mechanism by which the redox cycling of paraquat leads to lung damage is still subject to considerable speculation. Emphasis has been placed on the ability of radical species of oxygen to react with lipid membranes and cause lipid peroxidation (Bus *et al.* 1974, 1975, 1976). Peroxidation of membranes is considered to lead to their dysfunction and hence damage to the cell. Both direct and indirect evidence for this mechanism has been produced. However, the *in vitro* evidence for the production of lipid peroxides is contradictory (Shu *et al.* 1979; Steffen & Netter 1979). The system to measure *in vitro* lipid peroxidation is usually one that uses microsomal preparations from either the liver or lung together with NADPH or an NADPH-generating system. Malondialdehyde production or diene conjugation is often used as a measure of lipid peroxidation in these systems. Several studies reported that paraquat does not cause lipid peroxidation *in vitro* (Illett *et al.* 1974; Steffen & Netter 1979; Kornbrust & Mavis 1980). However more recent studies by Trush *et al.* (1981) indicate that there is considerable *in vitro* stimulation of lipid peroxidation by paraquat in rat microsomes. It may be that in those studies where the *in vitro* evidence for lipid peroxidation was not obtained, NADPH became a limiting factor in the production of paraquat radical and hence superoxide anion.

Bus *et al.* (1976) have provided indirect evidence that lipid peroxidation is involved with the mechanism of paraquat toxicity *in vivo*. A mechanistic scheme incorporating their proposal is shown in figure 2 and using this scheme they have predicted the biochemical parameters that should be altered. Also, by interfering with some of the defence mechanisms shown in this scheme, the authors tested whether the toxicity of paraquat was altered. For example, the detoxification of lipid hydroperoxides, or indeed H_2O_2, via the enzyme systems glutathione peroxidase and glutathione reductase is known to require reduced glutathione and selenium as essential cofactors. Bus *et al.* (1975) demonstrated that paraquat toxicity was significantly enhanced in mice fed selenium-deficient diets. The studies with selenium, however, indicate the difficulty of this approach because selenium deficiency increases the toxic effect of paraquat to mouse liver, an organ that in mice is not usually damaged by paraquat.

In contrast to the evidence *in vitro*, there is little direct evidence that shows the generation of lipid peroxidation in the lungs of paraquat-treated animals. The problem of demonstrating lipid peroxidation *in vivo* is in many respects analogous to defining the amount of paraquat present in individual cell types. The inability to detect malondialdehyde or diene conjugates in the lungs of animals treated with paraquat may reflect the fact that only a relatively small proportion of the cell types are damaged. Thus, even if peroxidation does occur and is extensive enough to cause cellular damage, it may still not be large enough to be detected when the results are expressed per gram wet weight of whole lung. This problem is applicable to all investigations that attempt to demonstrate a critical biochemical event in a particular cell type which represents a small proportion of the total cell population. It is certainly no less applicable to

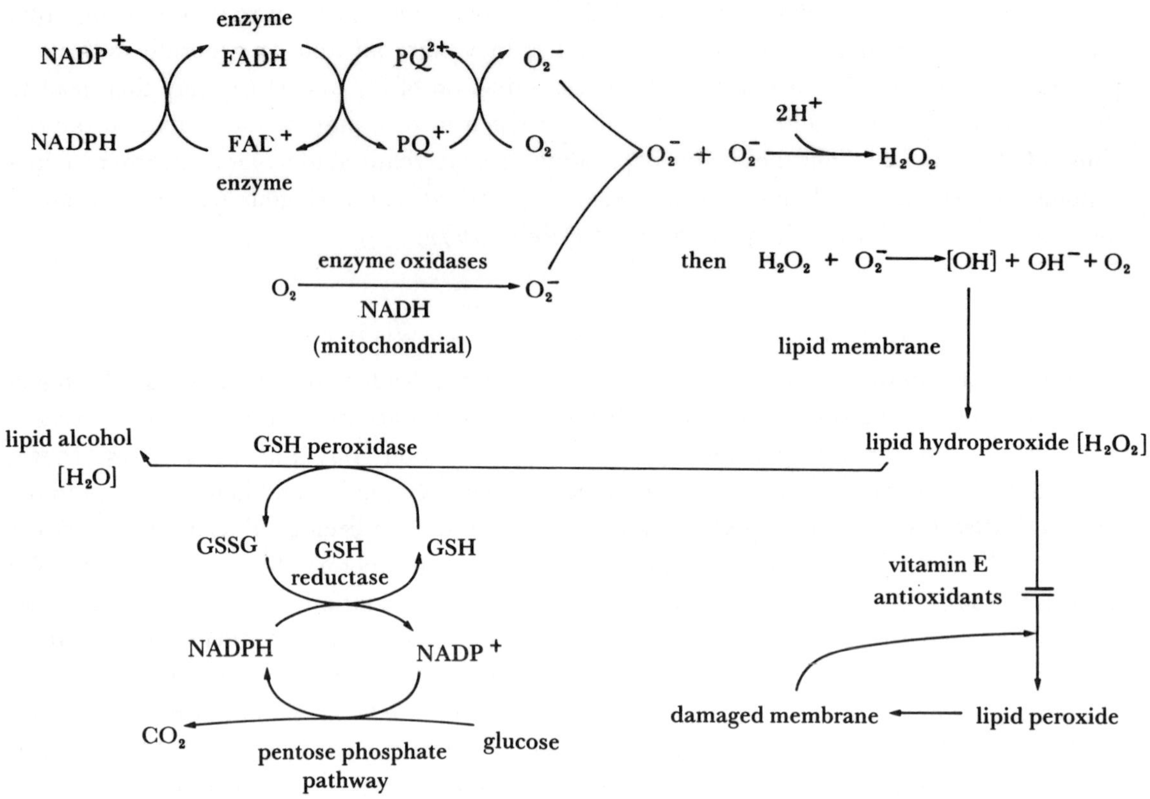

FIGURE 2. Mechanism of toxicity.

the suggestion that a critical biochemical event in the toxicity of paraquat is the depletion of NADPH levels in those cells in which paraquat is accumulated. (The sites of oxidation of NADPH are shown in figure 2.) This depletion would render the cells unable to perform essential physiological and biochemical functions. The first studies to indicate that this may be a possible explanation for the toxicity of paraquat were those of Fisher *et al.* (1975). They suggested that paraquat may shift the redox potential by oxidation of pyridine nucleotides and that the resulting scarcity of NADPH may then interfere with the synthesis of proteins and lipids. This hypothesis is supported by the evidence of Illett *et al.* (1974), who showed that NADPH oxidation was markedly increased in rat lung microsomes by paraquat. Fisher *et al.* (1975) also demonstrated that the oxidation of [1-^{14}C]glucose in lung slices was markedly enhanced in the presence of paraquat. They suggested this was consistent with a depletion in the levels of NADPH in the lung. They also demonstrated that lipid synthesis was inhibited in lung slices in the presence of paraquat, indicating that NADPH levels were reduced.

Witschi *et al.* (1977) measured the NADPH:NADP$^+$ ratio in the lungs of rats intravenously dosed with paraquat. They demonstrated that paraquat causes an oxidation of NADPH *in vivo*, although the relation of this effect to lung damage is not clear. More recently it has been shown that the pentose phosphate pathway (as measured by [1-^{14}C]glucose oxidation) is stimulated directly in proportion to the dose of paraquat given to rats and therefore the amount of paraquat present in the lung (figure 3). Also fatty-acid synthesis was inhibited in a dose-dependent manner in the lungs of treated rats (figure 4) and the degree of stimulation of the pentose phosphate pathway was directly related to the degree of inhibition of fatty-acid

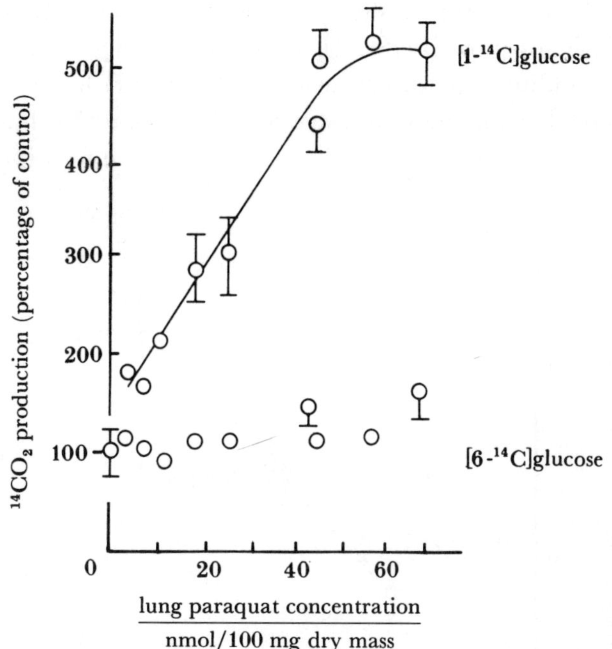

FIGURE 3. Relation between the production of $^{14}CO_2$ from [1-^{14}C]- or [6-^{14}C]glucose and lung paraquat concentration 2 h after dosing. Lung slices were prepared (0.6 mm thick) and $^{14}CO_2$ production from [1-^{14}C]- or [6-^{14}C]glucose was measured for 1 h. Results are expressed as a percentage of control. Points on the graph represent the mean ± s.e. with five rats per determination. Control values were 10102 ± 1304 d.p.m. (disintegrations per minute) $^{14}CO_2$ per 100 mg wet mass [1-^{14}C]glucose and 2851 ± 287 d.p.m. $^{14}CO_2$ per 100 mg wet mass [6-^{14}C]glucose. (From Keeling *et al.* 1982.)

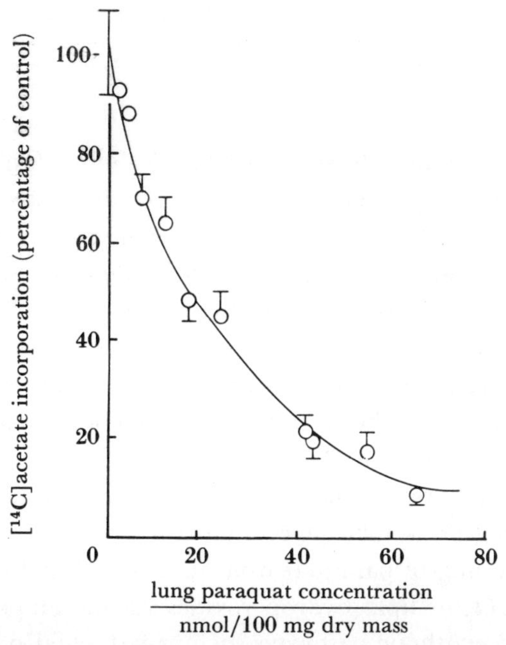

FIGURE 4. Relation between [^{14}C]acetate incorporation into fatty acids and lung paraquat concentration 2 h after dosing. Lung slices were prepared (0.6 mm thick) and [^{14}C]acetate incorporation into fatty acids was measured for 1.5 h. Results are expressed as a percentage of control. Points on graph represent the mean ± s.e. with five rats per determination. Control values were 10222 ± 580 d.p.m. ^{14}C per mg fatty acid. (From Keeling *et al.* 1982.)

synthesis (figure 5). Because both pathways appear to be equally affected (figure 5), this indicates that a single factor (probably NADPH) is affected, which in turn controls the activity of both pathways. In addition the direct measure of NADPH levels in the lung showed that this cofactor is reduced following the administration of paraquat (Keeling & Smith 1982). These data are consistent with the results of Witschi *et al.* (1975) in that the ratio of NADPH to $NADP^+$

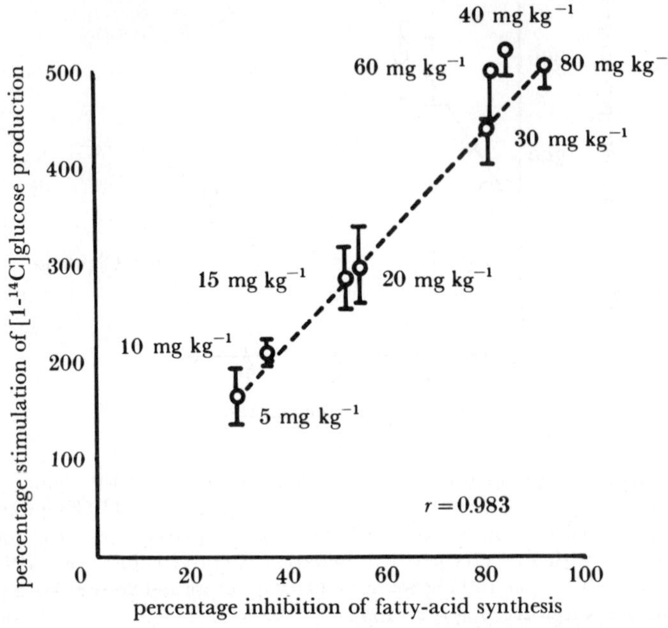

FIGURE 5. Relation between reduced [^{14}C]acetate incorporation into fatty acids and stimulated $^{14}CO_2$ production from [1-^{14}C]glucose in lung slices taken from rats 2 h after dosing subcutaneously with paraquat. Lung slices were prepared (0.6 mm thick) and $^{14}CO_2$ production from [1-^{14}C]glucose was measured for 1 h as d.p.m. $^{14}CO_2$ per 100 mg wet mass. [^{14}C]acetate incorporation into fatty acids was measured for 1.5 h as d.p.m. ^{14}C per milligram fatty acid. Results are expressed as percentage of control. Points on graph represent the mean of five rats per determination. (From Keeling *et al.* 1982.)

in the lung was reduced by paraquat. However Keeling & Smith (1982) demonstrated that it was a loss of NADPH that caused the change in the ratio rather than an alteration in both the NADPH and $NADP^+$ levels. The significance of these effects to an understanding of the mechanism of paraquat toxicity can best be demonstrated if they are considered in the context of the time course of events. The stimulation in pentose phosphate pathway activity and inhibition in fatty acid synthesis in the rat lung occurs within 2 h of administering an LD_{50} of paraquat (Keeling *et al.* 1982). These biochemical changes occurred at the time when it is not possible to demonstrate changes in the alveolar epithelium of the lung by using the electron microscope, nor can a significant fall in the NADPH content of the lung be measured (Keeling & Smith 1982). Thus it can be argued that when paraquat is accumulated into the lung, compensatory biochemical reactions take place that attempt to regenerate NADPH. The NADPH is utilized by the cycling of paraquat from its reduced and oxidized form and also in an attempt to detoxify H_2O_2 or lipid hydroperoxides via the glutathione peroxidase and reductase enzyme systems. Biosynthetic pathways such as fatty-acid synthesis that are highly dependent on NADPH (Wakil 1962) are inhibited owing to the rapid oxidation of NADPH. Only when these compensatory processes are no longer able to maintain NADPH levels in the lung can decreases in the amount of NADPH present be detected.

That there are early significant biochemical changes in the redox state of paraquat-treated lungs is shown by the studies of Keeling *et al.* (1982). These studies demonstrated that following the administration of paraquat the amount of protein SH groups in the lung is reduced and there is an increase in protein SS forms. These protein SS forms are mostly found in the form of mixed disulphides where low molecular mass SH compounds (such as glutathione and cysteamine) react with protein. Keeling *et al.* (1982) found a direct correlation between the amount of mixed disulphides generated and the stimulation of the pentose phosphate pathway or indeed inhibition of fatty-acid synthesis. The formation of mixed disulphides between protein and glutathione may be part of a significant regulatory mechanism in rat tissues that maintains the integrity of cell membranes during oxidative stress (Isaacs & Brinkley 1977*a*, *b*). Increases in lung-mixed disulphides may be a factor that is responsible for regulating the pentose phosphate pathway and fatty-acid synthesis. It seems likely that the redox stress caused by paraquat in target cells results in an increase in glutathione oxidation (in an attempt to detoxify lipid hydroperoxides and H_2O_2). Any excess GSSG formed will react with protein sulphydryl to form mixed disulphides, perhaps mediated by a thiol transferase activity. As sulphydryl groups of enzymes react with oxidized glutathione to form mixed disulphides the enzyme activity of these proteins is altered (Isaacs & Brinkley 1977*a*, *b*; Moron *et al.* 1979), presumably by conformational changes in their structure. In this way, as GSSG reacts with the protein sulphydryl, the coincident activation of the pentose phosphate pathway would generate NADPH that in turn would lead to the reduction of GSSG to GSH. Thus the activation of the pentose phosphate pathway by GSSG in the cell would contribute to the regulation of the normal redox state in the lung cells. It is of interest that a greater understanding of the regulation of intermediary metabolism in the lung has come about from an attempt to understand the mechanism of paraquat toxicity.

It seems clear that the primary reaction involved in the mechanism of paraquat toxicity in the lung is the cycling from its cationic form to reduced form with the production of superoxide anion and consumption of electrons. Those authors who favour lipid peroxidation as the mechanism of paraquat damage suggest that the production of superoxide anion directly or indirectly, through the formation of more reactive species of oxygen such as hydroxyl radical, leads to the peroxidation of vital lipid membranes and hence cell death. On balance the evidence *in vitro* favours the formation of lipid peroxides by lung microsomes whereas the *in vivo* data is more tenuous. Those authors who have emphasized the importance of NADPH depletion in paraquat toxicity suggest that the depletion of this vital cofactor renders the lung unable to perform essential biosynthetic processes, and in turn is more susceptible to lipid peroxidation. There is no reason why these hypotheses need to be mutually exclusive.

In conclusion, therefore, if the glutathione peroxidase or reductase systems together with the endogenous levels of antioxidant are inadequate to defend against peroxidation, then it is possible that cell damage may occur. The depletion of NADPH by the redox cycling of paraquat will lead to additional utilization of the remaining NADPH in an attempt to provide reduced glutathione for the glutathione peroxidase enzyme activity. Thus NADPH will be depleted both in the production of superoxide anion and in a defence to it. Consequently, the pentose phosphate pathway is stimulated in an attempt to regenerate NADPH. The pathways dependent on NADPH appear to be inhibited either as a compensatory mechanism or because NADPH is depleted within the cell. With further generation of NADPH this cofactor is available either to regenerate reduced glutathione or it may be used for the further reduction of paraquat and cycling with oxygen. On balance, the evidence suggests that the concentration of paraquat in lung cells causes a severe redox stress that leads eventually to a sustained depletion in NADPH levels. This by itself, or in combination with lipid peroxidation, initiates the cascade of biochemical events that lead to cell death.

References

Baldwin, R. C., Pasi, A., MacGregor, J. T. & Hine, C. H. 1975 The rates of radical formation from the dipyridyl herbicides, paraquat, diquat and morphamquat in homogenates of rat lung, kidney and liver. *Toxic. appl. Pharmac.* **32**, 298–304.

Bronkhurst, F. B., Van Daal, J. M. & Tan, H. D. 1968 Fatal poisoning with paraquat. *Ned. Tijdschrgeneesk* **112**, 310–313.

Bus, J. S., Aust, S. D. & Gibson, J. E. 1974 Superoxide- and singlet oxygen-catalysed lipid peroxidation as a possible mechanism for paraquat toxicity. *Biochem. biophys. Res. Commun.* **58**, 749–755.

Bus, J. S., Aust, S. D. & Gibson, J. E. 1975 Lipid peroxidation: a possible mechanism for paraquat toxicity. *Res. Commun. chem. Path. Pharmac.* **11**, 31–38.

Bus, J. S., Cagen, S. Z., Olgaard, M. & Gibson, J. E. 1976 A mechanism of paraquat toxicity in mice and rats. *Toxic. appl. Pharmac.* **35**, 501–513.

Bus, J. S. & Gibson, J. E. 1984 In *Drug metabolism and drug toxicity* (ed. J. R. Mitchell & M. G. Horning), pp. 21–32. New York: Raven Press.

Clarke, D. G., McElligott, T. F. & Weston-Hurst, E. 1966 The toxicity of paraquat. *Br. J. ind. Med.* **23**, 126–132.

Dodge, A. D. 1971 The mode of action of the bipyridylium herbicides, paraquat and diquat. *Endeavour* **30**, 130–135.

Fairshter, R. D., Rosen, S. M., Smith, W. R., Glauser, F. L., McRae, D. M. & Wilson, A. F. 1976 Paraquat poisoning: new aspects of therapy. *Q. Jl. Med.* **45**, 551–565.

Farrington, J. A., Ebert, M., Land, E. J. & Fletcher, K. 1973 Bipyridylium quaternary salts and related compounds: pulse radiolysis studies on the reaction of paraquat radical with oxygen. *Biochim. biophys. Acta* **314**, 372–381.

Fennelly, J. J., Fitzgerald, M. X. & Fitzgerald, O. 1971 Recovery from severe paraquat poisoning following forced diuresis and immunosuppressive therapy. *J. Ir. Med. Ass.* **64**, 69–71.

Fisher, H. K., Clements, J. A., Tierney, D. F. & Wright, R. R. 1975 Pulmonary effects of paraquat in the first day after injection. *Am. J. Physiol.* **228**, 1217–1223.

Fletcher, K. 1974 Paraquat poisoning. In *Forensic toxicology* (ed. B. Ballentyne), pp. 86–98. Bristol: John Wright and Sons Ltd.

Gage, J. C. 1968 The action of paraquat and diquat on the respiration of liver cell fractions. *Biochem. J.* **109**, 757–761.

Gordonsmith, R. H., Brooke-Taylor, S., Smith, L. L. & Cohen, G. M. 1983 Structural requirements of compounds to inhibit pulmonary diamine accumulation. *Biochem. Pharmac.* **32**, 3701–3709.

Illett, K. F., Stripp, B., Menard, R. H., Reid, N. D. & Gillette, J. R. 1974 Studies on the mechanism of the lung toxicity of paraquat. *Toxic. appl. Pharmac.* **28**, 216–226.

Isaacs, J. T. & Brinkley, F. 1977*a* Cyclic AMP-dependent control of the rat hepatic glutathione disulphide-sulphydryl ratio. *Biochim. biophys. Acta* **498**, 29–38.

Isaacs, J. T. & Brinkley, F. 1977*b* Glutathione dependent control of protein disulphide-sulphydryl content by subcellular fractions of hepatic tissue. *Biochim. biophys. Acta* **497**, 192–204.

Keeling, P. L. & Smith, L. L. 1982 Relevance of NADPH depletion and mixed disulphide formation in the rat lung to the mechanism of cell damage following paraquat administration. *Biochem. Pharmac.* **31**, 3243–3249.

Keeling, P. L., Smith, L. L. & Aldridge, W. N. 1982 The formation of mixed disulphides in rat lung following paraquat administration. Correlation with changes in intermediatory metabolism. *Biochim. biophys. Acta* **716**, 249–257.

Kornbrust, D. J. & Mavis, R. D. 1980 The effect of paraquat on microsomal lipid peroxidation *in vitro* and *in vivo*. *Toxic. appl. Pharmac.* **53**, 323–332.

Lock, E. A., Smith, L. L. & Rose, M. S. 1976 Inhibition of paraquat accumulation in lung slices by a component of rat plasma and a variety of drugs and endogenous amines. *Biochem. Pharmac.* **25**, 1769–1772.

Michaelis, L. & Hill, E. S. 1933 Potentiometric studies on semiquinones. *J. Am. chem. Soc.* **55**, 1481–1494.

Moron, M. S., DePierre, J. W. & Mannervik, B. 1979 Levels of glutathione, glutathione reductase and glutathione S-transferase activities in rat lung and liver. *Biochim. biophys. Acta* **582**, 67–78.

Nagi, A. H. 1970 Paraquat and adrenal cortical necrosis. *Br med. J.* **2**, 669.

Nienhaus, H. & Ehrenfeld, M. 1971 Pathogenesis of lung disease in paraquat poisoning. *Beitr. Path.* **142**, 244–267.

Oreopoulos, D. G., Soyannwao, M. A. O., Sinniah, R., Fenton, S. S. A., McGeown, M. G. & Bruce, J. H. 1968 Acute renal failure in case of paraquat poisoning. *Br. med. J.* **1**, 749–750.

Rose, M. S., Lock, E. A., Smith, L. L. & Wyatt, I. 1976 Paraquat accumulation: tissue and species specificity. *Biochem. Pharmac.* **25**, 419–423.

Rose, M. S., Smith, L. L. & Wyatt, I. 1974 Evidence for the energy-dependent accumulation of paraquat into rat lung. *Nature, Lond.* **252**, 314–315.

Sharp, C. W. M., Ottolengi, A. & Posner, H. S. 1972 Correlation of paraquat toxicity with tissue concentration and weight loss in the rat. *Toxic. appl. Pharmac.* **22**, 241–251.

Shu, H., Talcott, R. E., Rice, S. A. & Wei, E. T. 1979 Lipid peroxidation and paraquat toxicity. *Biochem. Pharmac.* **28**, 327–331.

Smith, L. L. 1982 The identification of an accumulation system for diamines and polyamines into the lung and its relevance to paraquat toxicity. *Arch. Toxicol. Suppl.* **5**, 1–14.

Smith, L. L. & Rose, M. S. 1977 A comparison of the effects of paraquat and diquat on rat lung. *Toxicology* **8**, 223–230.

Smith, L. L., Wright, A. F., Wyatt, I. & Rose, M. S. 1974 Effective treatment of paraquat poisoning in rats and its relevance to the treatment of paraquat poisoning in man. *Br. med. J.* **4**, 569–571.

Smith, L. L. & Wyatt, I. 1981 The accumulation of putrescine into slices of rat lung and brain and its relevance to the accumulation of paraquat. *Biochem. Pharmac.* **30**, 1053–1058.

Smith, L. L., Wyatt, I. & Cohen, G. M. 1982 The accumulation of diamines and polyamines into rat lung slices. *Biochem. Pharmac.* **31**, 3029–3033.

Smith, L. L., Wyatt, I. & Rose, M. S. 1981 Factors affecting the efflux of paraquat from rat lung slices. *Toxicology* **19**, 197–207.

Smith, P. & Heath, D. 1976 Paraquat. *C.R.C. Crit. Rev. Toxicol.* **4**, 411–445.

Sorokin, S. P. 1970 The cells of the lungs. *Conference on Morphology of Experimental Respiratory Carcinogenesis, May 1970.* (ed. P. Nettesheim, M. G. Hanna & J. W. Deatherage). U.S. Atomic Energy Commission.

Steffen, C. & Netter, K. J. 1979 Mechanism of paraquat action on microsomal oxygen reduction and its relation to lipid peroxidation. *Toxicol. appl. Pharmac.* **47**, 99–103.

Trush, M. A., Mimnaugh, E. G., Ginsburg, E. & Gram, T. E. 1981 *In vitro* stimulation by paraquat of reactive oxygen-mediated lipid peroxidation in rat lung microsomes. *Toxicol. appl. Pharmac.* **60**, 279–286.

Vijeyaratnam, G. S. & Corrin, B. 1971 Paraquat poisoning: a histological and electron-optical study of changes in the lung. *J. Path.* **103**, 123–129.

Waddell, W. J. & Marlowe, C. 1980 Tissue and cellular disposition of paraquat in mice. *Toxical. appl. Pharmac.* **56**, 127–140.

Wakil, S. J. 1962 Lipid metabolism. *A. Rev. Biochem.* **31**, 369–406.

Weidel, H. & Rosso, M. 1882 Studien uber das pyridin. *Monatsh. Chem.* **3**, 850–885.

Witschi, H., Kacew, S., Hirai, K. I. & Cote, M. G. 1977 *In vivo* oxidation of reduced nicotinamide adenine dinucleotide phosphate by paraquat and diquat in rat lung. *Chem. Biol. Interact.* **19**, 143–160.

Phil. Trans. R. Soc. Lond. B **311**, 659–671 (1985)
Printed in Great Britain

Metal ions and oxygen radical reactions in human inflammatory joint disease

By B. Halliwell[1], J. M. C. Gutteridge[2] and D. Blake[3]

[1] *Department of Biochemistry, King's College London, Strand, London WC2R 2LS, U.K.*
[2] *National Institute for Biological Standards and Control, Holly Hill, Hampstead, London NW3 6RB, U.K.*
[3] *Rheumatology Research Wing, University Medical School, Birmingham B15 2TJ, U.K.*

Activated phagocytic cells produce superoxide (O_2^-) and hydrogen peroxide (H_2O_2); their production is important in bacterial killing by neutrophils and has been implicated in tissue damage by activated phagocytes. H_2O_2 and O_2^- are poorly reactive in aqueous solution and their damaging actions may be related to formation of more reactive species from them. One such species is hydroxyl radical ($OH^{\bullet}$), formed from H_2O_2 in the presence of iron- or copper-ion catalysts. A major determinant of the cytotoxicity of O_2^- and H_2O_2 is thus the availability and location of metal-ion catalysts of $OH^{\bullet}$ formation. Hydroxyl radical is an initiator of lipid peroxidation. Iron promoters of $OH^{\bullet}$ production present *in vivo* include ferritin, and loosely bound iron complexes detectable by the 'bleomycin assay'. The chelating agent Desferal (desferrioxamine B methanesulphonate) prevents iron-dependent formation of $OH^{\bullet}$ and protects against phagocyte-dependent tissue injury in several animal models of human disease. The use of Desferal for human treatment should be approached with caution, because preliminary results upon human rheumatoid patients have revealed side effects. It is proposed that $OH^{\bullet}$ radical is a major damaging agent in the inflamed rheumatoid joint and that its formation is facilitated by the release of iron from transferrin, which can be achieved at the low pH present in the micro-environment created by adherent activated phagocytic cells. It is further proposed that one function of lactoferrin is to protect against iron-dependent radical reactions rather than to act as a catalyst of $OH^{\bullet}$ production.

Introduction

There has been considerable interest in the role of oxygen-derived species such as superoxide (O_2^-) and hydrogen peroxide (H_2O_2) as agents of tissue damage by activated phagocytic cells in a number of disorders, including the adult respiratory distress syndrome, autoimmune disease, chronic joint inflammation, reperfusion injury after ischaemic cardiac damage, and cancer. In this paper recent developments of our knowledge of the role of oxygen-derived species in human rheumatoid arthritis will be discussed.

Production of oxygen-derived species by phagocytic cells

Reviewing the massive literature on the production of oxygen-derived species by phagocytes is not a task that can be attempted here, but the authors believe that the present state of the subject can be summarized as follows.

(i) It is well established that neutrophils, monocytes, macrophages and eosinophils produce superoxide radical (O_2^-) and hydrogen peroxide (H_2O_2) upon activation of the respiratory burst.

(ii) The products of the respiratory burst play a key role in bacterial killing by neutrophils, as shown by the greatly impaired killing seen in patients suffering from chronic granulomatous disease.

(iii) The myeloperoxidase system in neutrophils plays a lesser role in killing because inborn defects in myeloperoxidase activity rarely produce clinically observable defects in bacterial killing (see, for example, Nauseef *et al.* 1983).

(iv) There is no clear evidence that activated phagocytes produce singlet oxygen (see, for example, Kanofsky *et al.* 1984). Indeed, hypochlorous acid (HOCl) itself, and the products of its reaction with various amines, are toxic (Albrich *et al.* 1981); there is no need to invoke singlet O_2 to explain the toxicity of the myeloperoxidase–H_2O_2– halide system. HOCl and its reaction products may mediate some of the damage done by activated phagocytic cells to their environment (Test *et al.* 1984).

Studies of the chemistry of oxygen-derived products (for reviews see, for example, Halliwell & Gutteridge 1984, 1985; di Guiseppi & Fridovich 1984) also lead to the following conclusions. Pure O_2^- and pure H_2O_2 are poorly reactive in *aqueous* solution, yet biological or chemical O_2^--generating systems can damage many biomolecules. This damage is often prevented or decreased by addition of superoxide dismutase (SOD), or catalase, or both, to the reaction mixtures. Superoxide and H_2O_2 themselves do not peroxidize membrane lipids or degrade DNA. In addition, H_2O_2 can cross cell membranes whereas O_2^- cannot unless there is a specific channel for it (for example, the 'anion channel' in the erythrocyte membrane).

It therefore follows that the damage done by activated neutrophils that has been attributed to oxygen radicals, such as hyaluronic acid degradation (Greenwald & Moy 1980) or DNA fragmentation (Weitberg *et al.* 1983; Phillips *et al.* 1984), is unlikely to be a result of direct reactions of O_2^- or H_2O_2 themselves. A suggestion that O_2^- contributes to formation of a 'chemotactic factor' that attracts more neutrophils into a site of inflammation, has recently been challenged. Indeed, although copper–zinc superoxide dismutase has an anti-inflammatory effect, there is increasing doubt about whether this is entirely because of removal of O_2^- radical (Baret *et al.* 1984).

Perhaps, then, the damage done is due to some 'reactive factor' that O_2^- and H_2O_2 produce. Suggestions have included the following.

(i) Singlet oxygen. There is no clear evidence for the production of this by phagocytic cells.

(ii) $HO_2^{\cdot}$ radical. The protonated form of O_2^- is a more powerful oxidizing agent than is O_2^- itself and reacts with unsaturated fatty acids. Formation of $HO_2^{\cdot}$ is favoured at low pH values (its pK_a is *ca.* 4.8) but the pH beneath adherent activated macrophages has been reported to be 5 or less (Etherington *et al.* 1981) and so it may be biologically relevant.

(iii) Hydroxyl radical. Hydroxyl radical, $OH^{\cdot}$, is produced when water is exposed to high-energy ionizing radiation, and so its properties have been well documented by radiation chemists. Hydroxyl radical is highly reactive, and any produced *in vivo* will react at or close to its site of formation. The nature of the direct damage done by $OH^{\cdot}$ *in vivo* would therefore depend on what its site of formation was; so production of $OH^{\cdot}$ close to DNA could lead to strand breakage (Mello Filho & Meneghini 1984), whereas production close to an enzyme molecule that is present in excess in the cell, such as lactate dehydrogenase, might have no biological consequences. It should also be remembered that reaction of $OH^{\cdot}$ with a biomolecule will produce another radical, usually of lower reactivity (because of the extremely high reactivity of $OH^{\cdot}$). Such less-reactive radicals can cause their own problems, because they can diffuse away from their site of formation and attack specific biomolecules. For example, uric acid reacts with $OH^{\bullet}$ radical; it protects lactate dehydrogenase against inactivation by $OH^{\bullet}$ but *accelerates* inactivation of alcohol dehydrogenase (Kittridge & Willson 1984). The urate-derived radicals are less reactive overall, so more of them survive to reach the sensitive sites on alcohol dehydrogenase that they can react with. Perhaps the best example of the importance of secondary radicals is the ability of $OH^{\bullet}$ to initiate lipid peroxidation by hydrogen-atom abstraction, with subsequent formation of peroxy radicals (for reviews see, for example, Halliwell & Gutteridge 1984, 1985).

Hydroxyl radical is produced when H_2O_2 comes into contact with iron ions. 'Free' iron ions cannot exist *in vivo*, but complexes of iron salts with phosphate esters (such as ATP and GTP), organic acids (such as citrate and lactate) and DNA are all effective in decomposing H_2O_2 (Floyd 1981, 1983; Flitter *et al.* 1983). Copper ions also promote the decomposition of H_2O_2 to $OH^{\bullet}$ (Samuni *et al.* 1981; Rowley & Halliwell 1983). The DNA damage done by H_2O_2 to human fibroblasts and mouse cells in culture can be minimized by metal chelating agents (Mello Filho & Meneghini 1984), suggesting that the H_2O_2 crosses the plasma membrane and combines with metal ions (probably iron) bound at or close to DNA.

Hence a major determinant of the actual toxicity of O_2^- and H_2O_2 to cells is the availability and location of metal-ion catalysts of $OH^{\bullet}$ radical formation. If, for example, iron salts are bound to DNA or membrane lipids, introduction of H_2O_2 and O_2^- can fragment the DNA and peroxidize the membrane lipids. The site of attack of the $OH^{\bullet}$ radicals will be determined by the site of the bound metal ion; such 'site-specific' damage is rarely prevented by 'scavengers' of $OH^{\bullet}$ (Czapski 1984) and the nature of the damage will probably not even resemble that done by $OH^{\bullet}$ radical generated in free solution and attacking the target randomly (Gutteridge & Halliwell 1982; Samuni *et al.* 1981; Gutteridge 1984*a*; Czapski 1984). The postulated 'crypto-$OH^{\bullet}$ radical' may represent formation and localized reactivity of 'real' $OH^{\bullet}$ radical (Youngman 1984; Moorhouse *et al.* 1985). For example, exposure of proteoglycan or mucus preparations to metal-dependent systems that generate $OH^{\bullet}$ seems to produce specific patterns of fragmentation rather than just random attack (Creeth *et al.* 1983; Cross *et al.* 1984), perhaps because the metal catalysts bind readily to specific sites.

There is much debate about the role played by the $OH^{\bullet}$ radical in initiating lipid peroxidation in systems containing iron salts (Aust & Svingen 1982). It is known that $OH^{\bullet}$ radicals, such as those produced radiolytically, can initiate peroxidation by abstracting hydrogen atoms, yet laboratory experiments in which peroxidation of liposomes or microsomal fractions is 'initiated' by iron-salt–ascorbate mixtures or NADPH–iron-salt–ADP mixtures often show little inhibition of peroxidation on addition of $OH^{\bullet}$ radical scavengers, even though formation of $OH^{\bullet}$ radicals can be demonstrated in the reaction mixtures (Aust & Svingen 1982; Gutteridge 1984*b*). There have been suggestions that perferryl, ferryl or Fe^{II}–Fe^{III}–O_2 complexes are the true initiators of peroxidation. This may be correct, but other possibilities exist. Those $OH^{\bullet}$ radicals important in initiation are formed by metal ions attached to the membranes; such site-specific production of $OH^{\bullet}$ could not be prevented by scavengers in bulk solution (Czapski 1984). Alternatively, once the peroxidation process has begun, initiation is continued by the alkoxy and peroxy radicals formed upon iron-dependent decomposition of lipid peroxides; $OH^{\bullet}$ radicals are not necessary under these conditions. Most lipid systems contain some pre-formed lipid hydroperoxide (Gutteridge 1984*b*, *c*), and it is merely necessary to fragment these peroxides to continue the chain reaction.

Chelating agents and oxidative tissue damage

If $OH^{\cdot}$ radicals are formed by low molecular mass metal complexes in free solution and then have to diffuse a short distance to attack a target, protection by added $OH^{\cdot}$ scavengers should be seen and has been reported in several *in vivo* systems (see, for example, Cohen 1978; Fox 1984). On the other hand, site-specific $OH^{\cdot}$ radical formation is much more difficult to protect against in this way (Czapski 1984; Moorhouse *et al.* 1985). Protection here can be achieved by (i) superoxide dismutase and H_2O_2-removing enzymes, to stop O_2^- and H_2O_2 ever reaching the bound metal complexes, and (ii) chelating agents that pull the metals away from sensitive sites and render them inactive. Complexes of iron salts with diethylenetriaminepenta-acetic acid and bathophenanthroline show diminished reactivity in $OH^{\cdot}$ production (Halliwell 1978*a*, *b*), but the chelating agent most effective in preventing iron-dependent $OH^{\cdot}$ production is desferrioxamine (Gutteridge *et al.* 1979), which is also a powerful inhibitor of iron-dependent lipid peroxidation (Wills 1969). Desferrioxamine, available from Ciba-Geigy as its mesylate salt (Desferal) is widely used to prevent iron overload in thalassaemia and other conditions requiring repeated blood transfusion. Desferal cannot be given by mouth. Doses of 50–60 mg per kilogram of body mass (administered subcutaneously) appear safe in the treatment of iron overload, but high doses have been associated with ocular abnormalities and patients receiving Desferal for any purpose should be periodically checked for both ocular (Davies *et al.* 1983) and auditory changes, especially in diabetics (Arden *et al.* 1984).

EDTA renders copper ions less reactive in $OH^{\cdot}$ production (Samuni *et al.* 1981), but iron–EDTA chelates still react with O_2^- and H_2O_2 (Butler & Halliwell 1982). Hence EDTA has a number of potential effects in biological systems.

(i) It could minimize damage by pulling iron ions off sensitive sites and causing the $OH^{\cdot}$ formed by iron–EDTA chelates in free solution to react 'randomly' with less-important targets. Such an effect may account for its ability to inhibit DNA degradation by bleomycin–iron complexes (Gutteridge *et al.* 1981*a*).

(ii) It could make damage worse by pulling iron off 'safe sites'; i.e. the 'random attack' does more damage than the site-specific attack (see, for example, Wong *et al.* 1981). It thus 'decompartmentalizes' iron in a harmful way (Willson 1978).

(iii) It will, in any case, change the nature of the damage and increase its susceptibility to inhibition by $OH^{\cdot}$ 'scavengers' (Gutteridge 1984*a*), because, in the presence of EDTA, the attack will be from $OH^{\cdot}$ generated close to the target 'in free solution' rather than from $OH^{\cdot}$ formed on the target.

EDTA is thus an interesting experimental tool, but for physiological purposes the authors use desferrioxamine, which tightly binds iron(III) *and* renders it inactive as a catalyst.

Hydroxyl radical formation by phagocytic cells

Direct measurement of $OH^{\cdot}$ radical formation *in vivo* is difficult because of its high reactivity, but recent studies in the authors' laboratories with the use of the technique of aromatic hydroxylation (Richmond *et al.* 1981; Grootveld & Halliwell 1985; Moorhouse *et al.* 1985) look promising. Conversion of dimethylsulphoxide (DMSO) into methane has been used as an assay method to show $OH^{\cdot}$ production in whole cells (Repine *et al.* 1979, 1981).

Hydroxyl radical formation by isolated phagocytic cells has been observed by using a wide range of techniques, including e.s.r. spin trapping (Green *et al.* 1979), conversion of DMSO into methane (Repine *et al.* 1979), benzoate decarboxylation (Sagone *et al.* 1980) and ethene formation (Weiss *et al.* 1977). DMSO penetrates into cells and probably measures $OH^{\bullet}$ production both extra- and intracellularly, but the other techniques are probably detecting only external $OH^{\bullet}$ generation. This external $OH^{\bullet}$ generation is important in considering the damage done by activated phagocytes to their surroundings (see Halliwell 1982*a*; Fox 1984) and to themselves; an increased availability of catalytic iron complexes can decrease phagocyte activity by damaging the phagocytes (Sweder van Asbeck *et al.* 1984*a*, *b*). However, as far as bacterial killing is concerned, it is the formation of $OH^{\bullet}$ inside the phagocytic vacuole that matters. Again there are two possibilities, not mutually exclusive. First, that the phagocytes produce $OH^{\bullet}$ from O_2^- and H_2O_2 in the phagocytic vacuole and it attacks the bacteria from the outside. Some evidence consistent with this is available (Johnston *et al.* 1975; Repine *et al.* 1984). Secondly, that H_2O_2 generated in the phagocytic vacuole penetrates into the bacteria and produces $OH^{\bullet}$ inside them. Several bacteria contain iron complexes capable of catalysing radical reactions (Gutteridge & Wilkins 1984) and the killing of *S. aureus* cells by H_2O_2 becomes more effective if the internal iron content of the bacteria is increased (Repine *et al.* 1981).

Hydroxyl radical formation inside and outside phagocytic cells: what is the physiological metal catalyst?

Although both iron and copper salts mediate radical reactions, almost all the work concerning $OH^{\bullet}$ radical formation *in vivo* has been done with iron salts. Discussion will therefore be largely confined to these, although the development of an assay that measures available copper complexes (Gutteridge 1984*d*) should increase our knowledge of their importance. What iron catalysts are available *in vivo*?

Simple iron chelates

The first possibility is low molecular mass iron chelates such as iron–ATP, –GTP, or –citrate. Small 'transit' pools of these iron complexes are present within cells (Jacobs 1977; Halliwell 1982*b*), and the killing of fibroblasts by H_2O_2 requires intracellular metal ions, probably bound to DNA (Mello Filho & Meneghini 1984; Floyd 1981).

Gutteridge *et al.* (1981*a*, 1982*a*) have developed an assay for complexes of iron capable of accelerating radical reactions. This so-called 'bleomycin assay' shows that such complexes are not present in human serum or plasma, except in some cases of iron overload secondary to idiopathic haemochromatosis (Gutteridge *et al.* 1985*a*). They are present in sweat (Gutteridge *et al.* 1985*b*), cerebrospinal fluid (Gutteridge *et al.* 1982*b*) and in the knee-joint synovial fluid of many patients suffering from rheumatoid arthritis. Not all rheumatoid synovial fluids contain bleomycin-detectable iron, but many do and the concentrations present correlate positively with both clinical and laboratory parameters of disease activity (Rowley *et al.* 1984). The precise molecular form of bleomycin-detectable iron has not been established, but it is, at least partly, ultrafilterable and may represent iron chelated to organic acids, phosphate esters or urate, or perhaps iron *weakly* attached to such proteins as albumin, so that it can easily detach during ultrafiltration (Gutteridge *et al.* 1985*b*). The reasons for the presence of bleomycin-detectable iron in synovial fluid are considered later.

Evidence for the importance *in vivo* of iron-mediated radical reactions comes from studies with desferrioxamine. Desferal blocks the haemolytic action of several radical-generating drugs in mice (Clark & Hunt 1983) and decreases acute lung vascular injury in animals after complement activation due to infusion of cobra venom factor or to severe burn injury (Ward *et al.* 1983; Fligiel *et al.* 1984). It also has a beneficial effect on the course of allergic encephalomyelitis in rats (Bowern *et al.* 1984).

Ferritin

Ferritin is often regarded as a safe 'storage form' of iron, yet ferritin stimulates both lipid peroxidation (Gutteridge *et al.* 1983) and the formation of OH˙ radicals from O_2^- and H_2O_2 (Bannister *et al.* 1984). The protein shell can be attacked by lipid peroxides, causing liberation of the iron, which is probably why ferritin stimulates lipid peroxidation (Gutteridge 1985). Superoxide seems to mobilize iron from ferritin, leading to subsequent OH˙ production (Biemond *et al.* 1984). Haemosiderin also promotes hydroxyl radical formation but is less effective than ferritin on a unit-iron basis (O'Connell & Halliwell, in preparation).

Lactoferrin and transferrin

Ambruso & Johnston (1981) reported that iron-loaded lactoferrin (two moles of Fe^{III} per mole of protein) is an efficient catalyst of OH˙ formation from O_2^- and H_2O_2. Bannister *et al.* (1982*a*) confirmed this with the use of a different assay system, although Winterbourn (1983) has pointed to some artefacts in the assays used and has concluded that iron-replete lactoferrin is, if effective at all, a poor catalyst compared with simple iron chelates; a view supported by the authors (Gutteridge *et al.* 1981*b*) and by Baldwin *et al.* (1984). Similarly, the activity of transferrin-bound iron in promoting OH˙ production has been reported as good (McCord & Day 1978), moderate (Bannister *et al.* 1982*b*), poor (Motohashi & Mori 1983) and zero (Maguire *et al.* 1982; Baldwin *et al.* 1984).

At physiological pH, transferrin and lactoferrin seem to release iron much less readily than does ferritin; unlike ferritin-bound iron, iron bound to the former proteins is ineffective in stimulating lipid peroxidation (Gutteridge *et al.* 1981*b*). One possible explanation of the above discrepancies is that, when the fully iron-loaded proteins are studied, there is sometimes iron bound to non-specific sites on the protein molecule that becomes detached during the assay and is the true catalyst of OH˙ production. It seems difficult to imagine that iron(III) embedded in the two specific binding sites of transferrin or lactoferrin can give rise to OH˙ radical that escapes into free solution without reacting with the protein. The authors find it especially hard to understand the claims that iron on lactoferrin (Ambruso & Johnston 1981) or transferrin (McCord & Day 1978) equals iron–EDTA in its efficiency of forming OH˙ radicals.

In normal human plasma, the transferrin present is only partly loaded with iron, i.e. few molecules have two Fe^{III} ions bound to them. Similarly, the lactoferrin released by phagocytic cells contains little iron and, indeed, its release has often been suggested to represent an antibacterial mechanism by binding the iron that bacteria require. These observations would make it even less likely that native transferrin and lactoferrin are significant catalysts of OH˙ radical production *in vivo*. Even if lactoferrin is a catalyst of OH˙ production by phagocytic cells, it is not the only one, because lactoferrin-deficient phagocytes still produce OH˙ (Boxer *et al.* 1982; Newburger & Tauber 1982).

Lactoferrin and transferrin are similar in many respects, but a major difference in their properties is that iron is released from transferrin at pH values of 5.6 and below, whereas lactoferrin holds on to its iron down to pH values of 2 or less (Lonnerdal *et al.* 1981).

Iron and rheumatoid disease

Iron has long been known to play a part in the pathology of rheumatoid disease; in active human disease there is increased deposition of iron in the synovial membranes, a drop in blood haemoglobin and often the presence of iron complexes catalytic for radical reactions in synovial fluid (Blake *et al.* 1981*a*, 1984; Rowley *et al.* 1984). The extensive deposition of ferritin iron (Muirden 1970) within synovial membranes in the inflamed rheumatoid joint would increase its sensitivity to radical reactions; H_2O_2 generated by phagocytes could easily penetrate into the synovial cells and react with iron mobilized from ferritin to form $OH^{\bullet}$ radicals. The synovial fluid of rheumatoid knee joints also contains increased ferritin concentrations (Blake *et al.* 1980) but this ferritin contains little or no iron and appears abnormal in its properties. Idiopathic haemochromatosis is often associated with joint inflammation (Schumacher 1982), which resolves when the iron overload is controlled. Infusion of iron–dextran into rheumatoid patients can aggravate the synovial inflammation (Blake *et al.* 1985*a*). The bleomycin-detectable iron present in synovial fluid can accelerate lipid peroxidation and $OH^{\bullet}$ formation in *in vitro* experiments (Gutteridge *et al.* 1982*a*). There is as yet no direct evidence that $OH^{\bullet}$ is formed in the inflamed rheumatoid joint, although aromatic hydroxylation as an *in vivo* method is being used to study this in our laboratories. However, $OH^{\bullet}$ radical formation could account for some or all of the hyaluronic acid depolymerization (Gutteridge *et al.* 1979) and cartilage degradation (Dean *et al.* 1985) seen in the rheumatoid joint.

Treatment with Desferal

Blake *et al.* (1983) observed that low doses of Desferal aggravated acute rat models of inflammation, but larger doses were anti-inflammatory. In Glynn-Dumond synovitis in guinea pigs, Desferal (100 mg per kilogram body mass) aggravated the acute phase of the inflammation, but repeated administration depressed the chronic phase. A similar effect is seen in the rat allergic air-pouch model of acute to chronic inflammation (Yoshino *et al.* 1984). It is not clear why Desferal may make inflammation worse during the acute phase; by stopping $OH^{\bullet}$ production it may protect neutrophils against self-destruction (Sweder van Asbeck *et al.* 1984), or perhaps some $OH^{\bullet}$ production is necessary for effective regulation of inflammation, for example, by inactivation of arachidonic acid metabolites (Henderson & Klebanoff 1983).

The suppressive action of Desferal on chronic inflammation was sufficiently encouraging for preliminary trials with rheumatoid patients to be performed. Giordano *et al.* (1984) injected 1 g of Desferal intramuscularly and observed an abrupt rise in haemoglobin. No fall in the acute phase response was observed, but the speed of the change in haemoglobin suggests that this effect was mediated by suppressing inflammation. No ill effects were reported. Of seven rheumatoid patients given larger doses (up to 3 g d^{-1} for 5 days each week, for 1–3 weeks), four developed ocular abnormalities that reversed on Desferal withdrawal. Two patients who received prochlorperazine to combat nausea during Desferal therapy became unconscious for 48–72 h (Blake *et al.* 1985*b*), possibly because this combination of drugs mediates iron transfer across the blood–brain barrier and achieves removal of iron essential to the functioning of the nervous system (Gutteridge *et al.* 1982*b*). It is clear that doses of Desferal suitable for treatment of iron overload are not necessarily safe in other disease states, and the combination of Desferal with phenothiazine drugs should be strictly avoided. Although Desferal is of potential use in a number of human diseases (on the basis of animal experiments) caution should be employed in its use, and patients receiving it should be carefully checked for ocular and auditory abnormalities.

Why is bleomycin-detectable iron present in rheumatoid fluids?

As discussed earlier, samples of knee-joint synovial fluid from rheumatoid patients often contain bleomycin-detectable iron. Addition of excess commercial (Sigma) apotransferrin to the assay renders the iron in synovial fluid undetectable, i.e. it is present in a form that can bind to added transferrin under the conditions of the bleomycin assay.

Bleomycin-detectable iron does not seem to be an artefact of the collection or handling of body fluids. It is not present in serum, handled and stored in an identical manner, except in some cases of idiopathic haemochromatosis where the normal iron-binding capacity of transferrin is at or close to saturation (Gutteridge *et al.* 1985*b*). Inflammatory exudates taken from the chronic allergic air-pouch model of inflammation in rats (Yoshino *et al.* 1984) never show bleomycin iron. Synovial fluids taken from patients and assayed at once still often contain bleomycin-detectable iron; in those that do not, such iron does not appear on storage at 4 °C or on freeze–thawing.

Is bleomycin-detectable iron present because the transferrin in the fluid is fully iron loaded? This may be so in cerebrospinal fluid (Gutteridge *et al.* 1982*b*), but it is certainly not in synovial fluid. Our (unpublished) data show a low percentage degree of saturation of transferrin in rheumatoid synovial fluids, calculated by comparing the total iron content of the fluids with the amount of immunologically determined transferrin protein. Addition of iron–NTA chelates to the fluids directly demonstrates this unsaturation; there is substantial iron-binding capacity. Why then is bleomycin-detectable iron present in the synovial fluid? A possible explanation could be that as macrophages and other phagocytes 'slide over' cell or cartilage surfaces, a micro-environment sealed off from the bulk synovial fluid is created (Wright & Silverstein 1984). The pH in this micro-environment can fall to 5 or less (Etherington *et al.* 1981). Oxygen radicals, proteolytic enzymes, myeloperoxidase (from neutrophils) and other products are released into this micro-environment. The low pH facilitates damage to collagen, cartilage and cell membranes by first favouring formation of $HO_2^{\bullet}$ from O_2^-, and secondly, by causing release of iron from transferrin, assisted by the presence of ascorbic acid in synovial fluid. The released iron is kept in the iron(II) state by $HO_2^{\bullet}$, O_2^-, or ascorbate or both and thus cannot re-bind to transferrin or bind to lactoferrin (from neutrophils). Hydroxyl radicals are formed and attack the surface to which the phagocytes are attached (as well as the phagocytes themselves) and also inactivate antiproteases in the micro-environment, causing further damage.

When the phagocytic cells detach or move away, the micro-environment is 'opened' to the bulk synovial fluid. All that can then be seen in this fluid is a slightly lower pH than normal, an oxidation of ascorbic acid (Blake *et al.* 1981*b*), partial inactivation of 'bulk' antiprotease activity (Lewis *et al.* 1984) and the presence of micromolar concentrations of iron available to bleomycin. This iron will not re-bind easily to transferrin or lactoferrin because ascorbate keeps it in the reduced state, and because transferrin does not bind iron easily from some physiological iron complexes (Bates & Schlabach 1973).

We further suggest that one function of lactoferrin released from neutrophils, far from acting as a catalyst of $OH^{\bullet}$ formation, is to minimize damage by binding some iron, because it can hold iron(III) at much lower pH values than can be achieved even in the postulated micro-environment. The fact that macrophages do not synthesize lactoferrin may facilitate damage by iron-dependent OH˙ radical formation from the extracellular H_2O_2 that they produce.

We are grateful to the Arthritis and Rheumatism Council, Wellcome Trust, British Technology Group, West Midlands Health Authority and Ciba-Geigy Pharmaceuticals for research support. BH is a Lister Institute Research Fellow. BH is grateful to Professor Roger Dean for helpful discussions.

References

Albrich, J. M., McCarthy, C. A. & Hurst, T. K. 1981 Biological reactivity of hypochlorous acid: implications for microbicidal mechanism of leukocyte myeloperoxidase. *Proc. natn. Acad. Sci. U.S.A.* **78**, 210–214.

Ambruso, D. R. & Johnston, R. B. 1981 Lactoferrin enhances hydroxyl radical production by human neutrophils, neutrophil particulate fractions and an enzymatic generating system. *J. clin. Invest.* **67**, 352–360.

Arden, G. B., Wonke, B., Kennedy, C. & Huehns, E. R. 1984 Ocular changes in patients undergoing long-term desferrioxamine treatment. *Br. J. Ophthal.* **68**, 873–877.

Aust, S. D. & Svingen, B. A. 1982 The role of iron in enzymatic lipid peroxidation. In *Free radicals in biology*, vol. 5 (ed. W. A. Pryor), pp. 1–28. New York: Academic Press.

Baldwin, A., Jenny, E. R. & Aisen, P. 1984 The effect of human serum transferrin and milk lactoferrin on hydroxyl radical formation from superoxide and hydrogen peroxide. *J. biol. Chem.* **259**, 13391–13394.

Bannister, J. V., Bannister, W. H., Hill, H. A. O. & Thornalley, P. J. 1982*a* Enhanced production of hydroxyl radicals by the xanthine–xanthine oxidase system in the presence of lactoferrin. *Biochim. biophys. Acta* **715**, 116–120.

Bannister, J. V., Bannister, W. H. & Thornalley, P. J. 1984 The effect of ferritin iron loading on hydroxyl radical production. *Life Chem. Rep.* suppl. 2, pp. 64–72.

Bannister, J. V., Bellavite, P., Davoli, A., Thornalley, P. J. & Rossi, F. 1982*b* The generation of hydroxyl radicals following superoxide production by neutrophil NADPH oxidase. *FEBS Lett.* **150**, 300–302.

Baret, A., Jadot, G. & Michelson, A. M. 1984 Anti-inflammatory properties in the rat of CuZnSOD from various species. *Life Chem. Rep.* suppl. 2, pp. 417–421.

Bates, G. W. & Schlabach, M. R. 1973 The reaction of ferric salts with transferrin. *J. biol. Chem.* **248**, 3228–3232.

Biemond, P., van Eijk, H. G., Swaak, A. J. G. & Koster, J. F. 1984 Iron mobilization from ferritin by superoxide derived from stimulated polymorphonuclear leucocytes. *J. clin. Invest.* **73**, 1576–1579.

Blake, D. R., Bacon, P. A., Eastham, E. J. & Brigham, K. 1980 Synovial fluid ferritin in rheumatoid arthritis. *Br. med. J.* **281**, 715–716.

Blake, D. R., Gallagher, P. J., Potter, A. R., Bell, M. J. & Bacon, P. A. 1984 The effect of synovial iron on the progression of rheumatoid disease. *Arthritis Rheum.* **27**, 495–501.

Blake, D. R., Hall, N. D., Bacon, P. A., Dieppe, P. A., Halliwell, B. & Gutteridge, J. M. C. 1981*a* The importance of iron in rheumatoid disease. *Lancet ii*, 1141–1144.

Blake, D. R., Hall, N. D., Bacon, P. A., Dieppe, P. A., Halliwell, B. & Gutteridge, J. M. C. 1983 Effect of a specific iron-chelating agent on animal models of inflammation. *Ann. Rheum. Dis.* **42**, 89–93.

Blake, D. R., Hall, N. D., Treby, D. A., Halliwell, B. & Gutteridge, J. M. C. 1981*b* Protection against superoxide and hydrogen peroxide in synovial fluid from rheumatoid patients. *Clin. Sci.* **61**, 483–486.

Blake, D. R., Lunec, J., Ahern, M., Ring, E. F. J., Bradfield, J. & Gutteridge, J. M. C. 1985*a* The effect of intravenous iron-dextran on rheumatoid synovitis. *Ann. Rheum. Dis.* **44**, 183–188.

Blake, D. R., Winyard, P., Lunec, J., Williams, A., Good, P. A., Gutteridge, J. M. C., Rowley, D. & Halliwell, B. 1985*b* Cerebral and ocular toxicity induced by desferrioxamine. *Q. Jl Med.* (In the press.)

Bowern, N., Ramshaw, I. A., Clark, I. A. & Doherty, P. C. 1984 Inhibition of autoimmune neuropathological process by treatment with an iron-chelating agent. *J. exp. Med.* **160**, 1532–1543.

Boxer, L. A., Coates, T. D., Haak, R. A., Wolach, J. B., Hoffstein, S. & Baehner, R. L. 1982 Lactoferrin deficiency associated with altered granulocyte function. *New Engl. J. Med.* **307**, 404–410.

Butler, J. & Halliwell, B. 1982 Reaction of iron–EDTA chelates with the superoxide radical. *Archs Biochem. Biophys.* **218**, 174–178.

Clark, I. A. & Hunt, N. H. 1983 Evidence for reactive oxygen intermediates causing haemolysis and parasite death in malaria. *Infect. Immun.* **39**, 1–6.

Cohen, G. 1978 The generation of hydroxyl radicals in biologic systems: toxicological aspects. *Photochem. Photobiol.* **28**, 669–675.

Creeth, J. M., Cooper, B., Donald, A. S. R. & Clamp, J. R. 1983 Studies on the limited degradation of mucus glycoproteins. *Biochem. J.* **211**, 323–332.

Cross, C. E., Halliwell, B. & Allen, A. 1984 Antioxidant protection: a function of tracheobronchial and gastrointestinal mucus. *Lancet*, i, pp. 1328–1330.

Czapski, G. 1984 On the use of OH˙ scavengers in biological systems. *Israel J. Chem.* **24**, 29–32.

Davies, S. C., Marcus, R. E., Hungerford, J. L., Miller, M. H., Arden, G. B. & Huehns, E. R. 1983 Ocular toxicity of high-dose intravenous desferrioxamine. *Lancet* i, pp. 181–184.

Dean, R. T., Roberts, C. R. & Forni, L. 1985 Oxygen-centred free radicals efficiently degrade the polypeptide of proteoglycans in cartilage. *Biosci. Rep.* (In the press.)

Etherington, D. J., Pugh, G. & Silver, I. A. 1981 Collagen degradation in an experimental inflammatory lesion: studies on the role of the macrophage. *Acta biol. med. germ.* **40**, 1625–1631.

Fligiel, S. E. G., Ward, P. A., Johnson, K. J. & Till, G. O. 1984 Evidence for a role of hydroxyl radical in immune-complex-induced vasculitis. *Am. J. Path.* **115**, 375–382.

Flitter, W., Rowley, D. A. & Halliwell, B. 1983 Superoxide-dependent formation of hydroxyl radicals in the presence of iron salts. *FEBS Lett.* **158**, 310–312.

Floyd, R. A. 1981 DNA–ferrous iron catalysed hydroxyl free radical formation from hydrogen peroxide. *Biochem. biophys. Res. Commun.* **99**, 1209–1215.

Floyd, R. A. 1983 Direct demonstration that ferrous ion complexes of di- and triphosphate nucleotides catalyse hydroxyl free radical formation from hydrogen peroxide. *Archs Biochem. Biophys.* **225**, 263–270.

Fox, R. B. 1984 Prevention of granulocyte-mediated oxidant lung injury in rats by a hydroxyl radical scavenger, dimethylthiourea. *J. clin. Invest.* **74**, 1456–1465.

Giordano, N., Fioravanti, A., Sancasciani, S., Marcolongo, R. & Borghi, C. 1984 Increased storage of iron and anaemia in rheumatoid arthritis: usefulness of desferrioxamine. *Br. med. J.* **289**, 961–962.

Green, M. R., Hill, H. A. O., Okolow-Zubkowska, M. J. & Segal, A. W. 1979 The production of hydroxyl and superoxide radicals by stimulated human neutrophils – measurements by e.p.r. spectroscopy. *FEBS Lett.* **100**, 23–26.

Greenwald, R. A. & Moy, W. W. 1980 Effect of oxygen-derived free radicals on hyaluronic acid. *Arthritis Rheum.* **23**, 455–463.

Grootveld, M. & Halliwell, B. 1985 An aromatic hydroxylation assay for hydroxyl radicals utilizing high-performance liquid chromatography (HPLC). Use to investigate the effect of EDTA on the Fenton reaction. *Free radical research communications.* (In the press.)

di Guiseppi, J. & Fridovich, I. 1984 The toxicology of molecular oxygen. *CRC Crit. Rev. Toxicol.* **12**, 315–342.

Gutteridge, J. M. C. 1984*a* Reactivity of hydroxyl and hydroxyl-like radical discriminated by release of thiobarbituric-acid-reactive material from deoxyribose, nucleosides and benzoate. *Biochem. J.* **224**, 761–767.

Gutteridge, J. M. C. 1984*b* Lipid peroxidation initiated by superoxide-dependent hydroxyl radicals using complexed iron and hydrogen peroxide. *FEBS Lett.* **172**, 245–250.

Gutteridge, J. M. C. 1984*c* Ferrous iron–EDTA-stimulated phospholipid peroxidation. A reaction changing from alkoxyl radical to hydroxyl-radical-dependent initiation. *Biochem. J.* **224**, 697–701.

Gutteridge, J. M. C. 1984*d* Copper–phenanthroline-induced site-specific oxygen radical damage to DNA. *Biochem. J.* **218**, 983–985.

Gutteridge, J. M. C. 1985 Age pigments and free radicals: fluorescent lipid complexes formed by iron- and copper-containing proteins. *Biochim. biophys. Acta.* **834**, 144–148.

Gutteridge, J. M. C. & Halliwell, B. 1982 The role of superoxide and hydroxyl radicals in the degradation of DNA and deoxyribose induced by a copper-phenanthroline complex. *Biochem. Pharmacol.* **31**, 2801–2805.

Gutteridge, J. M. C., Halliwell, B., Treffry, A., Harrison, P. M. & Blake, D. 1983 Effect of ferritin-containing fractions with different iron loading on lipid peroxidation. *Biochem. J.* **209**, 557–560.

Gutteridge, J. M. C., Paterson, S. K., Segal, A. W. & Halliwell, B. 1981*b* Inhibition of lipid peroxidation by the iron-binding protein lactoferrin. *Biochem. J.* **199**, 259–261.

Gutteridge, J. M. C., Richmond, R. & Halliwell, B. 1979 Inhibition of the iron-catalysed formation of hydroxyl radicals from superoxide and of lipid peroxidation by desferrioxamine. *Biochem. J.* **184**, 469–472.

Gutteridge, J. M. C., Rowley, D. A. & Halliwell, B. 1981*a* Superoxide-dependent formation of hydroxyl radicals in the presence of iron salts. Detection of 'free' iron in biological systems by using bleomycin-dependent degradation of DNA. *Biochem. J.* **199**, 263–265.

Gutteridge, J. M. C., Rowley, D. A., Griffiths, E. & Halliwell, B. 1985*b* Low molecular mass iron complexes and oxygen radical reactions in idiopathic haemochromatosis. *Clin. Sci.* **68**, 463–467.

Gutteridge, J. M. C., Rowley, D. A. & Halliwell, B. 1982*a* Superoxide-dependent formation of hydroxyl radicals and lipid peroxidation in the presence of iron salts. *Biochem. J.* **206**, 605–609.

Gutteridge, J. M. C., Rowley, D. A., Halliwell, B., Cooper, D. F. & Heeley, D. M. 1985*a* Copper and iron complexes catalytic for oxygen radical reactions in sweat from human athletes. *Clinica chim. Acta.* **145**, 267–273.

Gutteridge, J. M. C., Rowley, D. A., Halliwell, B. & Westermarck, T. 1982*b* Increased non-protein-bound iron and decreased protection against superoxide radical damage in cerebrospinal fluid from patients with neuronal ceroid lipofuscinoses. *Lancet* ii, pp. 459–461.

Gutteridge, J. M. C. & Wilkins, S. 1984 Non-protein-bound iron within bacterial cells and the action of bleomycin. *Biochem. Int.* **8**, 89–93.

Halliwell, B. 1978*a* Superoxide-dependent formation of hydroxyl radicals in the presence of iron chelates. *FEBS Lett.* **92**, 321–326.

Halliwell, B. 1978*b* Superoxide-dependent formation of hydroxyl radicals in the presence of iron salts. *FEBS Lett.* **96**, 238–242.

Halliwell, B. 1982*a* Production of superoxide, hydrogen peroxide and hydroxyl radicals by phagocytic cells: a cause of chronic inflammatory disease? *Cell Biol. Int. Rep.* **6**, 529–541.

Halliwell, B. 1982*b* Superoxide-dependent formation of hydroxyl radicals in the presence of iron salts is a feasible source of hydroxyl radicals *in vivo*. *Biochem. J.* **205**, 461–462.

Halliwell, B. & Gutteridge, J. M. C. 1984 Oxygen toxicity, oxygen radicals, transition metals and disease. *Biochem. J.* **219**, 1–14.

Halliwell, B. & Gutteridge, J. M. C. 1985 *Free radicals in biology and medicine*. Oxford University Press.

Henderson, W. R. & Klebanoff, S. J. 1983 Leukotriene B_4, C_4, D_4 and E_4 inactivation by hydroxyl radicals. *Biochem. biophys. Res. Commun.* **110**, 266–272.

Jacobs, A. 1977 Iron overload – clinical and pathological aspects. *Semin. Hematol.* **14**, 89–113.

Johnston, R. B., Keele, B. B., Misra, H. P., Lehmeyer, J. E., Webb, L. S., Baehner, R. L. & Rajagopalan, K. V. 1975 The role of superoxide anion generation in phagocytic bactericidal activity. *J. clin. Invest.* **55**, 1357–1372.

Kanofsky, J. R., Wright, J., Miles-Richardson, G. E. & Tauber, A. I. 1984 Biochemical requirements for singlet oxygen production by purified human myeloperoxidase. *J. clin. Invest.* **74**, 1489–1495.

Kittridge, K. J. & Willson, R. L. 1984 Uric acid substantially enhances the free radical-induced inactivation of alcohol dehydrogenase. *FEBS Lett.* **170**, 162–164.

Lewis, D. A., Parrott, D. P., Bird, J., Cosh, J. A. & Ring, F. 1984 Inactive forms of plasma antiproteases in arthritic synovial fluids. *Int. Res. Commun. Syst. Med. Sci.* **12**, 304–305.

Lonnerdal, B., Keen, C. L. & Hurley, L. S. 1981 Iron, copper, zinc and manganese in milk. *Ann. Rev. Nutr.* **1**, 149–174.

McCord, J. M. & Day, E. D. 1978 Superoxide-dependent production of hydroxyl radical catalysed by iron–EDTA complex. *FEBS Lett.* **86**, 139–142.

Maguire, J. T., Kellogg, E. W. & Packer, L. 1982 Protection against free radical formation by protein-bound iron. *Toxic. Lett.* **14**, 27–34.

Mello Filho, A. C. & Meneghini, R. 1984 *In vivo* formation of single-strand breaks in DNA by hydrogen peroxide is mediated by the Haber–Weiss reaction. *Biochim. biophys. Acta* **781**, 56–63.

Modell, B., Letsky, E. A., Flynn, D. M., Peto, R. & Weatherall, D. J. 1982 Survival and desferrioxamine in thalassaemia major. *Br. med. J.* **284**, 1081–1084.

Moorhouse, C. P., Halliwell, B., Grootveld, M. & Gutteridge, J. M. C. 1985 Cobalt (II) ion as a catalyst of hydroxyl radical and possible 'crypto hydroxyl' radical formation. Differential effects of hydroxyl radical scavengers. *Biochim biophys. Acta* (In the press.)

Motohashi, N. & Mori, I. 1983 Superoxide-dependent formation of hydroxyl radical catalysed by transferrin. *FEBS Lett.* **157**, 197–199.

Muirden, K. D. 1970 The anaemia of rheumatoid arthritis. The significance of iron deposits in the synovial membrane. *Aust. Ann. Med.* **2**, 97–104.

Nauseef, W. M., Root, R. K. & Malech, H. L. 1983 Biochemical and immunological analysis of hereditary myeloperoxidase deficiency. *J. clin. Invest.* **71**, 1297–1307.

Newburger, P. E. & Tauber, A. I. 1982 Heterogeneous pathways of oxidizing radical production in human neutrophils and the HL-69 cell line. *Pediat. Res.* **16**, 856–860.

Phillips, B. J., James, T. E. B. & Anderson, D. 1984 Genetic damage in CHO cells exposed to enzymically generated active oxygen species. *Mutat. Res.* **126**, 265–271.

Repine, J. E., Eaton, J. W., Anders, M. W., Hoidal, J. R. & Fox, R. B. 1979 Generation of hydroxyl radical by enzymes, chemicals and human phagocytes *in vitro*. *J. clin. Invest.* **64**, 1642–1651.

Repine, J. E., Fox, R. B. & Berger, E. M. 1981 Hydrogen peroxide kills *Staphylococcus aureus* by reacting with staphylococcal iron to form hydroxyl radical. *J. biol. Chem.* **256**, 7094–7096.

Repine, J. E., Johansen, K. S. & Berger, E. M. 1984 Hydroxyl radical scavengers produce similar decreases in the chemiluminescence responses and bactericidal activities of neutrophils. *Infect. Immun.* **43**, 435–438.

Richmond, R., Halliwell, B., Chauhan, J. & Darbre, A. 1981 Superoxide-dependent formation of hydroxyl radicals; detection of hydroxyl radicals by the hydroxylation of aromatic compounds. *Analyt. Biochem.* **118**, 328–335.

Rosen, H. & Klebanoff, S. J. 1981 Role of iron and EDTA in the bactericidal activity of a superoxide anion-generating system. *Archs Biochem. Biophys.* **208**, 512–519.

Rowley, D. A., Gutteridge, J. M. C., Blake, D., Farr, M. & Halliwell, B. 1984 Lipid peroxidation in rheumatoid arthritis: thiobarbituric-acid-reactive material and catalytic iron salts in synovial fluid from rheumatoid patients. *Clin. Sci.* **66**, 691–695.

Rowley, D. A. & Halliwell, B. 1983 Superoxide-dependent and ascorbate-dependent formation of hydroxyl radicals in the presence of copper salts: a physiologically-significant reaction? *Archs Biochem. Biophys.* **225**, 279–284.

Sagone, A. L., Decker, M. A., Wells, R. M. & Democko, C. 1980 A new method for the detection of hydroxyl radical production by phagocytic cells. *Biochim. biophys. Acta* **628**, 90–102.

Samuni, A., Chevion, M. & Czapski, G. 1981 Unusual copper-induced sensitization of the biological damage due to superoxide radical. *J. biol. Chem.* **256**, 12632–12635.

Schumacher, H. R. 1982 Articular cartilage in the degenerative arthropathy of haemochromatosis. *Arthrit. Rheum.* **25**, 1460–1467.

Sweder van Asbeck, B., Marx, J. J. M., Struyvenberg, A., van Kats, J. H. & Verhoef, J. 1984*a* Effect of iron(III) in the presence of various ligands on the phagocytic and metabolic activity of human polymorphonuclear leukocytes. *J. Immun.* **132**, 851–856.

Sweder van Asbeck, B., Marx, J. J. M., Struyvenberg, A., van Kats, J. H. & Verhoef, J. 1984*b* Deferoxamine enhances phagocytic function of human polymorphonuclear leukocytes. *Blood* **63**, 714–718.

Test, S. T., Lampert, M. B., Ossanna, P. J., Thoene, J. G. & Weiss, S. J. 1984 Generation of nitrogen-chlorine oxidants by human phagocytes. *J. clin. Invest.* **74**, 1341–1349.

Ward, P. A., Till, G. O., Kunkel, R. & Beauchamp, C. 1983 Evidence for role of hydroxyl radical in complement and neutrophil-dependent tissue injury. *J. clin. Invest.* **72**, 789–801.

Weiss, S. J., King, G. W. & Lo Buglio, A. F. 1977 Evidence for hydroxyl radical generation by human monocytes. *J. clin. Invest.* **60**, 370–376.

Weitberg, A. B., Weitzman, S. A., Destrempes, M., Latt, S. A. & Stossel, T. P. 1983 Stimulated human phagocytes produce cytogenetic changes in cultured mammalian cells. *New Engl. J. Med.* **308**, 26–30.

Wills, E. D. 1969 Lipid peroxide formation of microsomes. The role of non-haem iron. *Biochem. J.* **113**, 325–332.

Willson, R. L. 1978 Iron, zinc, free radicals and oxygen in tissue disorders and cancer control. In *Iron metabolism: CIBA Foundation Symposium* **65**, pp. 331–354. Amsterdam: Elsevier.

Winterbourn, C. C. 1983 Lactoferrin-catalysed hydroxyl radical production. *Biochem. J.* **210**, 15–19.

Wong, S. F., Halliwell, B., Richmond, R. & Skowroneck, W. R. 1981 The role of superoxide and hydroxyl radicals in the degradation of hyaluronic acid induced by metal ions and by ascorbic acid. *J. inorg. Biochem.* **14**, 127–135.

Wright, S. D. & Silverstein, S. C. 1984 Phagocytosing macrophages exclude proteins from the zones of contact with opsonised targets. *Nature, Lond.* **309**, 359–360.

Yoshino, S., Blake, D. R. & Bacon, P. A. 1984 The effect of desferrioxamine on antigen-induced inflammation in the rat air pouch. *J. Pharm. Pharmacol.* **36**, 543–545.

Youngman, R. J. 1984 Oxygen activation: is the hydroxyl radical always biologically relevant? *Trends biochem. Sci.* **9**, 280–284.

Discussion

Roberta J. Ward (*Division of Clinical Cell Biology, M.R.C., Clinical Research Centre, Harrow*). What percentage of the total synovial fluid iron is non-protein bound iron? In our studies at Northwick Park Hospital, we have shown a significantly lower mean total iron concentration (7.5 ± 3 μmol l^{-1}, $n = 10$) in the synovial fluid of rheumatoid arthritis patients with severe disease activity compared with rheumatoid arthritis patients with mild disease activity (16.0 ± 6 μmol l^{-1}, $n = 10$); $P > 0.02$. Therefore if the non-protein bound iron was a consistent percentage of the total iron, Dr Halliwell's results of a linear relation between non-protein iron and TBA-reacting material might indicate that there should be higher concentrations of TBA-reacting material in the synovial fluid of patients with mild disease activity. (TBA is thiobarbituric acid.)

B. Halliwell. Bleomycin-detectable iron is only a small percentage of the total iron present in synovial fluid and shows no consistent relation to the total iron present, either in synovial fluid or in plasma from patients suffering idiopathic haemochromatosis. Our results are thus not necessarily inconsistent with those of Dr Ward. Iron bound to lactoferrin or transferrin will not accelerate lipid peroxidation or register in the bleomycin assay, yet will show up in 'total iron' determinations.

P. M. May (*Department of Applied Chemistry, UWIST, Cardiff*). Dr Halliwell suggests that chelating agents might be used to sequester iron from the low molecular mass pool in cells. However, it is important to remember the competitive effects of other metal ions in the biological medium. For example, has he actually observed the reactions of EDTA that he discusses? I would expect them to be entirely inhibited by the binding of endogenous Ca^{II} or Mg^{II} to the chelating agent.

B. Halliwell. Because EDTA promotes iron-dependent formation of hydroxyl radicals we do not recommend it for use *in vivo*. Desferrioxamine has a binding constant for Fe^{III} several orders of magnitudes greater than that for any other metal ion.

EDTA administered to animals in other experiments is often given as its calcium chelate.

H. Sies (*Institut für Physiologische Chemie I, Universität Düsseldorf, F.R.G.*). I might comment on some clinical observations in West Germany. Superoxide dismutase has been introduced as a drug in 1982 under the name of Peroxinorm, largely in the field of orthopedic surgery. As far as I know, about 600000 patients have been treated for disorders such as gonarthritis, coxarthritis and other inflammations, and the results have been positive.

B. Halliwell. These observations are encouraging in showing the role of oxygen radicals in inflammation *in vivo*, but are the beneficial effects of SOD any better than those of steroids, non-steroidal anti-inflammatory drug therapy or other drugs? It must not be assumed that the ability of SOD to scavenge O_2^- necessarily accounts for its anti-inflammatory effect (Baret *et al.* 1984).

Reference

Baret, A., Jadot, G. & Michelson, A. M. 1984 Anti-inflammatory properties in the rat of CuZnSOD from various species. *Life Chem. Rep.* suppl. 2, pp. 417–421.

Phil Trans. R. Soc. Lond. B **311**, 673–677 (1985)
Printed in Great Britain

Oxidative stress studied in intact mammalian cells

By S. Orrenius

Department of Toxicology, Karolinska Institutet, P.O. Box 60400, *S*-104 01 *Stockholm, Sweden*

Exposure of isolated rat hepatocytes to toxic doses of menadione (2-methyl-1,4-naphthoquinone) results in enhanced formation of active oxygen species, depletion of cellular glutathione and protein thiols, and perturbation of intracellular calcium ion homeostasis. An increase in cytosolic Ca^{2+} concentration, resulting from inhibition of the plasma membrane Ca^{2+} translocase by menadione metabolism, appears to be critically involved in the development of cytotoxicity.

Introduction

The toxicological implications of the formation of active oxygen species have attracted growing interest in recent years. Under aerobic conditions oxygen radicals are normal cellular metabolites that are inactivated by various enzyme systems, including superoxide dismutases, catalase, and glutathione peroxidase. The production of oxygen radicals may, however, be greatly stimulated by the presence of various redox-active compounds. Under certain conditions, this stimulation of oxygen radical formation may be so great as to overwhelm the cellular defence systems and cause oxidative stress and toxicity.

In a series of studies we have used freshly isolated rat hepatocytes incubated with menadione (2-methyl-1,4-naphthoquinone) to investigate the sequence of events involved in the development of oxidative cell injury. The results of our studies, which are briefly summarized in this presentation, suggest that depletion of glutathione (GSH) and protein thiols and perturbation of calcium ion homeostasis are important early events in the development of cytotoxicity.

Metabolism of quinones

The metabolism of quinones by flavoenzymes can occur by either one- or two-electron reduction routes, which often differ greatly in cytotoxicity. As illustrated schematically in figure 1, one-electron reduction results in the formation of semiquinone radicals which can rapidly reduce dioxygen, forming the superoxide anion radical, O_2^-, and regenerating the quinone. Dismutation of O_2^- and production of other highly reactive species of oxygen quickly lead to conditions of oxidative stress and toxicity as redox cycling of the quinone continues (Hassan & Fridovich 1979). However, quinones can also undergo two-electron reduction, forming hydroquinones without production of free semiquinone intermediates. This reaction is catalysed by DT-diaphorase (Ernster 1967), and may serve an important protective function for the cell by competing with the single-electron pathway, because hydroquinones are often less reactive, and more easily excreted by the cell, than semiquinone radicals.

Although menadione can be released from liver cells as the free hydroquinone (Thor *et al.*, 1982), conjugates with glucuronic acid, sulphate and glutathione appear to constitute the predominant metabolites. Whereas conjugation with glucuronic acid and sulphate occurs with

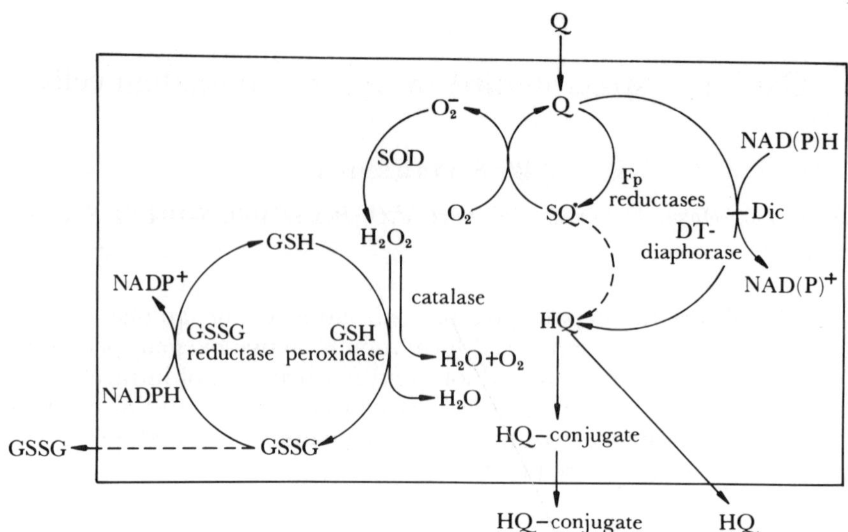

FIGURE 1. Schematic representation of quinone metabolism in isolated hepatocytes. Q, quinone; SQ˙, semiquinone radical; HQ, hydroquinone; Fp, flavoprotein; Dic, dicoumarol; SOD, superoxide dismutase; GSH, reduced glutathione; GSSG, glutathione disulphide.

menadiol, it is not yet clear to what extent the formation of glutathione conjugates also is preceded by reduction of the quinone. Quinones are able to form thiol conjugates non-enzymatically, and a glutathione conjugate of menadione has been reported to contribute to the overall redox cycling of the quinone in the perfused liver (Wefers & Sies 1983).

Metabolism and toxicity of menadione in intact hepatocytes

In isolated liver cells, menadione metabolism involves both one- and two-electron reduction pathways (Thor *et al.* 1982). The relative contribution of the two routes depends on menadione concentration, and can be manipulated by selective induction of either NADPH–cytochrome-P_{450} reductase or DT-diaphorase. Thus, hepatocytes isolated from phenobarbital-treated rats exhibit increased redox cycling and toxicity when exposed to menadione, whereas the reverse is true for hepatocytes from 3-methylcholanthrene-treated rats.

As mentioned above, the cytotoxicity of menadione and other quinones has been related to the oxidative stress caused by the redox cycling of these agents in their target cells (Hassan & Fridovich 1979; Thor *et al.* 1982). O_2^- is produced during the one-electron reduction of menadione to the semiquinone radical and its subsequent reoxidation by molecular oxygen, as well as by the direct interaction of menadione with intracellular thiols (Thor *et al.* 1982; Di Monte *et al.* 1984*b*). Dismutation of O_2^- yields H_2O_2 which can be further metabolized to H_2O by the glutathione peroxidase system at the expense of GSH and NADPH (Wendel 1980).

Exposure of hepatocytes to toxic concentrations of menadione is associated with extensive formation of O_2^- and H_2O_2, and the oxidation of glutathione and pyridine nucleotides (Thor *et al.* 1982). Treatments which promote one-electron reduction and redox cycling of the quinone potentiate menadione toxicity. Conversely, toxicity is diminished when resynthesis of glutathione is facilitated, and in hepatocytes from selenium-deficient rats in which menadione-induced GSH depletion occurs more slowly than in normal cells because of the decreased glutathione peroxidase activity (Orrenius *et al.* 1984). It thus appears that glutathione depletion is a critical event in the initiation of menadione toxicity in isolated hepatocytes, and that the metabolism of large amounts of H_2O_2 by glutathione peroxidase is responsible for the loss of the major fraction of reduced glutathione.

A recent analysis has shown that menadione metabolism by isolated hepatocytes affects the status of both soluble (GSH) and protein thiols (Di Monte *et al.* 1984*b*). In addition to oxidation of GSH to GSSG, formation of mixed disulphides and glutathione conjugate contribute to GSH depletion during menadione metabolism. Following the decrease in soluble thiols there is also a time- and dose-dependent loss of protein sulphydryl groups which is a result of both oxidation and arylation of the protein thiols, oxidation being the quantitatively more important mechanism (Di Monte *et al.* 1984*a*). This loss of protein thiols correlates very closely with the onset of toxicity and invariably precedes cell death. It therefore seems that the depletion of protein thiols, chiefly through oxidative mechanisms, is a critical event which follows GSH depletion in oxidative cell injury caused by menadione.

Perturbation of intracellular Ca^{2+} homeostasis during oxidative stress

We have previously reported that menadione and several other toxic agents cause the formation of numerous small blebs on the surface of isolated hepatocytes (Thor *et al.* 1982; Jewell *et al.* 1982). The same type of surface blebbing can be induced by the calcium ionophore A23187, suggesting that the alterations in surface structure caused by menadione and other toxins are a result of a redistribution of intracellular Ca^{2+} (Jewell *et al.* 1982). This hypothesis is further supported by the observation that GSH depletion is followed by a loss of cell Ca^{2+} from hepatocytes incubated with menadione (Thor *et al.* 1982). Detailed analysis has shown that this effect is due to inhibition of Ca^{2+} sequestration in both the mitochondria and endoplasmic reticulum as result of menadione metabolism (Thor *et al.* 1982: Jewell *et al.* 1982). Whereas the effect of menadione on mitochondrial Ca^{2+} sequestration seems to be related to the oxidation of pyridine nucleotides (Bellomo *et al.* 1982), the inhibitory effect on endoplasmic reticular Ca^{2+} sequestration appears to be a result of oxidation–arylation of thiol group(s) critical for Ca^{2+}-ATPase activity (Jones *et al.* 1983). The effects of menadione metabolism on regulation of Ca^{2+} compartmentation in hepatocytes are illustrated schematically in figure 2.

Because of the inhibition of Ca^{2+} sequestration in the endoplasmic reticulum and mitochondria, incubation of hepatocytes with menadione results in a release of Ca^{2+} sequestered in these compartments into the cytosol. Normally, this would probably only cause a transient rise in the cytosolic Ca^{2+} concentration, because the plasma membrane Ca^{2+} pump would remove this Ca^{2+} from the cell, and the cytosolic Ca^{2+} concentration would return to its usual very low level. Recent studies in our laboratories have shown, however, that the hepatic plasma membrane Ca^{2+}-ATPase is inhibited by agents which oxidize protein thiol groups, including menadione (Bellomo *et al.* 1983).

The redox cycling of menadione in hepatocytes could therefore inhibit all three Ca^{2+} translocases present in the mitochondria, endoplasmic reticulum and plasma membrane. This would undoubtedly lead to a sustained rise in cytosolic Ca^{2+} level, which could cause surface blebbing by altering microfilament organization. Obviously, measurements of cytosolic Ca^{2+} concentration following exposure of hepatocytes to menadione are required to substantiate this hypothesis, but this is very difficult in practice. We have therefore used phosphorylase *a* activity to monitor alterations in cytosolic free-Ca^{2+} concentration in hepatocytes subjected to oxidative

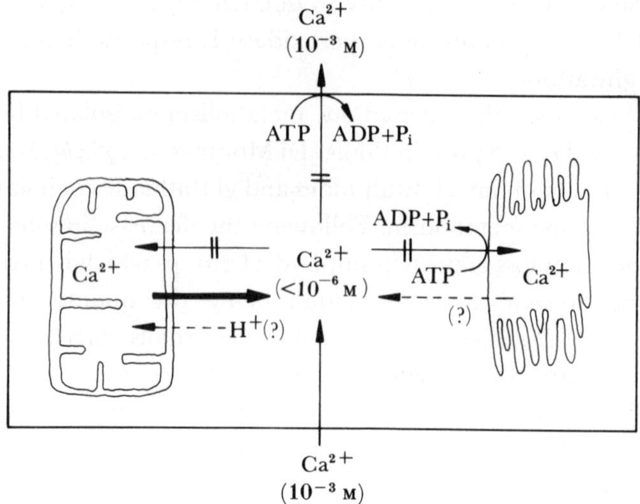

FIGURE 2. Schematic representation of regulation of Ca^{2+} compartmentation in hepatocytes. Effects of menadione metabolism. This illustrates how menadione metabolism causes the release of Ca^{2+} sequestered in the mitochondria and inhibition of the Ca^{2+} translocases located in the endoplasmic reticular and plasma membranes.

stress. Phosphorylase *a* has previously been demonstrated to be a valid indicator of fluctuations in cytosolic Ca^{2+} level because, under appropriate experimental conditions, its activity is strictly dependent on a Ca^{2+}-requiring phosphorylase kinase (Malencik & Fisher 1982). Using this approach we have found that exposure of hepatocytes to toxic levels of menadione causes prolonged phosphorylase activation (Bellomo *et al.* 1984; Di Monte *et al.* 1984*a*). Thus, it appears that menadione-induced oxidative stress in hepatocytes can cause depletion of GSH and protein thiols, and a perturbation of Ca^{2+} homeostasis that may lead to an uncontrollable rise in cytosolic free-Ca^{2+} concentration, resulting in cell death. A schematic representation of these events is given in figure 3.

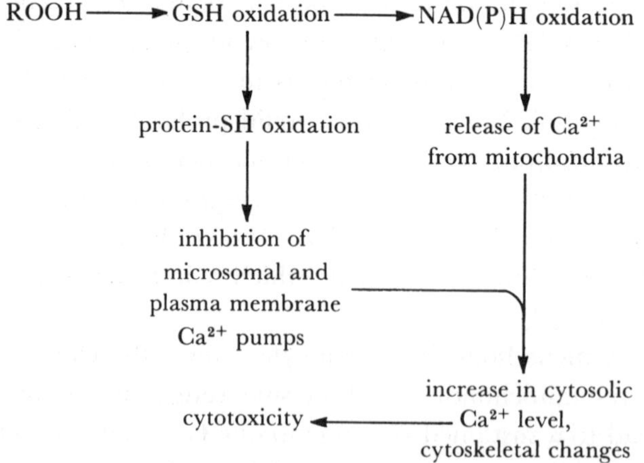

FIGURE 3. Schematic representation of the possible relation between hydrogen peroxide formation, glutathione and pyridine nucleotide oxidation, perturbation of Ca^{2+} homeostasis and cell toxicity during menadione metabolism in isolated hepatocytes.

Concluding remarks

Our studies have shown that exposure of hepatocytes to toxic levels of menadione results in extensive production of oxygen radicals, oxidation of cellular thiols and pyridine nucleotides, and perturbation of intracellular Ca^{2+} homeostasis. The latter is associated with characteristic blebbing of the plasma membrane, which appears to be an early morphologic sign of cell injury. Agents that promote one-electron reduction of menadione, and thereby its redox cycling with molecular oxygen, potentiate menadione toxicity.

References

Bellomo, G., Jewell, S. A. & Orrenius, S. 1982 The metabolism of menadione impairs the ability of rat liver mitochondria to take up and retain calcium. *J. biol. Chem.* **257**, 11558–11562.

Bellomo, G., Mirabelli, F., Richelmi, P. & Orrenius, S. 1983 Critical role of sulfhydryl group(s) in ATP-dependent Ca^{2+} sequestration by the plasma membrane fraction from rat liver. *FEBS Lett.* **163**, 136–139.

Bellomo, G., Thor, H. & Orrenius, S. 1984 Increase in cytosolic Ca^{2+} concentration during *t*-butyl hydroperoxide metabolism by isolated hepatocytes involves NADPH oxidation and mobilization of intracellular Ca^{2+} stores. *FEBS Lett.* **168**, 38–42.

Di Monte, D., Bellomo, G., Thor, H., Nicotera, P. & Orrenius, S. 1984*a* Menadione-induced cytotoxicity is associated with protein thiol oxidation and alteration in intracellular Ca^{2+} homeostasis. *Arch. Biochem. Biophys.* **235**, 343–350.

Di Monte, D., Ross, D., Bellomo, G., Eklöw, L. & Orrenius, S. 1984*b* Alterations in intracellular thiol homeostasis during the metabolism of menadione by isolated rat hepatocytes. *Arch. Biochem. Biophys.* **235**, 334–342.

Ernster, L. 1967 DT-Diaphorase. *Meth. Enzym.* **10**, 309–317.

Hassan, H. M. & Fridovich, J. 1979 Intracellular production of superoxide radical and of hydrogen peroxide by redox active compounds. *Arch. Biochem. Biophys.* **196**, 385–395.

Jewell, S. A., Bellomo, G., Thor, H., Orrenius, S. & Smith, M. T. 1982 Bleb formation in hepatocytes during drug metabolism is caused by alterations in intracellular thiol and Ca^{2+} homeostasis. *Science, Wash.* **217**, 1257–1259.

Jones, D. P., Thor, H., Smith, M. T., Jewell, S. A. & Orrenius, S. 1983 Inhibition of ATP-dependent microsomal Ca^{2+} sequestration during oxidative stress and its prevention by glutathione. *J. biol. Chem.* **258**, 6390–6393.

Malencik, D. A. & Fisher, E. H. 1982 Structure, function and regulation of phosphorylase kinase. In *Calcium and Cell Function*, vol. III (ed. W. Y. Cheung), pp. 161–168. New York: Academic Press.

Orrenius, S., Thor, H. & Di Monte, D. 1984 Metabolic activation and inactivation – a critical balance in toxicity. In *Foreign compound metabolism* (ed. J. Caldwell & G. D. Paulson), pp. 235–255. London: Taylor & Francis.

Thor, H., Smith, M. T., Hartzell, P., Bellomo, G., Jewell, S. A. & Orrenius, S. 1982 The metabolism of menadione (2-methyl-1,4-naphthoquinone) by isolated hepatocytes. A study of the implications of oxidative stress in intact cells. *J. biol. Chem.* **257**, 12419–12425.

Wefers, H. & Sies, H. 1983 Hepatic low-level chemiluminescence during redox cycling of menadione and the menadione-glutathione conjugate. Relation to glutathione and NAD(P)H: quinone reductase (DT-diaphorase) activity. *Archs. Biochem. Biophys.* **224**, 568–578.

Wendel, A. 1980 Glutathione peroxidase. In *Enzymatic basis of detoxication*, vol. I (ed. W. B. Jakoby), pp. 333–348. New York: Academic Press.

Discussion

H. Sies (*Institut für Physiologische Chemie I, Universität Düsseldorf, F.R.G.*). There were several interesting points in Professor Orrenius's lecture. Just for one: does he think that the thioether formation of menadione with protein-bound thiols might be crucial for the expression of quinone toxicity?

S. Orrenius. Although this mechanism may contribute to the toxicity, our present results suggest that the oxidation of protein thiols is more important.

Phil. Trans. R. Soc. Lond. B **311**, 679–689 (1985)
Printed in Great Britain

Radicals and melanomas

By P. A. Riley

Department of Biochemical Pathology, University College London, University Street, London WC1E 6JJ, U.K.

The synthesis of melanin involves the oxidation of phenolic substrates by the enzyme tyrosinase. In vertebrates tyrosinase is present only in specialized cells (melanocytes), where it catalyses the oxidation of tyrosine and certain diphenolic intermediate products to quinones which polymerize to give rise to melanin. This specialized metabolic pathway provides a possible approach to the specific chemotherapy of malignant tumours of pigment cells (malignant melanoma). Certain analogues of tyrosine are oxidized by tyrosinase generating reactive orthoquinones with cytotoxic potential. One such analogue, 4-hydroxyanisole, has been investigated as a possible specific melanocytotoxic precursor. The parent compound inhibits DNA synthesis but exhibits little general toxicity, while the tyrosinase oxidation products are highly toxic to cells. The mechanism of this toxicity may involve semiquinone radicals. Encouraging initial results have been obtained from clinical pilot studies using intra-arterial infusion of hydroxyanisole in patients with localized recurrences of malignant melanoma.

Introduction

Melanins

One of the most widespread surface pigments in the animal kingdom is 'melanin', a term which is thought to have been first used by Berzelius to refer to dark pigments without specific chemical implications other than those of relative insolubility. Subsequent attempts to derive a more specific chemical definition have foundered because the detailed structures of most melanins are unknown, but they are formed by the copolymerization of quinones to give rise to structures in which there is sufficient conjugation between aromatic rings to lower the quantal energies required for light absorption: an effect sometimes termed 'bathochromicity'. The highly conjugated structure of melanins permits electron movement and electron exchanges with neighbouring molecules to take place relatively easily (Sealy *et al.* 1980). Commoner *et al.* (1954) were the first to demonstrate the presence of free radicals in melanins, and it was subsequently shown that the free-radical property of melanins is a result of semiquinones stabilized by resonance in the highly conjugated polymer and by steric restrictions on internal radical annihilation reactions (Mason *et al.* 1960).

Melanogenic pathway

Melanins are generally classified according to their predominant component, which in most cases is an oxidation product of the amino acid tyrosine. The metabolic pathway was first described by Raper (1928) and subsequent studies have established that the enzyme tyrosinase catalyses several oxidative steps leading to the formation of an indole-containing polymer (scheme 1). The evolutionary importance of this pathway is not clear (Riley 1977). Quinones

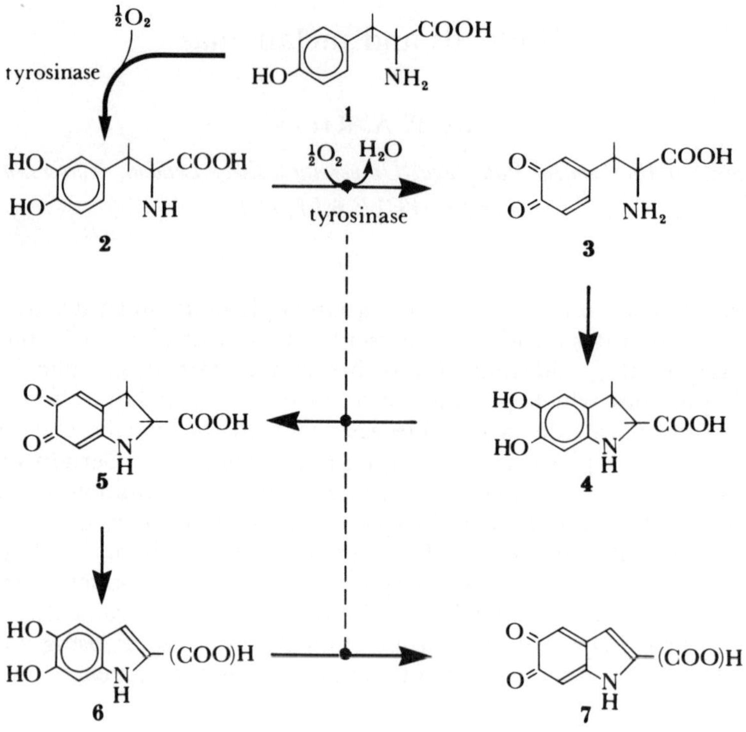

SCHEME 1. Generalized scheme of the melanogenic pathway illustrating the formation of melanin precursors from tyrosine. The scheme shows that tyrosinase catalyses the hydroxylation of tyrosine (**1**) to 3,4-dihydroxyphenylalanine (**2**) and the dehydrogenation of dopa to dopaquinone (**3**). Spontaneous formation of the indolene ring **4** is followed by an oxidation step that is either a result of direct enzymatic action or to a redox exchange with dopaquinone being reduced to dopa and subsequently reoxidized by the enzyme. Dopachrome (**5**) undergoes spontaneous reduction with rearrangement to give 5,6-dihydroxyindole-2-carboxylic acid (**6**), which is reoxidized to indole-2-carboxylic acid 5,6-quinone (**7**) by either the direct or indirect mechanism described above. The possible linked reactions are indicated by the broken line. In some cases spontaneous decarboxylation leads to dihydroxyindole formation, and this is indicated by the brackets in structures **6** and **7**.

are found in the defensive secretions of several arthropods and millipedes, and are present in the surface secretions of many small insects. The protein-tanning effect of quinones seems to be of importance in the hardening of insect cuticles and in sclerotization reactions in other invertebrates. As pigments, the melanins have importance in the protective coloration of many species, and in man, melanins appear to be important in reducing radiation damage to exposed structures such as the skin and the retina.

Melanocytes

In adult mammals, melanogenesis takes place exclusively in specialized pigment cells (melanocytes) that are embryologically derived from the neural crest. These cells are located in the eye, the skin, the hair bulbs and other sites, and synthesize melanin which is donated to surrounding cells. Melanocytes generate melanin in specialized intracellular organelles known as melanosomes, which contain the active form of the enzyme tyrosinase. The melanin is deposited on a protein matrix within these organelles to give rise to pigment granules that are transferred to surrounding cells by a process in which portions of the melanocyte cytoplasm containing the melanin granules are phagocytosed.

Melanomas

Tumours of melanocytes are known as melanomas and the malignant variety of melanoma is an extremely aggressive cancer. There is evidence that the incidence of malignant melanoma is increasing and affects a relatively young age group. The increased incidence of malignant melanoma is thought to be related to sun exposure on the basis of epidemiological evidence (Lee & Strickland 1980). Many malignant melanomas generate melanin, or melanin precursors, in large quantities and the tumour load may be estimated from the excretion of urinary melanogens. Therefore, it has for many years been the aim of rational chemotherapy for these tumours to use the melanogenic pathway as a means of targeting cytotoxic therapy.

Depigmentation

Phenolic antioxidants

Our approach to this problem was prompted by investigations of the mechanisms of action of certain depigmenting agents. It was found by Oettel (1936) that dosing black cats with hydroquinone caused their hair to turn grey within six to eight weeks, and some time later it was shown that the monobenzyl ether of hydroquinone **8**, which was used as an antioxidant

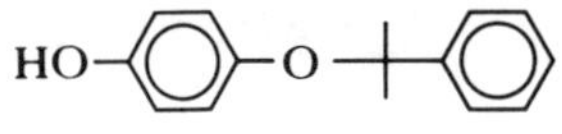

HO–C₆H₄–O–CH₂–C₆H₅

8

in rubber gloves, caused severe and extensive depigmentation of the skin of black workmen in a tanning factory (Oliver *et al.* 1940). The depigmenting effects of phenolic compounds in man have been reviewed elsewhere (Searle & Riley 1979). Peck & Sabotka (1941) performed experiments on the depigmentation caused by the monobenzyl ether of hydroquinone when applied to the skin of guinea pigs and humans and showed that the microscopic depigmentation possessed the features of vitiligo. Extensive studies of depigmentation were made by Brun (1961), who showed that a number of antioxidants caused loss of pigmentation of the areolar skin of guinea pigs. Of the agents tested the most effective long-term depigmenting agent was 4-hydroxyanisole (**9**, 4-HA).

Structural requirements

The structural requirements for the depigmenting activity were investigated for a range of phenolic compounds applied topically to the ear skin of black guinea pigs. In these experiments it was found that hydroxylation in the *para* position relative to a non-polar side chain was an optimal requirement for the depigmenting activity (Riley 1969*a*). For the structural isomers of hydroxyanisole (**9**, **10**, **11**), this was shown to be independent of their rates of penetration

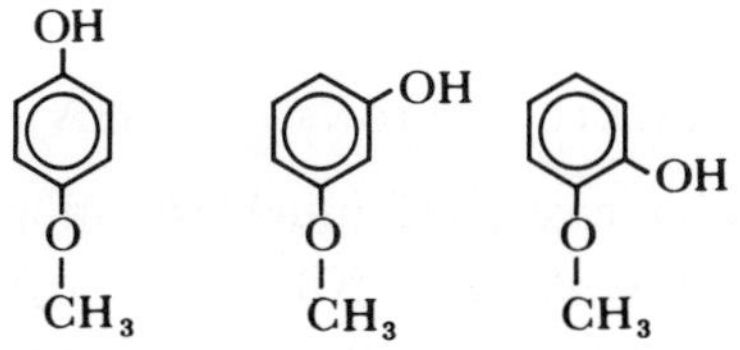

OH, O, CH_3 (**9**) — OH, O, CH_3 (**10**) — OH, O, CH_3 (**11**)

into the skin. Depigmenting compounds with an ether link in the side chain were shown to be more effective than those with hydrocarbon chains.

4-Hydroxyanisole oxidation by tyrosinase

The structural resemblance to tyrosine initially suggested the possibility that loss of pigmentation was caused by competitive inhibition of tyrosinase, but it was shown that the depigmenting phenols were substrates for tyrosinase (Riley 1969*b*). Studies on cultured melanocytes showed that tritiated 4-HA is selectively incorporated into melanogenic cells in culture and transferred with pigment granules to keratocytes, suggesting that the substance is oxidized by tyrosinase and incorporated into melanin (Riley 1970). A stepwise reaction resembling the oxidation of tyrosine was suggested by spectrophotometric observations (figure 1) and the formation of a free-radical semiquinone intermediate was postulated (Riley 1969*b*),

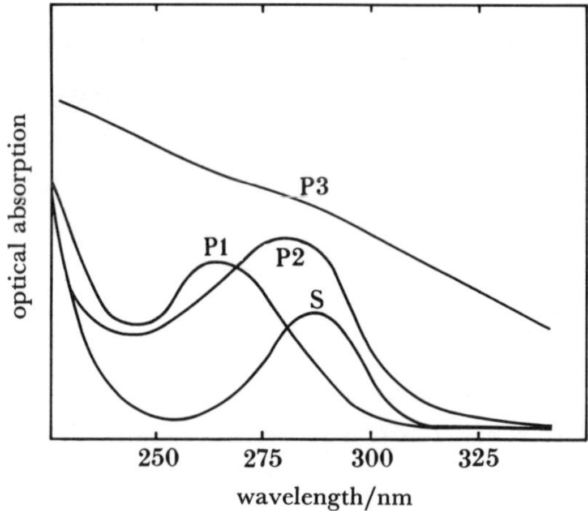

FIGURE 1. Ultraviolet absorption spectra of 4-hydroxyanisole and its tyrosinase oxidation products. The successive spectra are indicated as follows: S, substrate; P1, initial product; P2, secondary product; P3, spectral absorbance of final reaction mixture.

consistent with a new electron spin resonance signal generated in guinea pig skin depigmented by local application of 4-HA (Riley 1970). Erythrocyte suspensions exposed to a reaction mixture of mushroom tyrosinase and 4-HA showed evidence of membrane damage (Riley 1969*b*), and it was suggested that the cytotoxic action of 4-HA might be exerted as a result of the initiation of lipid peroxidation by a free-radical oxidation product (Riley *et al.* 1975). Melanocyte damage is visible in skin treated with depigmenting phenols, both at the light- and electron-microscopic levels (Bleehen *et al.* 1968; Riley 1969*a*) and this evidence makes it probable that the depigmenting action of 4-HA and related phenols is brought about by a reactive oxidation product that is able to initiate cellular damage in melanocytes.

Toxic actions of 4-HA

The cytotoxic actions of 4-HA have received much attention (Passi & Nazarro-Porro 1981; Breathnach *et al.* 1981; Breathnach *et al.* 1983; Nilges *et al.* 1984; Riley 1984). It appears that 4-HA has at least three main modes of action, two of which appear not to require prior metabolism or activation of the compound. These direct actions include a phase-specific effect on the cell cycle (Galpine & Dewey 1984), probably by interference with the synthesis of DNA by inhibiting ribonucleotide reductase (Elford 1984), and an inhibition of mitochondrial oxidation (Passi *et al.* 1984). The third action depends on the oxidation of 4-HA by tyrosinase. There is now good evidence (Naish *et al.* 1985) that the initial product (P1) of tyrosinase oxidation of 4-HA is the corresponding anisyl-3,4-quinone (**12**). This may, under appropriate conditions, polymerize rapidly to melanoid products (scheme 2). Electron spin resonance

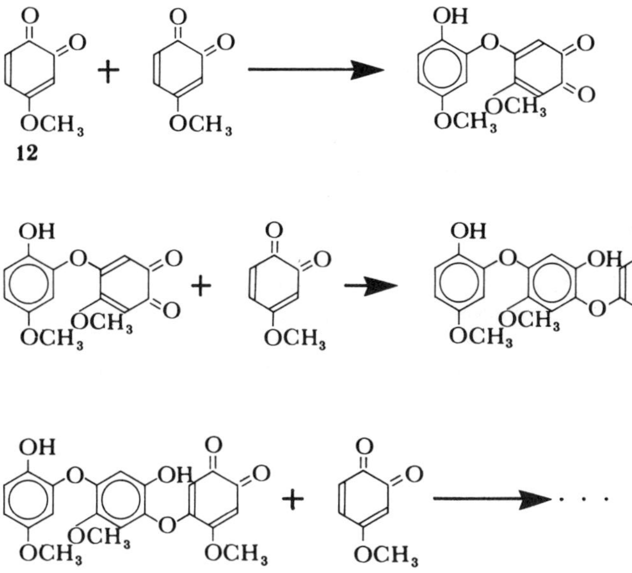

SCHEME 2. Schematic outline of proposed reactions in the linear polymerization of anisyl-3,4-quinone.

evidence (Nilges *et al.* 1984; Nilges & Swartz 1984; Tomasi 1984) shows that under certain conditions the anisyl semiquinone is formed. Whether the ultimate cytotoxic product of 4-HA is the quinone, as proposed for analogous compounds (Wick 1980), or the semiquinone species remains to be determined.

Formation of semiquinone

Nilges & Swartz (1984) have demonstrated that the semiquinone radical signal is inversely related to the oxygen concentration and interpret this finding in terms of a cycle of reduction of anisyl quinone to the corresponding hydroquinone, with simultaneous oxidation of a postulated trihydric phenol intermediate (Nilges *et al.* 1984), with enzymatic reoxidation of the 3,4-hydroquinone to the quinone. It is argued that, if the semiquinone is formed by reverse dismutation of the anisyl-3,4-quinone and the corresponding quinol, the steady-state concentration of the semiquinone will remain low while oxygen is available because the hydroquinone concentration will be small. The influence of tyrosinase on the rate of disappearance of anisyl-3,4-quinone has been examined. While the rate of removal of the quinone is significantly diminished in the presence of tyrosinase this effect cannot be the result of reoxidation of the corresponding 3,4-hydroquinone because it is also observed in a nitrogen atmosphere (figure 2). Also, recent studies of the stoichiometry of the oxidation of 4-HA suggest that the oxygen used is accounted for by the oxidation of 4-HA to anisyl quinone (figure 3). Thus, the role

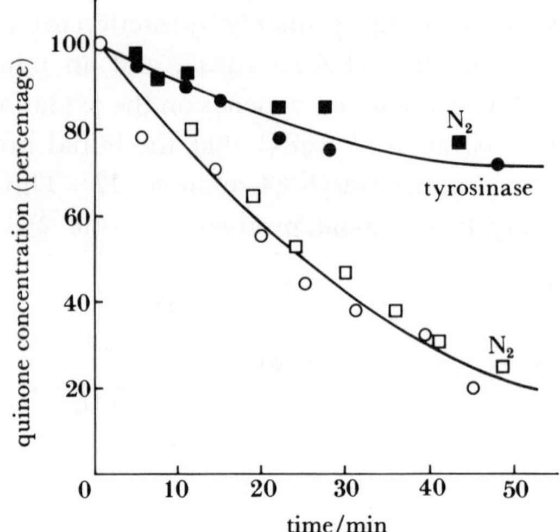

FIGURE 2. Rate of removal of anisyl-3,4-quinone in the presence and absence of tyrosinase (S. Naish, unpublished results). Anisyl-3,4-quinone (100 μM) was incubated at room temperature in 0.1 M phosphate-buffered saline (pH 7.4) in the presence (filled symbols) or absence (open symbols) of mushroom tyrosinase (15 μg ml^{-1}), and sampled at intervals. The quinone concentration was estimated by high-performance liquid chromatography, by a method essentially as described by Egan (1984). Incubations were performed in air (circles) and in nitrogen (squares).

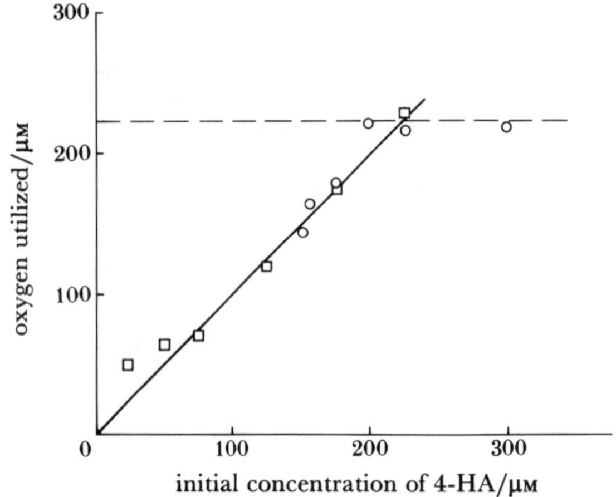

FIGURE 3. Stoichiometry of oxygen utilization as a function of 4-hydroxyanisole concentration in tyrosinase-catalysed oxidation (data of Dobrucki & Riley 1985). In two separate sets of experiments (○ and □), a range of 4-HA concentrations in 0.1 M phosphate buffer (pH 7.4) was incubated at 24 °C in the presence of mushroom tyrosinase (15 μg ml^{-1}), and the total oxygen utilized measured at the end of the reaction. The broken line indicates the saturating oxygen concentration in the incubation mixture (223 μM). With the possible exception of the lowest concentrations of 4-HA, the stoichiometry is close to a 1:1 molar ratio with oxygen.

of oxygen as a regulator of the semiquinone is obscure. The possibility that the concentration of semiquinone is diminished by the single-electron reduction of oxygen, with the implication that the generation of reactive oxygen species from the anisyl semiquinone might be a factor influencing the cytotoxicity of 4-hydroxyanisole, has not received support from recent e.s.r. studies (M. Davies, unpublished results), in which 5,5-dimethyl-1-pyrroline-*N*-oxide (DMPO) was used as a spin trap for oxygen radicals.

If the anisyl semiquinone is not formed by reverse dismutation of the quinone with the corresponding hydroquinone, an alternative pathway of generation needs to be postulated. A redox exchange between the anisyl-3,4-quinone and 4-hydroxyanisole in conditions where the substrate concentration exceeds the available oxygen concentration is one possibility, but to date no evidence has been obtained of the formation of the corresponding phenoxy radical. However, Tomasi (1984) has obtained e.s.r. spectra of radical species that may correspond to substituted phenoxyls, which could be derived by single-electron oxidation from oligomeric species of the class shown in scheme 2. Thus, redox-exchange reactions between the tyrosinase oxidation products of 4-HA may be the source of the free radicals detected, including anisyl semiquinone, as indicated in the speculative scheme (scheme 3). Whether this does, in fact, represent the processes occurring in the system, and to what extent the oxygen concentration influences the balance of reactions, remains to be determined.

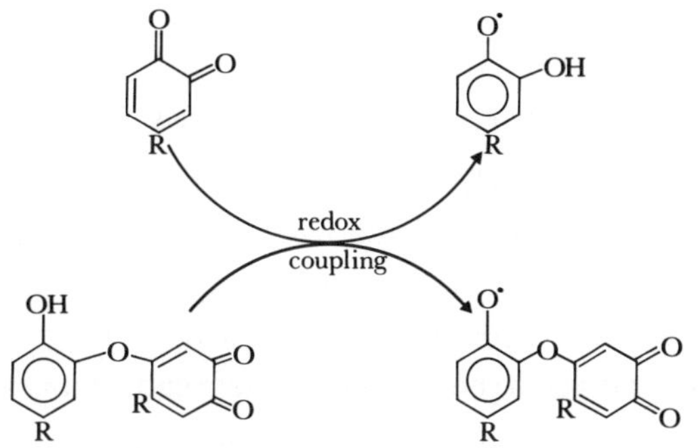

SCHEME 3. Speculative scheme of redox coupling between anisyl-3,4-quinone and its polymerization products. The polymers are represented by a dimerization product (R = OCH_3), which may lose an electron to the quinone, generating the semiquinone radical and a substituted phenoxy radical.

Cytotoxic damage by melanin precursors

The possibility that toxic damage to cells could occur from the action of oxidation products of the melanogenic pathway has been previously recognized (Hochstein & Cohen 1969), and it may be for this reason that a specialized organelle is necessary for pigment production to protect the cell from potentially damaging species (Slater & Riley 1966). It is a matter of some interest as to why, when the oxidation products of 4-HA cause cell damage, the natural intermediates of melanogenesis, such as dopaquinone, are not overtly toxic to cells. This is the case even in model systems (see table 1). It may be that this is a result of the ability of natural

TABLE 1. EFFECT ON PLATING EFFICIENCY OF NATURAL AND ANALOGUE SUBSTRATES OF TYROSINASE

(from Holden 1984)

substrate	LD_{63}/µM
L-tyrosine	700
L-Dopa	>1000
4-HA	11
p-ethoxyphenol	36
MBEH	31

Hamster lung fibroblast cultures (V79) were incubated in Ham's F12 medium in the presence of several concentrations of the agents, with mushroom tyrosinase (100 µg ml^{-1}), and the dose required to reduce survival to 37 % of controls was estimated from the plating efficiency data. Abbreviations used are as follows: Dopa = 3,4-dihydroxyphenylalanine; 4-HA = 4-hydroxyanisole; MBEH = monobenzyl ether of hydroquinone.

substrates to form an indolene ring (see scheme 1), a reaction that takes place rapidly owing to the proximity of the amino group on the side chain of dopaquinone, so that the relative persistence of dopaquinone is very brief. An alternative explanation for differences in apparent toxicity may be the comparative lipid solubilities of the quinones, which may permit the oxidation products of analogue substrates more readily to enter and penetrate phospholipid membranes. For example, of the compounds listed in table 1, tyrosine and dihydroxyphenylalanine have aqueous solubilities at 25 °C of 0.45 and 1.65 gl^{-1} respectively, but are insoluble in most organic solvents; and at the other end of the scale the monobenzyl ether of hydroquinone dissolves readily in alcohol, ether or benzene, but is insoluble in water. 4-hydroxy-anisole is soluble in both, with a partition coefficient of about 15 between octanol and 0.1 M Tris–HCl buffer at pH 7.4 (Cheeseman 1984). It is probable that the corresponding *ortho* quinones exhibit similar solubilities.

Melanoma therapy

Although it is now known that 4-HA has several actions, some of which may not depend on metabolism, the relative lack of generalized toxicity of 4-HA (Hodge *et al.* 1949) suggested the possibility of its systemic use as a treatment for malignant melanoma (Riley 1970). Encouraging results from animal experiments (Dewey *et al.* 1977) led to pilot clinical studies in patients suffering from local recurrences and widespread metastatic melanoma. A number of routes of parenteral administration have been used; intralesional injection, intravenous infusion and intra-arterial infusion (Morgan *et al.* 1981; Webster *et al.* 1984). Therapy by the intra-arterial route has shown encouraging results for local recurrences and 4-HA has now been given intra-arterially to twenty patients, of which nine cases have shown a response (table 2). In two patients the recurrences disappeared completely after treatment (Morgan 1984) and in one of these cases there is no evidence of disease after three years. The usual pattern of

Table 2. Summary of cases treated by intra-arterial infusion of 4-hydroxyanisole

patient	age	sex	site treated	total dose/g	response[a]
LT	83	F	cheek	20	R
MP	75	F	leg	11	R
BN	53	F	foot	68	R
MG	31	F	arm	6	N
MH	42	F	arm	2	N
ED	76	F	leg	17	N
AM	40	F	leg	16	R
HH	50	M	oral	98	N
TG	46	M	arm	5	N
AD	73	F	leg	74	R
FT	70	F	leg	184	R
LY	70	F	leg	184	R
VG	53	F	leg	7	N
ML	65	F	leg	190	R
HB	53	M	liver	104	N
LP	41	M	leg	224	N
MN	63	F	foot	56	N
AG	44	M	leg	224	N
HS	60	M	leg	108	R
MA	64	F	leg	112	N

[a] Responses are indicated as R (positive) and N (negative).

response has been for the intra-arterial infusion with 4-HA to halt the growth of tumour in the infused region, but for the disease to progress gradually elsewhere in the body. In general, it has been found that the larger metastases are more responsive to treatment (Riley *et al.* 1982). Although there are technical problems associated with the arterial route of administration of 4-HA, it is clear from the preliminary clinical results that a significant local response can be elicited in some patients.

Pharmacokinetics

Studies of the pharmacokinetics of 4-HA in mice (Holden & Dewey 1984) and some studies on patients receiving the drug by intra-arterial infusion (Morgan *et al.* 1984) have shown that the compound is rapidly cleared from the blood ($t_{\frac{1}{2}} = 9$ min in the human). Data from the human studies (Morgan *et al.* 1984) and investigations on rat-liver cells (Egan 1984) indicate that 4-HA is metabolized by the liver and that an inducible enzyme system is involved. The process of *o*-demethylation is apparently not implicated (Cheeseman 1984) and 4-HA is known to be a poor substrate for UDP-glucuronyl-transferase (Mulder & Van Doorn 1975). No recovery of 4-HA was observed after enzymatic deconjugation of the metabolic products formed from 4-HA in the rat-liver cell system (Egan 1984). It is possible that further ring hydroxylation is the preferred initial step in the metabolic pathway of 4-HA, with the subsequent possibilities of conjugation or oxidation to a quinone. Cummings & Prough (1983) have shown that the structurally related compound, BHA (E320) inhibits the mixed-function oxidase system through the formation of a quinone metabolite which diverts electrons from the microsomal electron-transport chain to the reduction of oxygen. In collaboration with other laboratories we are investigating modifications to 4-HA with the aim of improving its pharmacokinetic characteristics, by reducing hepatic metabolism without diminishing the tyrosinase affinity of the compound or its ease of oxidation by the enzyme. The key to systemic chemotherapy by using the tyrosinase analogue approach to the melanogenic pathway will be to solve the problem of the pharmacokinetics of 4-HA so that adequate doses can reach all the sites of metastasis of the tumour.

I am grateful to many colleagues for discussions, advice and collaboration; in particular Dr K. U. Ingold, Mr B. D. G. Morgan and Professor T. F. Slater. I thank C. J. Cooksey, M. Davies, J. Dobrucki, J. L. Holden and S. Naish for permission to mention unpublished results. The financial assistance of the Medical Research Council, the Wellcome Trust, the National Foundation for Cancer Research and the Association for International Cancer Research in supporting various phases of this work is gratefully acknowledged.

References

Bleehen, S. S., Pathak, M. A., Hori, Y. & Fitzpatrick, T. B. 1968 Depigmentation of skin with 4-isopropylcatechol, mercaptoamines and other compounds. *J. invest. Derm.* **50**, 103–117.

Breathnach, A. S., Diala, E. B., Gallagher, S. M., Nazzaro-Porro, M. & Passi, S. 1981 Ultrastructural observations on the effect of 4-hydroxyanisole on normal human melanocytes in tissue culture. *J. invest. Derm.* **77**, 292–296.

Breathnach, A. S., Robins, E., Ethridge, L., Bhasin, S., Gallagher, S., Passi, S. & Nazzaro-Porro, M. 1983 Ultrastructural and biochemical observations on the effect of 4-hydroxyanisole plus tyrosinase on normal human melanocytes and keratinocytes in tissue culture. *Br. J. Cancer* **47**, 813–822.

Brun, R. 1961 Contribution a l'étude de la depigmentation. *Bull. inst. Natu. Genevois.* **61**, 3–205.

Cheeseman, K. H. 1984 4-Hydroxyanisole and the rat liver microsomal drug metabolising enzymes. In *Hydroxyanisole: recent advances in anti-melanoma therapy* (ed. P. A. Riley), pp. 183–189. Oxford: I.R.L. Press.

Commoner, B., Townsend, J. & Pake, G. E. 1954 Free radicals in biological materials. *Nature, Lond.* **174**, 689.

Cummings, S. W. & Prough, R. A. 1983 Butylated hydroxyanisole-stimulated NADPH oxidase activity in rat liver microsomal fractions. *J. biol. Chem.* **258**, 12315–12319.

Dewey, D. L., Butcher, F. W. & Galpine, A. R. 1977 Hydroxyanisole-induced regression of the Harding–Passey melanoma in mice. *J. Path.* **122**, 117–127.

Dobrucki, J. & Riley, P. A. 1985 Paper in preparation.

Egan, C. 1984 Assay of 4-hydroxyanisole in culture medium by high performance liquid chromatography and some preliminary results in primary maintenance cultures of adult rat hepatocytes. In *Hydroxyanisole: recent advances in anti-melanoma therapy* (ed. P. A. Riley), pp. 177–182. Oxford: I.R.L. Press.

Elford, H. L. 1984 The effect of hydroxyanisole on mammalian ribonucleotide reductase. In *Hydroxyanisole: recent advances in anti-melanoma therapy* (ed. P. A. Riley), pp. 71–75. Oxford: I.R.L. Press.

Galpine, A. R. & Dewey, D. L. 1984 Inhibition of DNA synthesis and S-phase killing by 4-hydroxyanisole. In *Hydroxyanisole: recent advances in anti-melanoma therapy* (ed. P. A. Riley), pp. 119–127. Oxford: I.R.L. Press.

Hochstein, P. & Cohen, G. 1963 The cytotoxicity of melanin precursors. *Ann. N.Y. Acad. Sci.* **100**, 876–886.

Hodge, H. C., Sterner, J. H., Maynard, C. A. & Thomas, T. 1949 Short-term toxicity tests on the mono and dimethyl ethers of hydroquinone. *J. Ind. Hyg. Toxicol.* **31**, 79–92.

Holden, J. L. 1984 Toxicity of phenolic depigmenting agents. In *Gray Laboratory Annual Report*, pp. 96–97. London: Cancer Research Campaign.

Holden, J. L., Dewey, D. L. & Riley, P. A. 1984 A study of the pharmacokinetics of 4-hydroxyanisole. In *Hydroxyanisole: recent advances in anti-melanoma therapy* (ed. P. A. Riley), pp. 213–220. Oxford: I.R.L. Press.

Lee, J. A. H. & Strickland, D. 1980 Malignant melanoma: social status and outdoor work. *Br. med. J.* **41**, 747–763.

Mason, H. S., Ingram, H. E. & Allen, B. 1960 Free radical property of melanins. *Archs Biochem. Biophys.* **86**, 225–230.

Morgan, B. D. G. 1984 Recent results of a clinical pilot study of intra-arterial 4-HA chemotherapy in malignant melonoma. In *Hydroxyanisole: recent advances in anti-melanoma therapy* (ed. P. A. Riley), pp. 233–241. Oxford: I.R.L. Press.

Morgan, B. D. G., Holden, J. L., Dewey, D. L. & Riley, P. A. 1984 Human pharmacokinetics of 4-hydroxyanisole. In *Hydroxyanisole: recent advances in anti-melanoma therapy* (ed. P. A. Riley), pp. 221–226. Oxford: I.R.L. Press.

Morgan, B. D. G., O'Neill, T., Dewey, D. L., Galpine, A. R. & Riley, P. A. 1981 Treatment of malignant melanoma by intravascular 4-hydroxyanisole. *Clin. Oncol.* **7**, 227–234.

Mulder, G. J. & Van Doorn, A. B. D. 1975 A rapid NAD^+-linked assay for microsomal uridine diphosphate glucuronyl transferase of rat liver and some observations on substrate specificity of the enzyme. *Biochem. J.* **151**, 131–140.

Naish, S., Cooksey, C. J. & Riley, P. A. 1985 Paper in preparation.

Nilges, M. J., Swartz, H. M. & Riley, P. A. 1984 Electron spin resonance studies of reactive intermediates of tyrosinase oxidation of hydroxyanisole. *J. biol. Chem.* **259**, 2446–2451.

Nilges, M. J. & Swartz, H. M. 1984 Quinone and semiquinone intermediates in the reaction of 4-hydroxyanisole and tyrosinase. In *Hydroxyanisole: recent advances in anti-melanoma therapy* (ed. P. A. Riley), pp. 25–34. Oxford: I.R.L. Press.

Oettel, H. 1936 Die Hydrochinonvergiftung (Hydroquinone poisoning). *Arch. exp. Pathol. Pharmakol.* **183**, 319–326.

Oliver, E. A., Schwartz, L. & Warren, L. H. 1940 Occupational leukoderma. *Arch. Derm.* **42**, 993–1014.

Passi, S. & Nazzaro-Porro, M. 1981 Molecular basis of substrate and inhibitory specificity of tyrosinase: phenolic compounds. *Br. J. Derm.* **104**, 659–665.

Passi, S., Picardo, M. & Nazzaro-Porro, M. 1984 Effect of parahydroxyanisole on tyrosinase and mitochondrial oxidoreductases. In *Hydroxyanisole: recent advances in anti-melanoma therapy* (ed. P. A. Riley), pp. 57–70. Oxford: I.R.L. Press.

Peck, S. M. & Sabotka, H. 1941 Effect of monobenzyl hydroquinone on oxidase systems *in vivo* and *in vitro*. *J. invest. Derm.* **4**, 325–329.

Raper, H. S. 1928 The aerobic oxidases. *Physiol. Rev.* **8**, 245–282.

Riley, P. A. 1969*a* Hydroxyanisole depigmentation: *in vivo* studies. *J. Path.* **97**, 185–191.

Riley, P. A. 1969*b* Hydroxyanisole depigmentation: *in vitro* studies. *J. Path.* **97**, 193–206.

Riley, P. A. 1970 Mechanism of pigment cell toxicity produced by hydroxyanisole. *J. Path.* **101**, 163–169.

Riley, P. A. 1977 The Mechanism of melanogenesis. *Symp. zool. Soc. Lond.* **39**, 77–95.

Riley, P. A. 1984 (ed.) *Hydroxyanisole: recent advances in anti-melanoma therapy*. Oxford: I.R.L. Press.

Riley, P. A., Morgan, B. D. G., O'Neill, T., Dewey, D. L. & Galpine, A. R. 1982 Treatment of malignant melanoma with 4-hydroxyanisole. In *Free radicals, lipid peroxidation and cancer* (ed. D. C. H. McBrien & T. F. Slater), pp. 421–432. London: Academic Press.

Riley, P. A., Sawyer, B. & Wolff, M. A. 1975 The melanocytotoxic action of 4-hydroxyanisole. *J. invest. Derm.* **64**, 86–89.

Sealy, R. C., Felix, C. C., Hyde, J. S. & Swartz, H. M. 1980 In Free radicals in biology, vol. 4 (ed. W. Pryor), pp. 209–251. New York: Academic Press.

Searle, C. E. & Riley, P. A. 1979 Chemically-induced depigmentation of skin and hair. In *Hair and hair diseases* (ed. C. E. Orfanos), pp. 911–929. Stuttgart: Gustav Fischer Verlag.

Slater, T. F. & Riley, P. A. 1966 Photosensitisation and lysosomal damage. *Nature, Lond.* **209**, 151–153.
Tomasi, A. 1984 Radical activation of 4-hydroxyanisole by mushroom tyrosinase. In *Hydroxyanisole: recent advances in anti-melanoma therapy* (ed. P. A. Riley), pp. 15–24. Oxford: I.R.L. Press.
Webster, D. J. T., Whitehead, R. H., Tarr, M. J. & Hughes, L. E. 1984 A phase I study of 4-hydroxyanisole (4-HOA) in patients with advanced malignant melanoma. In *Hydroxyanisole: recent advances in anti-melanoma therapy* (ed. P. A. Riley), pp. 227–232. Oxford: I.R.L. Press.
Wick, M. M. 1980 An experimental approach to the chemotherapy of melanoma. *J. invest. Derm.* **74**, 63–65.

Discussion

R. Marris (793 *Harrow Rd, London NW10 5PA*). Three or four years ago at least two substances were referred to in the correspondence columns of the *Lancet* with rather similar effects upon a few malignant melanomas. Has Professor Riley tried them alone or in combination with 4-hydroxyanisole?

P. A. Riley. The substances Mr Marris is referring to are straight-chain dicarboxylic acids with 9–12 carbon atoms in the chain. Azelaic acid has been used successfully to treat patients with lentigo maligna, a lesion that is regarded as an early form of malignant melanoma. We have no experience of the use of these compounds in the treatment of advanced malignant melanoma.

R. Marris. There is a variety of domesticated dog that is especially subject to malignant melanoma in the mouth. It may be useful in investigations; grey horses and Angora goats, which have been used elsewhere, are expensive (Cheville 1983).

Reference

Cheville, N. F. 1983 *Cell pathology* (2 edn). Iowa State University Press.

P. A. Riley. Thank you.